COMPUTATIONAL METHODS IN MOLECULAR BIOLOGY

New Comprehensive Biochemistry

Volume 32

General Editor

G. BERNARDI
Paris

ELSEVIER
Amsterdam · Lausanne · New York · Oxford · Shannon · Singapore · Tokyo

Computational Methods in Molecular Biology

Editors

Steven L. Salzberg

*The Institute for Genomic Research,
9712 Medical Center Drive, Rockville, MD 20850, USA*

David B. Searls

*SmithKline Beecham Pharmaceuticals, 709 Swedeland Road,
P.O. Box 1539, King of Prussia, PA 19406, USA*

Simon Kasif

*Department of Electrical Engineering and Computer Science,
University of Illinois at Chicago, Chicago, IL 60607-7053, USA*

ELSEVIER

Amsterdam · Lausanne · New York · Oxford · Shannon · Singapore · Tokyo

Elsevier Science B.V.
P.O. Box 211
1000 AE Amsterdam
The Netherlands

First edition (Hardbound) 1998
Paperback edition 1999

Library of Congress Cataloging-in-Publication Data

Computational methods in molecular biology / editors, Steven L. Salzberg, David B. Searls, Simon Kasif.
    p.   cm. – – (New comprehensive biochemistry ; v. 32)
  Includes bibliographical references and index.
    ISBN 0-444-82875-3 (hardcover : alk. paper). – – ISBN 0-444-50204-1 (pbk. : alk. paper)
    1. Molecular biology– –Mathematics. I. Salzberg, Steven L., 1960– . II. Searls, David B. III. Kasif, Simon. IV. Series.
  QD415.N48 vo. 32
  [QH506]
  572 s– –dc21
  [572.8′01′51]   98–22957
                                    CIP

ISBN: 0 444 82875 3 (Hardbound)
ISBN: 0 444 50204 1 (Paperback)
ISBN: 0 444 80303 3 (series)

⊗ The paper used in this publication meets the requirements of ANSI/NISO Z39.48-1992 (Permanence of Paper).
Printed in the Netherlands

To Claudia and Annika – S.L.S.

To my parents – D.B.S.

To Orna and Mei – S.K.

Preface

The field of computational biology, or bioinformatics as it is often called, was born just a few years ago. It is difficult to pinpoint its exact beginnings, but it is easy to see that the field is currently undergoing rapid, exciting growth. This growth has been fueled by a revolution in DNA sequencing and mapping technology, which has been accompanied by rapid growth in many related areas of biology and biotechnology. No doubt many exciting breakthroughs are yet to come. All this new DNA and protein sequence data brings with it the tremendously exciting challenge of how to make sense of it: how to turn the raw sequences into information that will lead to new drugs, new advances in health care, and a better overall understanding of how living organisms function. One of the primary tools for making sense of this revolution in sequence data is the computer. Computational biology is all about how to use the power of computation to model and understand biological systems and especially biological sequence data.

This book is an attempt to bring together in one place some of the latest advances in computational biology. In assembling the book, we were particularly interested in creating a volume that would be accessible to biologists (as well as computer scientists and others). With this in mind, we have included tutorials on many of the key topics in the volume, designed to introduce biological scientists to some of the computational techniques that might otherwise be unfamiliar to them. Some of those tutorials appear as separate, complete chapters on their own, while others appear as sections within chapters. We also want to encourage more computer scientists to get involved in this new field, and with them in mind we included tutorial material on several topics in molecular biology as well. We hope the result is a volume that offers something valuable to a wide range of readers. The only required background is an interest in the exciting new field of computational biology.

The chapters that follow are broadly grouped into three sections. Loosely speaking, these can be described as an introductory section, a section on DNA sequence analysis, and a section on proteins. The introductory section begins with an overview by Searls of some of the main challenges facing computational biology today. This chapter contains a thought-provoking description of problems ranging from gene finding to protein folding, explaining the biological significance and hinting at many of the computational solutions that appear in later chapters. Searls' chapter should appeal to all readers. Next is Salzberg's tutorial on computation, designed primarily for biologists who do not have a formal background in computer science. After reading this chapter, biologists should find many of the later chapters much more accessible. The following chapter, by Fasman and Salzberg, provides a tutorial for the other main component of our audience, computational scientists (including computer scientists, mathematicians, physicists, and anyone else who might need some additional biological background) who want to understand the biology that underlies all the research problems described in later chapters. This tutorial introduces the non-biologist to many of the terms, concepts, and mechanisms of molecular biology and sequence analysis.

The second of the three major sections contains work primarily on DNA and RNA sequence analysis. Although the techniques covered here are not restricted to DNA sequences, most of the applications described here have been applied to DNA.

Krogh's chapter begins the section with a tutorial introduction to one of the hottest techniques in computational biology, hidden Markov models (HMMs). HMMs have been used for a wide range of problems, including gene finding, multiple sequence alignment, and the search for motifs. Krogh covers only one of these applications, gene finding, but he first gives a cleverly non-mathematical tutorial on this very mathematical topic.

The chapter by Overton and Haas describes a case-based reasoning approach to sequence annotation. They describe an informatics system for the study of gene expression in red blood cell differentiation. This type of specialized information resource is likely to become increasingly important as the amount of data in GenBank becomes ever larger and more diverse.

The chapter by States and Reisdorf describes how to use sequence similarity as the basis for sequence classification. The approach relies on clustering algorithms which can, in general, operate on whole sequences, partial sequences, or structures. The chapter includes a comprehensive current list of databases of sequence and structure classification.

Xu and Uberbacher describe many of the details of the latest version of GRAIL, which for years has been one of the leading gene-finding systems for eukaryotic data. GRAIL's latest modules include the ability to incorporate sequence similarity to the expressed sequence tag (EST) database and a nice technique for detecting potential frameshifts.

Burge gives a thorough description of how to model RNA splicing signals (donor and acceptor sites) using statistical patterns. He shows how to combine weight matrix methods with a new tree-based method called maximal dependence decomposition, resulting in a splice site recognizer that is state of the art. His technique is implemented in GENSCAN, currently the best-performing of all gene-finding systems.

Parsons' chapter includes tutorial material on genetic algorithms (GAs), a family of techniques that use the principles of mutation, crossover, and natural selection to "evolve" computational solutions to a problem. After the tutorial, the chapter goes on to describe a particular genetic algorithm for solving a problem in DNA sequence assembly. This description serves not only to illustrate how well the GA worked, but it also provides a case study in how to refine a GA in the context of a particular problem.

Salzberg's chapter includes a tutorial on decision trees, a type of classification algorithm that has a wide range of uses. The tutorial uses examples from the domain of eukaryotic gene finding to make the description more relevant. The chapter then moves on to a description of MORGAN, a gene-finding system that is a hybrid of decision trees and Markov chains. MORGAN's excellent performance proves that decision trees can be applied effectively to DNA sequence analysis problems.

Wei, Chang, and Altman's chapter describes statistical methods for protein structure analysis. They begin with a tutorial on statistical methods, and then go on to describe FEATURE, their system for statistical analysis of protein sequences. They describe several applications of FEATURE, including characterization of active sites, generation of substitution matrices, and protein threading.

Protein threading, or fold recognition, is essentially finding the best fit of a protein sequence to a set of candidate structures for that sequence. Lathrop, Rogers, Bienkowska, Bryant, Buturović, Gaitatzes, Nambudripad, White, and Smith begin their chapter with a tutorial section that describes what the problem is and why it is "hard" in the computer science sense of that word. This section should be of special interest to those who want to understand why protein folding is computationally difficult. They then describe their threading algorithm, which is an exhaustive search method that uses a branch-and-bound strategy to reduce the search space to a tractable (but still very large) size.

Jones' chapter describes THREADER, one of the leading systems for protein threading. He first introduces the general protein folding problem, reviews the literature on fold recognition, and then describes in detail the so-called double dynamic programming approach that THREADER employs. Jones makes it clear how this intriguing problem combines a wide range of issues, from combinatorial optimization to thermodynamics.

The chapter by Wolfson and Nussinov presents a novel application of geometric hashing for predicting the possibility of binding, docking and other forms of biomolecular interaction. Even when the individual structures of two molecules are accurately modeled, it remains computationally difficult to predict whether docking or binding are possible. Thus, this method naturally complements the work on structure prediction described in other chapters.

The chapter by Kasif and Delcher uses a probabilistic modeling approach similar to HMMs, but their formalism is known as probabilistic networks or Bayesian networks. These networks have slightly more expressive power and in some cases a more compact representation. For sequence analysis tasks, the probabilistic network approach allows one to model features such as motif lengths, gap lengths, long term dependencies, and the chemical properties of amino acids.

Finally, the end of the book contains some reference materials that all readers should find useful. The first appendix contains a list of Internet resources, including most of the software described in the book. This list is also available on a Web page whose address is given in the appendix. The Web page will be kept up to date long after the book's publication date. The second appendix contains an annotated bibliographical list for further reading on selected topics in computational biology. Some of these references, each of which contains a very short text description, point to more technical descriptions of the systems in the book. Others point to well-known or landmark papers in computational biology which would be of interest to anyone looking for a broader perspective on the field.

Steven Salzberg
David Searls
Simon Kasif
Baltimore, Maryland
October 1997

List of contributors*

Russ B. Altman   207
*Section of Medical Informatics, 251 Campus Drive, Room x-215,
Stanford University School of Medicine, Stanford, CA 94305-5479, USA*

Jadwiga Bienkowska   227
*BioMolecular Engineering Research Center, Boston University, 36 Cummington Street,
Boston, MA 02215, USA*

Barbara K.M. Bryant   227
Millennium Pharmaceuticals, Inc., 640 Memorial Drive, Cambridge, MA 02139, USA

Christopher B. Burge   129
*Center for Cancer Research, Massachusetts Institute of Technology,
40 Ames Street, Room E17-526a, Cambridge, MA 02139-4307, USA*

Ljubomir J. Buturović   227
Incyte Pharmaceuticals, Inc., 3174 Porter Drive, Palo Alto, CA 94304, USA

Jeffrey T. Chang   207
*Section of Medical Informatics, 251 Campus Drive, Room x-215,
Stanford University School of Medicine, Stanford, CA 94305-5479, USA*

Arthur L. Delcher   335
Computer Science Department, Loyola College in Maryland, Baltimore, MD 21210, USA

Kenneth H. Fasman   29
*Whitehead Institute/MIT Center for Genome Research, 320 Charles Street,
Cambridge, MA 02141, USA*

Chrysanthe Gaitatzes   227
*BioMolecular Engineering Research Center, Boston University, 36 Cummington Street,
Boston, MA 02215, USA*

Juergen Haas   65
*Center for Bioinformatics, University of Pennsylvania, 13121 Blockley Hall,
418 Boulevard, Philadelphia, PA 19104, USA*

---

* Authors' names are followed by the starting page number(s) of their contribution(s).

David Jones   285
Department of Biological Sciences, University of Warwick, Coventry CV4 7AL,
England, UK

Simon Kasif   335
Department of Electrical Engineering and Computer Science,
University of Illinois at Chicago, Chicago, IL 60607-7053, USA

Anders Krogh   45
Center for Biological Analysis, Technical University of Denmark,
Building 208, 2800 Lyngby, Denmark

Richard H. Lathrop   227
Department of Information and Computer Science, 444 Computer Science Building,
University of California, Irvine, CA 92697-3425, USA

Raman Nambudripad   227
Molecular Computing Facility, Beth Israel Hospital, 330 Brookline Avenue, Boston, MA
02215, USA

Ruth Nussinov   313
Sackler Inst. of Molecular Medicine, Faculty of Medicine,
Tel Aviv University, Tel Aviv 69978, Israel; and
Laboratory of Experimental and Computational Biology, SAIC, NCI-FCRDC,
Bldg. 469, rm. 151, Frederick, MD 21702, USA

G. Christian Overton   65
Center for Bioinformatics, University of Pennsylvania, 13121 Blockley Hall,
418 Boulevard, Philadelphia, PA 19104, USA

Rebecca J. Parsons   165
Department of Computer Science, University of Central Florida, P.O. Box 162362,
Orlando, FL 32816-2362, USA

William C. Reisdorf, Jr.   87
Institute for Biomedical Computing, Washington University in St. Louis,
700 South Euclid Avenue, St. Louis, MO 63110, USA

Robert G. Rogers Jr.   227
BioMolecular Engineering Research Center, Boston University, 36 Cummington Street,
Boston, MA 02215, USA

Steven Salzberg   11, 29, 187
The Institute for Genomic Research, 9712 Medical Center Drive,
Rockville, MD 20850, USA

David B. Searls 3
SmithKline Beecham Pharmaceuticals, 709 Swedeland Road, P.O. Box 1539,
King of Prussia, PA 19406, USA

Temple F. Smith 227
BioMolecular Engineering Research Center, Boston University, 36 Cummington Street,
Boston, MA 02215, USA

David J. States 87
Institute for Biomedical Computing, Washington University in St. Louis,
700 South Euclid Avenue, St. Louis, MO 63110, USA

Edward C. Uberbacher 109
Bldg. 1060 COM, MS 6480, Cumputational Biosciences Section,
Life Sciences Division, ORNL, Oak Ridge, TN 37831-6480, USA

Liping Wei 207
Section of Medical Informatics, 251 Campus Drive, Room x-215,
Stanford University School of Medicine, Stanford, CA 94305-5479, USA

James V. White 227
BioMolecular Engineering Research Center, Boston University, 36 Cummington Street,
Boston, MA 02215, USA; and TASC, Inc., 55 Walkers Brook Drive, Reading, MA
01867, USA

Haim Wolfson 313
Computer Science Department, Tel Aviv University,
Raymond and Beverly Sackler Faculty of Exact Sciences, Ramat Aviv 69978,
Tel Aviv, Israel

Ying Xu 109
Bldg. 1060 COM, MS 6480, Cumputational Biosciences Section,
Life Sciences Division, ORNL, Oak Ridge, TN 37831-6480, USA

Contents

I – Introduction and Tutorial Background

Chapter 6. Classification-based molecular sequence analysis

Chapter 7. Computational gene prediction using neural networks and similarity search

III – Protein Structure Modeling and Prediction

Chapter 13. THREADER: protein sequence threading by double dynamic programming

David Jones

Chapter 14. From computer vision to protein structure and association
Haim J. Wolfson and Ruth Nussinov 313

Chapter 15. Modeling biological data and structure with probabilistic networks
Simon Kasif and Arthur L. Delcher 335

IV – Reference Materials

Other volumes in the series

PART I

Introduction and Tutorial Background

S.L. Salzberg, D.B. Searls, S. Kasif (Eds.), *Computational Methods in Molecular Biology*
© 1998 Elsevier Science B.V.   All rights reserved

CHAPTER 1

Grand challenges in computational biology

David B. Searls

SmithKline Beecham Pharmaceuticals, 709 Swedeland Road, PO Box 1539, King of Prussia, PA 19406, USA
Phone: (610) 270-4551; Fax: (610) 270-5580; Email: David_B_Searls@sbphrd.com

1. Introduction

The notion of a "grand challenge" conjures up images ranging from crash engineering efforts involving mobilization on a national scale, along the lines of the moon landing, to the pursuit of more abstruse and even quixotic scientific goals, such as the proof of Fermat's conjecture. There are elements of both in the challenges now facing computational biology or, as it is often called, bioinformatics. This relatively new interdisciplinary science crucially underpins the movement towards "large scale biology", in particular genomics. As biology increasingly becomes an information-intensive discipline, the application of computational methods becomes not only indispensable to the management, understanding, and presentation of the data, but also interwoven in the fabric of the field as a whole.

One useful classification of the challenges facing computational biology is that which distinguishes the individual *technical* challenges, which to a large degree define the field, and what may be termed the *infrastructural* challenges faced by the field, qua scientific discipline. These latter "meta-challenges" are an inevitable result of the sudden ascendancy of bioinformatics. Unlike most other scientific fields, it has become an economic force before it has been thoroughly established as an academic discipline, with the exception of the protein structure aspects of the field. Computational biologists are in high demand in the pharmaceutical and biotechnology industries, as well as at major university laboratories, genome centers, and other institutional resources; yet, there were few established, tenured computational biologists in universities to begin with, and many of those have moved on to industry. This creates the danger of "lost generations" of would-be computational biologists who may have little opportunity for coordinated training in traditional settings and on traditional timetables.

A closely related infrastructural challenge centers on the interdisciplinary nature of bioinformatics. Many of those now in the field arrived by way of a convoluted path of career changes and retraining. It is now crucial to establish the training programs, and indeed the curricula, that will enable truly multidisciplinary education with appropriate attention to solid formal foundations. Even more importantly, the challenge facing the field as a whole is to establish itself with the *apparatus* of a scientific discipline: meetings, journals, and so on. To be sure, examples of these exist and have for some time, but only relatively recently have they begun to attain respectable levels of rigor and consistency. Largely in response to the current hiring frenzy in industry, government agencies have begun funding training programs more aggressively in recent years, though bioinformatics research funding has at best only kept pace with overall funding levels, where the trend is not encouraging. The "grand challenge" of establishing the field of bioinformatics in this sense should not be underestimated in importance.

2. Protein structure prediction

Turning to the technical challenges facing the field, it is easy to identify the most venerable of the "holy grails", that of ab initio protein structure prediction. Since the demonstration by Anfinsen that unfolded proteins can refold to their native three-dimensional conformation, or *tertiary structure*, strictly on the basis of their amino acid sequence, or *primary structure* [1], protein chemists have sought computational means of predicting tertiary structure from primary sequence [2]. Despite recent encouraging progress, however, a comprehensive solution to this problem remains elusive [3]. Without delving into the voluminous lore surrounding this field, it is perhaps instructive nonetheless to offer a superficial characterization of the nature of the challenge, which is essentially two-fold.

First, despite Anfinsen's result suggesting that the primary sequence plus thermodynamic principles should suffice to completely account for the folded state, the relevant aspects of those principles and the best way to apply them is still not certain. Moreover, it is also the case that there are exceptions to the thesis itself, due to eventualities such as covalent modification, interactions with lipid membranes, the necessity of cofactors, and the transient involvement of additional proteins to assist in folding [4]. This illustrates the first of several principles confounding solutions to many of the grand challenges in computational biology: that *there are no rules without exception in biology.*

Second, the search space of the problem is so daunting, because of the vast range of possible conformations of even relatively short polypeptides, that protein chemists have even been led to wonder how the molecule *itself* can possibly explore the space adequately on its progress to the folded state [5,6]. Whatever solution the efficient analog computer of the protein uses to prune this search space, the heuristic is not yet evident in computational approaches to the problem.

One obvious trick is to first solve the problem in some small, tractable subregion, and then to find a way to compose such intermediate results. In this regard, it is discouraging to find that, in the context of entire proteins, identical short stretches of amino acid sequence can exist in completely different conformations [7]. This illustrates a second confounding principle: *biological phenomena invariably have nonlocal components.* A recurring problem in sequence analysis in general is that, while strictly local effects may dominate any phenomenon (e.g. a consensus binding site), action at a distance always tends to throw a wrench into the works. In the case of protein folding, the reason is immediately evident, since folding tends to bring distant parts of the primary structure into apposition, and it is easy to imagine such interactions "overruling" any local tendency.

However, at times nature can also be benevolent to analysis, and it lends a helping hand in this instance by establishing some highly regular local themes that recur often enough to constitute useful structural cliches. These are the alpha-helices and beta-sheets that comprise *secondary structure*, and the bulk of amino acids in proteins can be identified as participating in one or the other of these characteristic arrangements.

The problem of predicting secondary structure features, which occur in relatively short runs of amino acids, is a minor variant of the protein structure prediction problem, which has been addressed only with decidedly asymptotic success. That is, the earliest empirical systems addressing this problem [8–10] gave accuracies of 56–60%, which proved not to be terribly useful in practice [11]. Since that time, by a combination of more sophisticated techniques and greater availability of homologous sequences, there has been a significant improvement, yet there is a sense that there is some sort of barrier at around 75% accuracy [12–14]. In all likelihood this is another manifestation of the nonlocality principle.

Still other simplified variants of the protein folding problem are yielding useful technologies, for example by reversing the problem in cases where the structure of a related protein is known [15–17]. Techniques of homology modeling and fold recognition ("threading") represent very active areas of research, and in fact have all been part of the CASP (Critical Assessment of techniques for protein Structure Prediction) contests that have been conducted for several years now to promote objective measurement and progress in the field [18]. That such an institution has been established as a focus for work in this area is itself indicative of the "grand challenge" nature of the overall endeavor.

3. Homology search

Another grand challenge facing computational biology, but one building rather more on success than frustration, is the detection of increasingly distant homologues, or proteins related by evolution from a common ancestor. The computational elucidation of such relationships stands as the fundamental operation and most pragmatic success of the field, because finding homologues is the shortest and surest path to determining the function of a newly discovered gene. Discovering closely related homologues, i.e. members of the same family of proteins or the corresponding genes in different related species, has been relatively straightforward since the development of efficient local alignment and similarity search algorithms [19,20]. However, the major challenge now resides in pushing further into the so-called "twilight zone" of much more distant similarities, where the signal begins to submerge in the noise. This challenge has stimulated both incremental improvement in the basic techniques of comparison, as well as the development of new methodologies, e.g. based on "profile" search, where essential features of a specific family of proteins are abstracted so as to afford more sensitive search [21]. Such improvements are delving into more and more sophisticated computer science, for example in the increasingly widespread use of hidden Markov model algorithms for profiles and other purposes [22,23].

Another challenge in this arena is that of algorithmic efficiency, a problem that can be formulated as a tradeoff (of a sort familiar to computer science): to a first approximation, the more thorough the search and/or comparison of sequences, the slower the algorithm. For example, the standard method for "full" local alignment of sequences, allowing for insertions and deletions to achieve the best fit, is intrinsically quadratic in the length of the sequences [19]; its efficient implementation in both time and space has inspired many clever speedups and parallel implementations over the years, but it is still not routinely used as a first choice. Rather, the famous BLAST algorithm, which achieves vastly more efficient search essentially by ignoring the possibility of gaps in alignments, is justifiably the single most important tool in the computational biologist's armamentarium [20]. Other algorithms, such as FASTA, strive for compromises that seek the best of both worlds [24]. These classes of algorithms are undergoing constant incremental enhancements, and at the same time the meaningfulness of the comparisons of protein sequences are being addressed, e.g. by improvements in the substitution matrices that serve to score putative alignments [25]. A related challenge in this arena centers on the statistical interpretation of sequence comparisons, a question that has been a concern for some time [26]; in this regard, the firm formal foundation that was an additional virtue of the BLAST algorithm is now being extended to a wider variety of analyses [27,28].

At the opposite extreme from the challenge of creating hyperefficient and statistically well-founded string search algorithms, is the challenge of injecting more and more elaborate domain models into alignment algorithms. The standard dynamic programming approach to determining minimal edit distance was early on adapted to a more "biological" model involving local alignment and so-called affine gaps (which allow the extension of an existing gap to cost much less than the initiation of a new gap) [29]. More recently, a spate of new dynamic programming algorithms have appeared that deal with messy biological facts such as the potential for frameshift errors in translating DNA to protein [30,31] and the spliced structure of genes in comparing genomic sequence [32,33]. It is evident that the possibility exists for even more elaborate domain modeling in alignment algorithms [34].

For increasingly distant relationships, sequence similarity signals must ultimately fade into the evolutionary noise in the face of even the cleverest algorithms, and for this reason it is widely held that similarity at the level of protein structure must take up the slack in detecting proteins of similar function. This leads us to a third confounding principle, that *problems in computational biology are perversely intertwined,* such that a complete solution to any one of them ultimately seems to require the solution to many if not all others. In this case, the principle is manifested by the growing realization that neither sequence, structure, nor indeed function can be studied in isolation. Thus, predicting how proteins fold, which is most tractable when a comparison can be made to a known structure based on alignment to its primary sequence, may be a necessary prelude to performing structural alignment when there is insufficient primary sequence similarity for conventional alignment. Making headway against this implicit conundrum is a challenging and active area of research [35,36].

4. Multiple alignment and phylogeny construction

The same confounding principle is at work in another pair of technical challenges (such complementary pairs of problems are known as *duals* in computer science), i.e. those of multiple alignment and phylogenetic tree reconstruction. The algorithms used for the latter problem depend on metrics of similarity derived by alignment of related sequences, but isolated pairwise alignments can be deceptive and it is far better first to align all related sequences together to determine the true evolutionary correspondences of sequence residues. Multiple alignment is hard enough that it requires approximate solutions for all but the smallest problems, and so has inspired a number of novel algorithms [37,38]. The most effective of these, however, take account of the degree of relatedness of the sequences being aligned, and therefore are most effective when an accurate phylogenetic tree is available – hence, a mutual dependency. Only recently have attempts been made to effectively combine these problems to develop algorithms that attempt to solve both at once, or in iterative fashion [39]. Both topics have not only produced many competing algorithms, but some controversy as well (one sure sign of a grand challenge), as between the camps favoring different principled approaches to such problems: maximum likelihood, parsimony, and distance-based methods [40].

Many of these challenges have benefitted from advances in theoretical computer science, and in fact biological problems seem to be a favorite source of inspiration for this community. In such cases there is always a concern that such theoretical studies be truly relevant to the domain, with the ultimate test being the development and widespread use of an application derived from, or pragmatically enabled by, a theoretical result or advance. To the extent that the development of BLAST was stimulated by advances in statistics, it would certainly satisfy the "widespread use" criterion. There are many other examples of highly promising theoretical work being done in topics such as phylogeny construction [41], protein folding [42], and physical mapping [43]. From the point of view of real-world bioinformatics, and quite apart from the intrinsic computational challenges in each area, the grand meta-challenge to the field is to reduce these fascinating advances to practice.

5. Genomic sequence analysis and gene-finding

The task of sequencing the human genome, and those of model organisms and microorganisms, is the canonical grand challenge in biology today, and it carries with it a set of computational challenges such as physical and genetic mapping algorithms, large-scale sequence assembly, direct support of sequencing data processing, database issues, and annotation. More than any other single factor, the sheer volume of data poses the most serious challenge – many problems that are ordinarily quite manageable become seemingly insurmountable when scaled up to these extents [44]. For this reason, it is evident that imaginative new applications of technologies designed for dealing with problems of scale will be required. For example, it may be imagined that data mining techniques will have to supplant manual search, intelligent database integration will be needed in place of hyperlink-browsing, scientific visualization will replace conventional interfaces to the data, and knowledge-based systems will have to supervise high-throughput annotation of the sequence data.

One traditional problem that constitutes a well-defined challenge, now of increasing importance as genomic data ramps up, is that of gene-finding and gene structure prediction from "raw" sequence data [45]. Great strides have been made from the days when open reading frames were evaluated on the basis of codon usage frequencies and perhaps a few signal sequences. A large number of statistical correlates of coding regions have been identified [46], and imaginative new frameworks for applying them [47–53].

Progress has also been made in the identification of particular signals related to gene expression [54]. As in the case of protein structure prediction, there have been efforts to comprehensively evaluate and compare the many competing methods [55].

This diverse topic serves to exemplify and thus summarize each of the confounding principles that have been elaborated. First, that *there are no rules without exception*: gene structure prediction depends on the uniformity of the basic gene architecture, but exceptions abound. In terms of signals, for example, there are exceptions to the so-called invariant dinucleotides found at intron boundaries [56]. Even reading frame is not inviolate, in cases such as translational recoding [57]. A well-known exception to the "vanilla" gene model is that of the immunoglobulin superfamily genes, whose architecture makes allowance for characteristic chromosomal rearrangements [58].

Second, that *every phenomenon has nonlocal components:* to wit, even a "syntactically correct" gene is not in fact a gene if it is never expressed. Whether a gene is ever transcribed, and details of the manner in which it is transcribed, are a function of its context, both on chromosomes and in tissues. Moreover, even as clean and regular a phenomenon as the genetic code may fall prey to this principle (as well as the first). It would appear to be strictly local in the sense that the ribosome seems to care only about the triplet at hand in selecting the amino acid to extend a nascent polypeptide chain. However, on occasion a certain stop codon can be translated as a selenocysteine residue, depending on flanking sequence and probably overall mRNA conformation [59,60].

Third, that *every problem is intertwined with others:* the detection of signals related to gene expression is greatly aided by the ability to identify coding regions, and vice versa. For example, if one knew the exact nature of the signals governing splicing, then introns and thus exons could be delineated, from first principles as it were. Conversely, if one could reliably and precisely determine coding sequences by their statistical properties, one would easily find the splicing signals at their boundaries. Neither indicators are sufficiently discriminatory on their own, though in this case the interdependence works in favor of the hybrid approach to gene structure prediction, that takes into account both statistical indicators and a framework model of gene structure to afford greater discrimination. Similarly, it has been shown that gene structures are much more effectively predicted when evidence from similarity search is also included [55], yet the identification of genes that might serve as homologues for future prediction is increasingly dependent on gene prediction algorithms, at least as a first step.

6. Conclusion

These confounding principles, together with the undeniable richness of the domain, contribute to the general air of a grand challenge surrounding computational biology. Undoubtedly these challenges will multiply as the field enters the so-called "post-genomic" era, where the problems of scale will be exacerbated by the need to efficiently bring to bear all manner of data on the problem of understanding genes and gene products in the global functional context of the cell and the organism. The intelligent integration of biological information to achieve true understanding will then constitute the grandest challenge of all.

References

[1] Anfinsen, C.B., Haber, E., Sela, M. and White Jr., F.H. (1961) Proc. Natl. Acad. Sci. USA 47, 1309–1314.

[2] Defay, T. and Cohen, F.E. (1995) Proteins 23(3), 431–445.

[3] Russell, R.B. and Sternberg, M.J.E. (1995) Curr. Biol. 5, 488–490.

[4] Gething, M.J. and Sambrook, J. (1992) Nature 355, 33–45.

[5] Creighton, T.E. (1990) Biochem. J. 270, 1–16.

[6] Creighton, T.E. (1992) Nature 356, 194–195.

[7] Kabsch, W. and Sander, C. (1984) Proc. Natl. Acad. Sci. USA 81, 1075–1078.

[8] Chou, P.Y. and Fasman, G.D. (1974) Biochemistry 13(2), 222–245.

[9] Lim, V.I. (1974) J. Mol. Biol. 88(4), 873–894.

[10] Garnier, J., Osguthorpe, D.J. and Robson, B. (1978) J. Mol. Biol. 120(1), 97–120.

[11] Kabsch, W. and Sander, C. (1983) FEBS Lett. 155(2), 179–182.

[12] Rost, B. and Sander, C. (1993) J. Mol. Biol. 232(2), 584–599.

[13] Salamov, A.A. and Solovyev, V.V. (1995) J. Mol. Biol. 247(1), 11–15.

[14] Frischman, D. and Argos, P. (1997) Proteins 27(3), 329–335.

[15] Taylor, W.R. (1989) Prog. Biophys. Mol. Biol. 54, 159–252.

[16] Barton, G.J. and Sternberg, M.J.E. (1990) J. Mol. Biol. 212, 389–402.

[17] Bowie, J.U., Luthy, R. and Eisenberg, D. (1991) Science 253, 164–170.

[18] Moult, J. (1996) Curr. Opin. Biotechnol. 7(4), 422–427.

[19] Smith, T.F. and Waterman, M.S. (1981) J. Mol. Biol. 147(1), 195–197.

[20] Altschul, S.F., Gish, W., Miller, W., Myers, E.W. and Lipman, D.J. (1990) J. Mol. Biol. 215(3), 403–410.

[21] Gribskov, M., McLachlan, A.D. and Eisenberg, D. (1987) Proc. Natl. Acad. Sci. USA 84, 4355–4358.

[22] Krogh, A., Brown, M., Mian, I.S., Sjolander, K. and Haussler, D. (1994) J. Mol. Biol. 235(5), 1501–1531.

[23] Eddy, S.R. (1996) Curr. Opin. Struct. Biol. 6(3), 361–365.

[24] Pearson, W.R. (1995) Protein Sci. 4(6), 1145–1160.

[25] Henikoff, S. (1996) Curr. Opin. Struct. Biol. 6(3), 353–360.

[26] Doolittle, R.F. (1981) Science 214, 149–159.

[27] Waterman, M.S. and Vingron, M. (1994) Proc. Natl. Acad. Sci. USA 91(11), 4625–4628.

[28] Altschul, S.F. and Gish, W. (1996) Methods Enzymol. 266, 460–480.

[29] Gotoh, O. (1982) J. Mol. Biol. 162(3), 705–708.

[30] Birney, E., Thompson, J.D. and Gibson, T.J. (1996) Nucleic Acids Res. 24, 2730–2739.

[31] Guan, X. and Uberbacher, E.C. (1996) Comput. Appl. Biosci. 12(1), 31–40.

[32] Hein, J. and Stovlbaek, J. (1994) J. Mol. Evol. 38(3), 310–316.

[33] Gelfand, M.S., Mironov, A.A. and Pevzner, P.A. (1996) Proc. Natl. Acad. Sci. USA 93, 9061–9066.

[34] Searls, D.B. (1996) Trends Genet. 12(1), 35–37.

[35] Gracy, J., Chiche, L. and Sallantin, J. (1993) Protein Eng. 6(8), 821–829.

[36] Orengo, C.A., Brown, N.F. and Taylor, W.R. (1992) Proteins 14(2), 139–167.

[37] Lipman, D.J., Altschul, S.F. and Kececioglu, J.D. (1989) Proc. Natl. Acad. Sci. USA 86(12), 4412–4415.

[38] Hirosawa, M., Totoki, Y., Hoshida, M. and Ishikawa, M. (1995) Comput. Appl. Biosci. 11(1), 13–18.

[39] Vingron, M. and Von Haeseler, A. (1997) J. Comput. Biol. 4(1), 23–34.

[40] Nei, M. (1996) Annu. Rev. Genet. 30, 371-403.

[41] Benham, C., Kannen, S., Paterson, M. and Warnow, T. (1995) J. Comput. Biol. 2(4), 515–525.

[42] Hart, W.E. and Istrail, S. (1997) J. Comput. Biol. 4(1), 1–22.

[43] Alizadeh, F., Karp, R.M., Weisser, D.K. and Zweig, G. (1995) J. Comput. Biol. 2(2), 159–184.

[44] Smith, R.F. (1996) Genome Res. 6(8), 653–660.

[45] Fickett, J.W. (1996) Trends Genet. 12(8), 316–320.

[46] Fickett, J.W. and Tung, C.S. (1992) Nucleic Acids Res. 20(24), 6441–6450.

[47] Dong, S. and Searls, D.B. (1994) Genomics 23, 540-551.

[48] Snyder, E.E. and Stormo, G.D. (1995) J. Mol. Biol. 248, 1–18.

[49] Salzberg, S., Chen, X., Henderson, J. and Fasman, K. (1996) ISMB 4, 201–210.

[50] Gelfand, M.S., Podolsky, L.I., Astakhova, T.V. and Roytberg, M.A. (1996) J. Comput. Biol. 3(2), 223–234.

[51] Uberbacher, E.C., Xu, Y. and Mural, R.J. (1996) Methods Enzymol. 266, 259–281.

[52] Zhang, M.Q. (1997) Proc. Natl. Acad. Sci. USA 94, 559–564.

[53] Burge, C. and Karlin, S. (1997) J. Mol. Biol. 268(1), 78–94.

[54] Bucher, P., Fickett, J.W. and Hartzigeorgiou, A. (1996) Comput. Appl. Biosci. 12(5), 361–362.

[55] Burset, M. and Guigo, R. (1996) Genomics 34(3), 353–367.

[56] Tarn, W.Y. and Steitz, J.A. (1997) Trends Biochem. Sci. 22(4), 132–137.

[57] Larsen, B., Peden, J., Matsufuji, T., Brady, K., Maldonado, R., Wills, N.M., Fayet, O., Atkins, J.F. and Gesteland, R.F. (1995) Biochem. Cell Biol. 73(11-12), 1123–1129.

[58] Hunkapiller, T. and Hood, L. (1989) Adv. Immunol. 44, 1–63.

[59] Engelberg-Kulka, H. and Schoulaker-Schwarz, R. (1988) Trends Biochem. Sci. 16, 419–421.

[60] Böck, A., Forchhammer, K., Heider, J. and Baron, C. (1991) Trends Biochem. Sci. 16, 463–467.

S.L. Salzberg, D.B. Searls, S. Kasif (Eds.), *Computational Methods in Molecular Biology*
© 1998 Elsevier Science B.V. All rights reserved

CHAPTER 2

A tutorial introduction to computation for biologists

Steven L. Salzberg

*The Institute for Genomic Research, 9712 Medical Center Drive, Rockville, MD 20850, USA
Phone: 301-315-2537; Fax: 301-838-0208; Email: salzberg@cs.jhu.edu or salzberg@tigr.org*

Department of Computer Science, Johns Hopkins University, Baltimore, MD 21218, USA

1. Who should read the tutorials?

This chapter and the one that follows provide two tutorials, one on computation and one on biological sequence analysis. Computational biology is an interdisciplinary field, and scientists engaged in this field inevitably have to learn many things about a discipline in which they did not receive formal training. These two chapters are intended to help bridge the gap and make it easier for more scientists to join the field, whether they come from the computational side or the biological side – or from another area entirely. The first tutorial chapter (this chapter) is intended to be a basic introduction to the computer science and machine learning concepts used in later chapters of this book. These concepts are used repeatedly in other chapters, and rather than introduce them many times, we have put together this gentle introduction. The presentation is designed for biologists unfamiliar with computer science, machine learning, or statistics, and can be skipped by those with more advanced training in these areas. Several of the later chapters have their own tutorial introductions, covering material specific to the techniques in those chapters, so readers with an interest in just one or two topics might go directly to the chapters covering those issues. If any of these chapters use unfamiliar terminology, then we suggest that you come back to this chapter for some review before proceeding.

The second tutorial chapter, immediately following this one, provides a brief introduction to sequence analysis and some of the underlying biology. Every chapter in this book touches on some of the topics covered in this tutorial. This second tutorial also tries to acquaint the reader with some of the many terms and concepts required to understand what biologists are talking about. (Biology is more than a science – it is also a foreign language.)

We have tried to keep the presentation in these tutorial chapters light and informal. For readers with a deeper interest in the computer science issues, the end of this chapter contains recommended introductory texts on artificial intelligence, algorithms, and computational molecular biology.

2. Basic computational concepts

Computational biology programs are designed to solve complex problems that may involve large quantities of data, or that may search for a solution in a very large space of possibilities. Because of this, we are naturally interested in creating programs that run as fast as possible. Ideally, the speed of our programs will be such that they are essentially "free" when compared with alternative laboratory methods of discovering the same information. Even the slowest computer program is usually a much faster way to get an answer than going to a wet lab and doing the necessary biological experiments. Of course, the computer's answer is frequently not as reliable as the lab results, but if we can get that answer rapidly, it can be extremely helpful.

In order to understand how much a computer program costs, we need a common language for talking about it. The language of computer science is not nearly as complex as that of biology, because there are not as many things to name, but it does require some getting used to. This chapter defines a few of the basic terms you need to navigate through a computer science discussion. We need to have a common language for talking about questions such as, how expensive is a computation? How does it compare to other computations? We want more than answers such as "this program takes 12 s on a Pentium Pro 200, while this other one takes 14 s," although this kind of answer is superficially very useful. Usually, though, the particulars of running time are determined more by the cleverness of the programmer than by the underlying algorithm. Therefore we need a way of discussing computation that is independent of any computer. Such abstract terms do not tell you exactly how long a system will take to run on your particular machine; instead, they give you a precise way of predicting the *relative* performance of one algorithm versus another. Armed with the proper tools for measuring computation, you can usually make such estimates accurately and confidently without worrying about what brand of computer will be used to run the algorithm.

2.1. What is an algorithm?

An *algorithm* is a precisely defined procedure for accomplishing a task. Algorithms are usually embodied as computer programs, but they can be any simple procedure, such as instructions on how to drive your car from one place to another, how to assemble a piece of furniture, or how to bake a chocolate cake. Algorithms are sometimes built directly into computer hardware, in order to make important programs run faster, but they are much more commonly implemented in software, as programs. It is important to understand the distinction between "algorithm" and "program": a program is the embodiment of an algorithm, and can take many forms. For example, the same algorithm could be implemented in C, Fortran, or Java, and it could be run on a Unix machine, a Macintosh, or a Windows PC. These programs might look completely different on the surface, but the underlying algorithm could still be the same. Computer scientists and computational biologists have many ways to describe their algorithms, and they use a great deal of shorthand in order to communicate their ideas in less space. For example, many algorithms require iteration (or loops), where an operation is repeated many times with only a slight change each time through the loop. One way to write this is to use a *for* loop such as this:

for $i = 1$ to N do

    Execute program P on item i

This is a small example of *pseudocode*, and it specifies that a sub-program called P is going to be run over and over again, N times in all. Each pass through the loop is called an *iteration*. The loop index, i, serves two functions: it counts how many times we have been through the loop, and it also specifies which object is going to be passed to the sub-program.

A *cycle* refers to the extremely small unit of time required to execute one machine instruction on a given CPU (central processing unit). Most computers today operate extremely fast, but they also have a very small number of low-level machine instructions. Therefore what looks like one instruction on a computer (one line of a program, for example) may require hundreds of CPU cycles. The clock speed of a computer is usually expressed in megahertz, or millions of cycles per second. For example, the fastest desktop PCs in 1997 run at just over 300 megahertz. In abstract discussions of algorithms, we rarely refer to these tiny units of time, but it is important to understand them when discussing applications.

The *memory* of a computer is a critical resource that all programs use. There are several kinds of memory, but the two most important for our purposes are *real* memory, known as RAM on most computers, and *virtual* memory. (RAM actually stands for random access memory, which is a historical artifact from the days when much memory was on tapes, which did not allow direct access to any item. With tapes, the tape itself had to be wound forward or backward to access each item stored there, so the amount of time to fetch a single item could vary enormously depending on where it was on the tape.) When a program is running, it uses *real* memory to do its computations and store whatever data it needs. If the RAM of the computer is smaller than the amount of memory needed by the program, then it has to "borrow" memory from the disk. This borrowed memory is what we call virtual memory, because it is not really there. So if your computer has 64 megabytes (MB) of RAM, you can still run programs that require more memory, because virtual memory allows the computer to borrow storage from the disk, which usually has much more memory available. However, this does slow everything down, because disk access is far slower than real memory access. Ideally, then, everything your program needs should fit in RAM. When a program has to go and fetch more memory and the RAM is already full, it exchanges part of the data stored in RAM with the virtual memory stored on the hard disk. This is called **swapping**, because the two pieces of memory are swapped between two locations. Swapping excessively can cause a program to slow down tremendously. Sometimes a program needs so much memory that it spends almost all its time swapping! (This is rather humorously known as thrashing.) If that happens, then the only good solution is to use a machine with more RAM.

2.2. How fast is a program?

When we measure how fast a program runs, we really only care about how much time passes on the clock while we are waiting for the program to finish. "Wall clock" time, as we call it, does not always tell us how fast the program is. For example, if you are using a shared computer, then other users might have most of the system's resources, and your program might only seem slow when in fact it is very fast. One way to time a program is to measure it on an unloaded machine; i.e. a machine where you are the only user.

14

This is usually quite reliable, but the time still depends on the processor speed and the amount of memory available.

A more general way to measure the speed of a computation is to figure out a reasonable unit for counting how many operations the computer has to do. We could use machine-level instructions as our primitive unit, but usually we do not know or care about that. In addition, the number of such instructions varies tremendously depending on the cleverness of the programmer. We could alternatively use some standard high-level operation, such as one retrieval from a database, but for some programs these operations will not correspond well to what the program has to do. So instead of providing a general answer to this question, we define operations differently depending on the algorithm. We try to use the most natural definition for the problem at hand. This may seem at first to be ad hoc, but in the history of computer science it has worked very well and very consistently. For example, consider programs that compare protein sequences, such as BLAST and FASTA. The most important operation for these programs involves comparing one amino acid to another, which might require fetching two memory locations and using them as indices into a PAM matrix. This operation will be repeated many times during the running of the program, and it is natural to use it as a unit of time. Then, by considering how many of these units are required by one program versus another, we can compare the running time of the two programs independently of any computer.

2.3. Computing time is a function of input size

The application-dependent notion of "operation" gives us an abstract way of measuring computation time. Next we need a way of measuring how run time changes when the program has different inputs. Here we can come up with a very standard measurement, based simply on the number of bytes occupied by the input. (One byte, which equals 8 bits, is sufficient to represent one character of input. "Bit" is shorthand for binary digit, either 0 or 1.) For example, with a sequence comparison program, the input will be two protein or DNA sequences, and the output will be an alignment plus a score for that alignment. We expect the program to take longer to compare long sequences than it will take for short ones. To express this more formally, we use the variable N to measure the size of the input. For our example, we might set N to be the sum of the lengths of the input sequences. Now we can really begin describe how long the program takes to run. A program might require N units of time, or $3N$, or N^2, or something else. Each of these time units corresponds to the application dependent notion of "operation" described above. This measurement is machine-independent, and it truly captures how much work the program needs to do. If two programs do the same thing and one requires less time using our abstract measurement scheme, then the one requiring less time is said to be more *efficient*.

Some of the later chapters in this book use notation such as O(N) to describe the running time of their algorithm. In most cases, it is okay to take this simply to mean that the running time is proportional to N. More precisely, this notation specifies the *worst case* running time, which means that the amount of time will never be more than kN for some constant k. It might be that the average time requirement is quite a bit less than the worst case, though, so this does not tell the whole story. For example, the Smith–

Waterman algorithm for local alignment of two sequences runs in $O(N^2)$ time; i.e. its time is proportional to the square of the length of the input sequences. (Actually, for Smith–Waterman we define N to be the length of the *longer* of the two input sequences.) The BLAST and FASTA algorithms, which also produce alignments, also require $O(N^2)$ time in the worst case – but in the vast majority of cases, BLAST and FASTA run much faster. This speed is what accounts (in part) for their popularity and widespread use.

2.4. Space requirements also vary with input size

The space requirements of a program are measured in much the same way as time is measured: we use the input size N to calibrate how much space is required. This is because most programs' use of memory is a function of the inputs, just as their running time is. For example, the Smith–Waterman sequence comparison algorithm builds a matrix using the two input sequences. The matrix is of size $N \times M$, where N and M are the sizes of the sequences. Each entry in the matrix contains a number plus a pointer, which requires a few bytes of storage. The space requirement of this algorithm is therefore $O(NM)$. In fact, with clever programming, you can implement Smith–Waterman using only $O(N)$ space (Waterman, 1995), at a cost of roughly doubling the running time. So if space is at a premium and time is not, then you might choose to implement the algorithm in a way that uses much less space but runs more slowly. This type of trade-off between space and time occurs frequently in algorithm design.

2.5. Really expensive computations

There is one class of problems that seem to take forever to solve, computationally speaking. These are the so-called *NP-Complete* problems. Computer scientists have been working for years to try to solve these problems, but the best algorithms they have devised take exponential time. This means that the time requirement is something on the order of 2^N (or even worse!), where the N appears in the exponent. These problems are impossibly difficult unless N is quite small. It is tempting to think that, since computers are getting so much faster every year, we will eventually be able to solve even these really hard problems with faster computers – but this just is not so. To illustrate, consider the most famous NP-Complete problem, the Traveling Salesman Problem. (This problem is so well known that computer scientists refer to it simply as "TSP.") The problem can be stated pretty simply: a traveling salesman has to visit N cities during a trip, starting and returning at the same city. He would like to find the shortest route that takes him to every city exactly once. The input for the problem here is the list of N cities plus all the distances between them. Now, suppose we have an algorithm that can find the shortest path in 2^N time (and in fact, such an algorithm has been described). If there are 20 cities, then we need roughly one million operations, which will probably take just a few seconds on a very fast computer. But 30 cities require a billion operations, which might take a few hours. When we get up to 60 cities (which is still not a very big problem), the number of operations is 2^{60}, a truly astronomical number. Even with computers a million times faster than today's fastest machine, it would take years to get a solution to our problem. The point is that computer speeds are not increasing exponentially, despite what the popular press would like us to believe. Therefore we can never hope to run exponential-time algorithms on our computers except for very small inputs.

Luckily, we can frequently design algorithms that give us good *approximate* solutions to NP-Complete problems, even though we cannot find exact solutions. These approximation algorithms may run very fast, and sometimes we can even run them over and over to get multiple different solutions. The solution we get may not be optimal, but often it is plenty good enough, and this provides us a way around the computational difficulties of NP-Complete problems.

It turns out that some of the most important problems in computational biology fall into the class of NP-Complete problems. For example, the protein threading problem that is described in the chapters by Jones and by Lathrop et al. is NP-Complete. As you will read in their chapters, both approximate and exact solutions can be found for the threading problem.

3. Machine learning concepts

Machine learning is an area of computer science that studies how to build computer programs that automatically improve their performance without programming. Of course, someone has to write the learning program, but after that it proceeds automatically, usually by analyzing data and using feedback on how well it is performing. Machine learning technology is turning out to be one of the most useful ways to solve some of the important problems in computational biology, as many of the chapters in this book illustrate. This section and the section that follows give some brief background material to prepare the reader for those later chapters.

3.1. Learning from data

One of the fundamental ideas in machine learning is that data can be the foundation for learning. This is especially true of biological sequence data, where in many cases the important knowledge can be summarized as a pattern in the sequence. An area of research known as "data mining," which overlaps to a large extent with machine learning, has recently arisen in response to the growing repositories of data of all types, especially on the Web. Data mining emphasizes the discovery of new knowledge by finding patterns in data. To cite just one example of how pattern discovery is important in biological sequence analysis, consider DNA promoter sequences. Promoter sequences are intensively studied in order to understand transcription and transcriptional regulation. Since most promoters are still unknown, a program might be able to discover new ones by examining the existing sequence databases and noticing common patterns in the sequences upstream from known genes. (Of course there are many people attempting to do this already, but so far it has been a difficult problem.)

Learning from data usually requires some kind of feedback as well. In the chapters on gene finding in this volume, we will see that learning programs can be trained to find genes in anonymous DNA. In order to accomplish this, they need to have a set of data for training in which the gene locations are already known. The knowledge of where the actual genes are is given to the programs as feedback so that they can adjust their pattern-finding behavior.

3.2. Memory-based reasoning

A long-established notion in the machine learning and pattern recognition communities is that a large memory of stored examples can serve as the basis for intelligent inference. This line of research takes advantage of the fact that memorization, while difficult for humans, is effortless for computers. By making the computer remember every example it encounters, we can give it the ability to respond more intelligently to new examples. The hard part is figuring out a clever way to retrieve the old examples, so that only the most relevant experiences are used for processing a new example. Thus most of the intelligence in memory-based algorithms is embodied in making these comparisons. Learning techniques that use a "knowledge base" of experiences go by a variety of names, including nearest neighbors, case-based reasoning (as discussed by Overton and Haas in chapter 5 of the present volume), and instance-based learning.

The nearest-neighbor algorithm is probably the oldest version of memory-based learning, dating back to the 1950s. The idea is simple: compute a distance from a new example to every example stored in memory, and use the nearest one to perform *classification*. A natural extension to this is to choose the k nearest examples, and combine them using a voting scheme or an interpolation method.

Classification is probably the most commonly performed task that a machine learning system does. This may seem somewhat limiting, but it is surprising how many different questions can be posed as classification problems. For example, the problem of recognizing a picture, which is usually regarded as a complex visual processing question, can also be considered as a problem of retrieving the most similar picture from a memory base. Since the most similar picture will usually be a picture of the same object, this visual recognition task is really just an image classification problem. In computational biology, once we have large enough databases, we should be able to use memory-based methods effectively for protein structure prediction and gene finding. Protein threading algorithms, which are described by Lathrop et al. (ch. 12) and Jones (ch. 13) in the present volume, are a very complicated form of memory-based reasoning: they use a database of previously seen structures as the basis for predicting the structure of a new protein sequence.

4. Where to store learned knowledge

Simply storing everything is not always the most effective strategy. Researchers in machine learning have designed many different data structures to capture the knowledge that is implicitly contained in data. In order to create these data structures, the experiences (the data) must be processed in advance to extract whatever generalizations they contain. This knowledge is then stored in a compact data structure, typically something much smaller than the data itself, in a form that can be used easily in the future. An introduction to all the different forms of learned information would take a book in itself, so here we

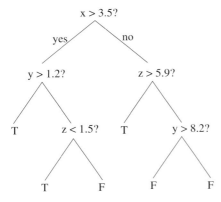

Fig. 1. A sample decision tree. Each internal node tests a variable or a set of variables, and if the test is satisfied, the program proceeds down the left branch. This procedure stops when it reaches the bottom of the tree. The bottom or "leaf" nodes contain class labels, shown here as either true (T) or false (F). In general, a tree may contain any number of classes.

will give just an introduction to two of the more popular data structures, both of which are used (and described in more detail) in later chapters.

4.1. Decision trees

A decision tree contains a set of *nodes* and *edges*, arranged in the form of a tree, as shown in Fig. 1. The nodes are either internal or leaf nodes, and the edges connect the nodes together. Decision trees are used most frequently for classification problems, which abound in scientific and medical domains. For example, diagnosis is a classification problem where the question is whether a patient should be classified as having a disease or not. Sequence analysis problems can easily be cast as classification problems: for example, is a DNA sequence part of a coding region or not? Is a protein sequence part of an alpha-helix, a beta-sheet, or some other structure? All these are classification questions.

A decision tree contains class labels at the *leaf* nodes, which are the nodes at the bottom of the tree in Fig. 1. To get to these nodes, an example (which can be a patient being diagnosed, a DNA sequence, or almost any other kind of object) must pass through a series of test nodes, beginning at the *root* node at the top of the tree. Each test node contains a question, or test, that is asked about the example. If the answer to the question is yes, then the example continues down to the left child of that node. Otherwise the example gets passed down to the right. In order to do this kind of processing on an example, it needs to be converted into a set of *features*, which are just attributes of the example that we find relevant to the classification task. In a medical diagnosis task, the features would be various measurements and tests that you could run on the patient. For protein sequence analysis, features might be statistical measures such as the frequency of various types of amino acids, or the presence of a series of hydrophobic residues, or perhaps the similarity to another protein sequence.

One advantage of decision trees is that they can distill a large quantity of data into a small, compact data structure that contains the essential information we need for classification. The machine learning community has developed several systems for automatically constructing decision trees from data. As we will see in chapter 10, these software systems can be used to build trees that do an excellent job at distinguishing

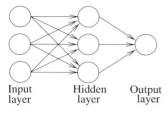

Input layer   Hidden layer   Output layer

Fig. 2. A simple 3-layer artificial neural network.

coding from non-coding DNA. The appendix on Internet resources describes how to get source codes for building decision trees over the Internet.

4.2. Neural networks

Neural networks are another important classification technique that are the subject of a large and active research area. (A more accurate name for these devices is "artificial neural networks", since their resemblance to actual neurons is quite superficial. We will use the shorter and more common name for convenience.) Like decision trees, neural nets are often used for classification, and it is in this context that we see them used most often in computational biology. The well-known gene-finding systems GRAIL, GeneParser, and Genie all use neural nets to do at least part of their processing.

Neural networks can help make decisions about examples in much the same way that decision trees can. Just as with decision trees, the examples need to be converted first to a set of feature values. These values are fed into the *input layer* of the network, as shown in Fig. 2. The most popular type of neural network is a 3-layer network, as shown in the figure, where the first layer holds the input features, and the third layer contains the output. The middle layer is called the *hidden* layer, because it is "hidden" from both the input and the output. This layer contains nodes that combine the input features in various ways to compute more complex functions. Most of the time these are linear functions, but other methods of combining the inputs have also been explored. The output of these nodes is fed to the third (output) layer, which combines them to come up with its decision. The output is usually a numeric value that gets converted into a classification. Sometimes this numeric value itself is used to indicate the confidence in the decision. For instance, a network might have an output value that ranges from 0 to 1, and its decision might be "yes" for all values above 0.5 and "no" otherwise. Alternatively, the value itself could be output, so the decision would appear to be more like a confidence value in the yes/no decision.

A detailed example of how neural networks can be used for sequence analysis is covered in chapter 7, which describes the GRAIL gene-finding system.

5. Search

Search is one of the big topics in computer science. Work on search has appeared in such diverse areas as theoretical computer science, artificial intelligence, and optimization. When computer scientists talk about search, they usually mean algorithms for searching in an abstract space. This space could be something like a database, or it could be a set of potential solutions, or it could be a data structure like a graph or a tree. Some of the systems described in this book are essentially just clever ways of searching a very large space to find the solution to a difficult problem.

5.1. Defining a search space

Before you can start searching, you have to define the space that your search algorithm will explore. Searching in a computer is analogous to a trial-and-error process such as you might perform in any scientific laboratory: you try many different combinations of ingredients, and eventually you find something that works well. You may even find something that works optimally, but you do not always need that. To keep the analogy simple, we will use the example of searching for a good recipe for a delicious cake.

When we search for a solution by computer, we usually want to do things systematically. Imagine that when trying to bake a cake, we start with the right ingredients (a helpful start!), but we do not know the proportions. If there are five ingredients, we might try mixing them first in equal proportions. This creates a big mess, so where do we go next? We could try mixing the ingredients in a series of different ratios, such as $1:2:3:4:5$. Obviously there are an awful lot of possibilities here, so this will take a long time. If we really had to do this for a physical process such as cooking, it would take much too long. However, suppose we can simply ask an oracle to give each ratio a score. Now we have something much more like computational search: we construct a solution (a set of ratios) and then evaluate it. If the answer comes back good enough, then we are done.

5.2. Search space size

Even here, though, the number of possibilities is huge – infinite, in fact. So we might add some constraints to make things easier. For example, since we know that cooking is not so precise, we might assume that the ratios of our ingredients can be expressed as integers. This certainly helps, but we still have infinitely many possibilities. So we might go further, using our knowledge of cooking, and assume that the ratio of the biggest to the smallest ingredient is no larger than 50. Now we have a finite space, because each of the five terms in our ratio is an integer from 1 to 50.

How big is this search space? Since we could in principle consider every ratio that satisfied our constraints, we have 50 possible values for each ingredient. With 5 ingredients, we have $50^5 = 3.125 \times 10^8$ possible combinations. Although this is a large number, we can run through all these combinations pretty quickly on a reasonably fast desktop computer. The limiting factor will be our oracle: if the oracle can score each combination in one millisecond, then we can check all combinations in about 86 hours of CPU time. But if the oracle takes one second instead of one millisecond, then we will have to wait many years to get an answer, obviously much too long.

5.3. Tree-based search

So far in our example we have only considered what is known as *exhaustive* search, where we consider all possibilities. Very often we can do better than that, and when we have a large space it might be essential that we figure out a way to avoid searching all of it. One way to get all the benefits of exhaustive search without incurring all the costs is to organize the search space in the form of a *tree*. This type of tree is not the same as a decision tree, but structurally it looks very similar. The difference is in how we use it.

For our cooking example, we could build a search tree as follows. The root of the tree will represent ingredient 1, and it will have 50 "branches" going down from it, one branch for each possible amount of that ingredient in our cake. The second level of the tree will represent ingredient 2, and so on. This tree will have 5 levels for the ingredients, and on the 6th level we will have the scores representing how good each combination is. Searching the space corresponds to searching this tree for the highest score.

Now let us create a new oracle that gives us some extra information about our recipe. This oracle will give us an *estimate* of how good our cake will taste as we make each decision, even before we have decided on all five ingredients. For example, if we decide that the ratio of ingredients 1 and 2 is 10 : 1, then the oracle might come back by saying "this will taste terrible!" In other words, we might know that certain proportions just can not work out. The question is, can we use this information? Well, it turns out that such information can be enormously useful. As Smith and Lathrop show in their chapter on protein folding, this type of oracle can be used in an algorithm called "branch and bound" that is tremendously fast compared to exhaustive search. Let us go a bit further to show how this might happen.

Once we have an estimate that the ratio 10 : 1 is terrible (let us give it a score of −100), we can temporarily stop searching down that branch of the tree. We could instead jump over to the next value of ingredient 2, and begin exploring there. Suppose our oracle tells us that 10 : 2 is still not good, but it is better than 10 : 1 (maybe it gets a score of −20). We can continue in this manner, considering all values of ingredient 2 and getting an estimate of how good they will be. We will eventually consider all combinations of ingredients 1 and 2 (2500 possibilities), getting estimates for all of them. Because we are scanning the entire breadth of the tree before going deeper, we call this "breadth-first" search.

Once we have got our estimates at level 2, we can proceed to level 3. But to save time, we can use our estimates to eliminate branches. Obviously, we should continue at level 3 by considering the highest scoring path we have found so far. Why start down a branch that is no better than −100 when we have other branches that might be much better? Remember that our estimates are all "upper bounds" that tell us how good the best solution will be; some paths below the 10 : 1 ratio for ingredients 1 and 2 might score worse than −100, but nothing will score better.

So the process will now work as follows: pick the node in the tree with the highest estimate, and "expand" that node by considering nodes one level below. Suppose our best node at level 2 was the ratio 30 : 7. We can expand this by considering all 50 values for ingredient 3 that go with 30 : 7. For each of these, we ask our oracle how good the resulting cake might taste, and we get back an upper bound estimate. The best of all these answers becomes our new estimate for the best score for 30 : 7. The idea is that as our solution becomes more complete, our oracle can give us more accurate estimates. For example, if $30:7$ had a score of $+100$ originally, we might find after expanding it that the best we can do is $+50$. Now we need to decide where to search next. If one of the other nodes on level 2 has a higher score, we should expand that one next. If not, then we can keep expanding node $30:7$, going down another level. As we search, we keep refining our estimate of the final solution quality.

Once we get to the bottom of the tree, we can calculate exact answers. This allows us to permanently eliminate some branches from the tree, saving huge amounts of computation. Suppose that we find a particular combination down the $30:7$ branch that gives a pretty tasty cake, maybe with a score of $+20$. Remember our estimate for the $10:1$ ratio of ingredients 1 and 2? The oracle told us that no matter what else we did, that cake would score no better than -100. But we know now that we can create a much tastier recipe, so there is no need to search the $10:1$ branch. Since there were three levels below this branch with 50 combinations each, this means we just eliminated $125\,000$ possibilities. As long as our oracle is accurate, we can be assured of still finding the optimal solution, but we can sometimes do it much faster.

6. Dynamic programming

"Dynamic programming" is a phrase that appears again and again in computational biology. This is an old name for a technique that appeared in the 1950s, before computer programming was an everyday term, so do not be fooled by the word "programming" here. Dynamic programming (DP) is an optimization technique: most commonly, it involves finding the optimal solution to a search problem. Many different algorithms have been called (accurately) dynamic programming algorithms, and quite a few important ideas in computational biology fall under this rubric. All these methods have a few basic principles in common, which we will introduce here. But the first thing to do, if you are not already familiar with dynamic programming (DP), is to ignore the name – it is misleading, because it has nothing to do with programming as we know it today.

The fundamental principle that all DP algorithms share is that they first produce optimal solutions to small pieces of the problem, and then glue them together to get a complete, and still optimal, solution. It is not always possible to break a problem up this way; sometimes gluing solutions together is too costly, and sometimes finding optimal solutions to sub-problems is just as hard as solving the whole problem. When a problem does have this property, though, it is said to satisfy the "principle of optimality."

An example will illustrate what we mean by this principle. Suppose we want to find the shortest path between two cities, A and B. We can represent shortest path problems as a graph, where each city is a node and edges between nodes are marked with the distances between cities. For all pairs of cities that are directly connected, we can keep a lookup table that contains their distances. We could also use cost instead of distance, in which case we had be finding the cheapest way to get from one city to another.

Now we use the principle of optimality as follows. The idea is to break up the problem into smaller pieces, solve the pieces, and then glue them together. With our path problem, we can break the problem into pieces by first finding the shortest path from A to X, then finding the shortest path from X to B. The length of the path A–X–B is simply the sum of the costs of the two sub-paths. Thus the "gluing" operation is very simple to perform.

But what if the shortest path from A to B does not pass through X at all? This is certainly possible, and if it happens then our algorithm will fail to give us the correct answer. We can solve this by running our algorithm for *all* cities in the graph. In other words, we will compute the shortest path from A to B that passes through every possible city. The shortest of these has to be the shortest overall.

Now we have two sub-problems to solve: find the shortest path from A to X, and find the shortest path from X to B. To solve the sub-problem, we simply call our dynamic programming algorithm itself! This recursive (self-invoking) call is also characteristic of dynamic programming: DP is almost always expressed recursively. So to find the shortest path from A to X, we will find the shortest path from A to Y, and from Y to X, and glue those together. We again repeat this for all values of Y, this time excluding B. (Note that by excluding B, we make the problem a little smaller.) As we continue making recursive calls, we eventually get to cities that are directly connected to each other. At this point we just look up the answer in our table.

For an example from computational biology, consider the problem of finding genes in DNA. Several later chapters describe dynamic programming approaches to this problem. For example, the hidden Markov models described in the chapter by Krogh use a special DP algorithm called Viterbi. In gene finding, the task is to produce the "optimal" interpretation of a DNA sequence by parsing it into a series of sub-sequences. Each sub-sequence is labeled according to its role in the genome; for example, intergenic DNA, exon, intron, and other labels. Each algorithm has its own way of scoring these sub-sequences, and the role of the dynamic program is to guarantee that, whatever the scoring method, the algorithm will find the optimal score. Of course, if the method does not produce accurate scores, then the overall output might be poor.

The dynamic programming version of gene finding says that to score or "parse" a sequence $S(1, n) = s_1, s_2, ..., s_n$, one can first find the optimal score for a *prefix* sequence $S(1, i) = s_1, s_2, ..., s_i$. (The notation $S(i, j)$ indicates the sub-sequence of S beginning with the base at position i and ending with j.) If this prefix is scored optimally, then you can "glue" it together with an optimal parse for the rest of the sequence $s_{i+1}, s_{i+2}, ..., s_n$. With this formulation, the algorithm can proceed in a left-to-right manner through any sequence, adding one base at a time. With each new base i, the algorithm computes the optimal score for the sequence up to i. Under the dynamic programming assumption, the optimal score for $S(1, i)$ must be the sum of the scores for $S(1, k)$ and $S(k + 1, i)$ for some k between 1 and i. Since we have already computed the optimal scores for all such k, we can find the optimal score for i itself by computing all the intermediate sums and choosing the maximum.

If this still seems a trifle obscure, then now is the time to skip ahead to one of the gene-finding chapters that uses dynamic programming, such as Xu and Uberbacher's description of GRAIL or my own description of MORGAN. These chapters put the dynamic program in the context of a larger problem where all the biological background and assumptions are clear.

7. Basic statistics and Markov chains

One of the most successful recent innovations in the world of computational biology is the use of hidden Markov models (HMMs) to solve a surprisingly wide range of different sequence analysis problems. Another, related technique is the Markov chain, which has been used successfully as the basis for bacterial gene finders and for modeling mRNA splicing, among other things. The success of these probabilistic methods is perhaps not surprising, since the biological world is highly stochastic in its behavior.

Krogh's chapter provides a thorough tutorial on HMMs, so this section is only intended to give a very basic background on the statistical underpinnings of Markov chains, HMMs, and other probabilistic methods. For a more thorough introduction to statistics, the reader should look up an introductory statistics text such as the one listed at chapter's end. For more details on HMMs, see Krogh's chapter and the references listed there.

To model a biological process statistically, you first must define the space of possible *events* in the world that you wish to model. For DNA sequences, we might want to model the next base that will occur in a sequence. In this case there are only four events, which we conveniently label A, C, G, and T. If we believe that the next base is going to be guanine (G) 25% of the time, then we can write that as $P(G) = 0.25$; i.e. the probability of G is 0.25. If we just count all the G's in a sequence and use that as the probability of guanine, then we call this the *prior* probability.

Frequently we can use conditional probabilities to get more accurate models. A conditional probability is the probability that one event will occur *given* that we already know that some other events have occurred. For example, if the previous two bases in a sequence are AT, then we might find that 30% of the time the next base is guanine. This can be true even if $P(G) = 0.25$, so we have to write it differently, as $P(G|AT) = 0.30$. If we are looking at a database comprising nothing but small regions around translational start sites (which almost always begin with ATG), then we might find that $P(G|AT)$ is much higher. The important point is that conditional probabilities capture the dependencies between bases at different locations in a sequence, and therefore they make for a better model of the data.

More generally, we might want to use probabilities to identify particular sequence types. *Markov chains* are an excellent statistical tool for modeling a sequence of any type, because a Markov chain is simply a sequence of events that occur one after another (in a chain, of course). The main restriction on a Markov chain is that the probability assigned to an event at any location in the chain can depend on only a fixed number of previous events. So, for example, the identity of a DNA base might depend on 2 previous bases, or 3, 4, or more, as long as you set a fixed limit.

Let us look at Markov chains as a method for scoring sequences, and in the process review Bayes' Law from statistics. Suppose we want to identify start codons, which in eukaryotes are almost always the trinucleotide ATG. Suppose that $P(A) = P(C) = P(G) = P(T) = 0.25$ in our data, so that all the priors are equal. We can model the start codon as a Markov chain with three states, as shown in Fig. 3. Each state is labeled with the base that it is mostly likely to "output," but this is just for convenience. More generally, we store a list of all possible outputs in each state, and assign a probability to each. In

a 0.91	a 0.03	a 0.03
c 0.03	c 0.03	c 0.03
g 0.03	g 0.03	g 0.91
t 0.03	t 0.91	t 0.03

(a) → (t) → (g)   Fig. 3. A simple Markov chain for a start codon.

our model, the first state outputs an "A" with probability 0.91 and outputs the other bases with probability 0.03.

To score a codon with this model, we have to compute the probability that the model will output the codon. In this trivial example, the model outputs ATG with probability $0.91^3 = 0.754$. In this Markov chain, each "state" outputs something without regard to what occurred previously, so we call this a 0th order Markov chain. In a first order chain, the outputs would depend on the one state immediately preceding each state, so for DNA sequences we would store 16 probabilities in each state.

Obviously, ATG is the most likely start codon in our model, but we might want to use it to score other codons as well. Here is where we must be careful to pose our questions in the proper manner, statistically speaking. If we are given a new sequence such as CTG, we could just run it through our model and get a probability, in this case $0.03 \times 0.91 \times 0.91 = 0.025$. This probability is the probability of CTG *given* our model. But our question is the reverse of this; i.e. what is the probability that we are looking at a start codon, given that the sequence is CTG? If we use M to represent our Markov chain, then we want to know $P(M|CTG)$. The number computed above is $P(CTG|M)$, the probability of CTG given that we are using our Markov chain. To answer our question about the probability that CTG is a start codon, we need to use Bayes' Law:

$$P(A|B) = \frac{P(B|A)\,P(A)}{P(B)}.$$

To use our Markov chain here, we compute

$$P(M|CTG) = \frac{P(CTG|M)\,P(M)}{P(CTG)}.$$

Since we said all bases are equally likely, we can compute $P(CTG) = 0.25^3 = 0.0156$. (This is actually another trivial Markov chain computation, in which we assume we can generate any random sequence by simply multiplying together the prior probabilities of each base in the sequence.) This gives us an answer of $16 \times P(M)$, but we do not know $P(M)$. What is the probability of our Markov model itself? Usually there is no good way of computing this, which may make it seem that we are stuck. Fortunately, though, in almost all uses of Markov chains and HMMs for biological sequence analysis (including this example), the probability of the model is ignored. This is because the typical use requires only that we pick the best pattern according to a model, so $P(M)$ is a constant for the purposes of comparison. To illustrate, suppose that we apply our Markov chain to a sequence of 10 000 bases in order to find the start codon. For every codon in the sequence, the same value for $P(M)$ appears in the computation, so we can ignore that when we try to choose the best codon.

A more realistic example of using Markov chains would be to score mRNA splice sites, where the identities of the bases in each position are crucially important to splicing but by no means easy to determine, and where they also have local dependencies resulting from their influence on the bond strength between the mRNA and the splicing machinery. Another example from current practice in computational biology is the Glimmer system for bacterial gene finding (http://www.cs.jhu.edu/labs/compbio/glimmer.html), which uses a special type of Markov chain called an IMM. This system finds bacterial genes by (in part) scoring all six possible reading frames in order to determine which one (if any) contains a gene. The system simply chooses the reading frame with the highest score to decide which one contains a gene. In both these examples, the prior probability of the model does not affect the outcome, and therefore it can be safely ignored.

8. Conclusion

This tutorial has been necessarily brief and rather biased. We attempted to cover primarily the topics that you might encounter elsewhere in this book, and of course many important topics in computer science were entirely omitted. Even within computational biology, there are additional topics worthy of coverage, but to try to be comprehensive would extend this chapter itself into a book. Instead, we recommend the following texts as one place to read further. Meanwhile, we hope the preceding discussion is enough to prepare most readers for the more technical chapters that follow.

9. Recommended introductory texts

We hope that the preceding gentle introduction will provide sufficient background to allow readers to get enough background to proceed through the chapters that follow. This chapter is clearly much too brief to give more than a superficial treatment to any of the topics, so for those with an interest in learning more, we recommend the texts below. These texts might also provide a good reference for further reading on some of the technical topics in the chapters ahead.

(1) D. Casey. Primer on Molecular Genetics. Available online at http://www.ornl.gov/hgmis/publicat/primer/intro.html. An excellent introduction to the basis of molecular biology, sequencing, and mapping.
(2) T. Cormen, T. Leiserson and R. Rivest (1995). *Introduction to Algorithms*. MIT Press, Cambridge, MA. A detailed presentation of a wide range of algorithms and data structures, including string-matching techniques (useful in sequence analysis), dynamic programming, and trees.
(3) J. Setubal and J. Meidanis (1997). *Introduction to Computational Molecular Biology*. PWS Publishing Company, Boston. A general textbook on algorithms for sequence analysis from a computer science perspective.
(4) S. Russell and P. Norvig (1995). *Artificial Intelligence*. Prentice-Hall, Upper Saddle River, NJ. A thorough survey of artificial intelligence, including search algorithms, probabilistic reasoning, and machine learning.

(5) M. Waterman (1995). *Introduction to Computational Biology: Maps, Sequences, and Genomes.* Chapman & Hall, London. A general textbook on computational biology, with an emphasis on statistical approaches.

S.L. Salzberg, D.B. Searls, S. Kasif (Eds.), *Computational Methods in Molecular Biology*
© 1998 Elsevier Science B.V. All rights reserved

CHAPTER 3

An introduction to
biological sequence analysis

Kenneth H. Fasman

*Whitehead Institute/MIT Center for Genome Research, 320 Charles Street, Cambridge, MA 02141, USA
and Astra Bioinformatics Center, 128 Sidney Street, Cambridge, MA 02139, USA*

Steven L. Salzberg

*Department of Computer Science, Johns Hopkins University, 3400 N. Charles St.,
Baltimore, MD 21218, USA and The Institute for Genomic Research,
9712 Medical Center Drive, Rockville, MD 20850, USA*

1. Introduction

Biological sequence analysis long predates the Human Genome Project (HGP). However, it has recently come into its own with the exponential growth in sequence information generated by the HGP. The idea of undertaking a coordinated international study of the entire human genetic complement arose from a series of meetings held between 1985 and 1987. The goal of the genome project, an informal worldwide scientific collaboration, is the fundamental characterization of the human genetic blueprint. Specifically, the HGP effort has been divided into two main parts: mapping, in which biologically important regions of the genome (especially genes) are localized on the chromosomes relative to a framework of landmarks; and sequencing, in which the genetic information encoded by the order of the DNA's chemical subunits is determined.

The ultimate goal of genomic mapping and sequencing is to associate specific human traits and diseases with one or more genes at precise locations on the chromosomes. The successful completion of the genome project will provide an unparalleled understanding of the fundamental organization of human genes and chromosomes. Consequently, it promises to revolutionize both therapeutic and preventive medicine by providing insights into the basic biochemical processes that underlie human disease.

This chapter provides a general introduction to biological sequence analysis. In order to keep the discussion brief, many topics have been omitted, and others have been only touched upon. The goal is to provide the reader with enough background to understand the biological underpinnings of the chapters that follow, along with some notion of the value of the analyses undertaken. More comprehensive introductions to nucleic acid and protein sequence analysis can be found in books such as Gribskov and Devereux (1991) [1] and Doolittle (1996) [2]. Similarly, a proper introduction to the fundamentals of molecular biology, genetics, and genomics is also outside the scope of this article. Good starting points for the newcomer include the publications by Berg and Singer (1992) [3], Casey (1996) [4], Lewin (1997) [5] and Watson et al. (1992) [6].

2. A little molecular biology

Sequences arise in biological research because of the polymeric nature of the major biological macromolecules, nucleic acids and proteins. For the purposes of this discussion, the fundamental processes of molecular biology can be summarized as follows:

- The human genetic code is captured in long molecules of DNA, which are packaged in the cell nucleus in the chromosomes. DNA is composed of a long chain of subunits; the nucleotides adenine, guanine, cytosine, and thymine (abbreviated A, G, C and T, respectively). DNA is usually found in the form of two paired chains, held together by hydrogen bonds between complementary pairs of nucleotides. A on one strand always pairs with T on the other, and similarly for G and C.
- Most of the important structural and functional components of a human cell are composed of proteins, which are long, folded chains composed of the 20 common amino acids, or peptides. (Thus, the frequent use of the term "polypeptide" for a protein.)
- Genes are individual stretches of the chromosomal DNA that encode the sequence of amino acids comprising a particular protein. The 64 possible nucleotide triplets (4^3), known as codons, are used to represent the 20 amino acids using a degenerate code. In higher organisms, portions of genes that code for proteins ("exons") are usually separated by noncoding regions ("introns").
- An intermediary form of nucleic acid called RNA *transcribes* a gene's DNA sequence in a form which can be directly *translated* by the cell's protein-making machinery (the ribosome). Prior to processing an RNA sequence, the intronic regions are spliced out to leave just the coding portions to be translated.

3. Frequency analysis

Perhaps the most basic analysis to perform on a sequence is to determine the frequency of occurrence of its component elements. In the case of biological sequences, the ratio of the four nucleotides or twenty amino acids in a given region of nucleic acid or protein sequence can give us important structural and functional clues. Similarly for the relative proportions of dimers, trimers, and higher order oligomers; i.e. pairs, triplets, etc. of nucleotides or amino acids.

For example, given the fact that A–T base pairing involves two hydrogen bonds, while G–C pairing involves three, DNA with a low G + C content will separate into its two component strands at lower temperatures than that with a higher G + C ratio. In fact, it has been demonstrated that thermophilic microorganisms living in hot springs and deep-sea vents tend to have genomes with higher G + C values than those living in cooler environments. Similarly, because of various steps that require the binding and subsequent dissociation of complementary DNA strands, genomes with a high G + C content are more difficult to sequence. Higher G + C content has also been associated with more gene-dense regions of mammalian genomes.

Oligomer frequency can be used to distinguish coding and noncoding regions. The coding regions (exons) in the DNA of higher organisms such as humans are typically short (averaging 150 nucleotides in length) and widely separated. Current estimates are that only 1–3% of human DNA is comprised of coding regions. In order to determine protein sequences based upon the DNA sequences of human genes, these coding regions need to be located. Staden and McLachlan [7] observed that since coding regions are translated in triplets (codons) while other regions are not, the codon frequency of these regions might provide a distinctive statistical signature. Since then, numerous programs have been written to utilize the 64 codon frequencies in methods for identifying coding regions [8]. As the amount of sequence data has grown, it has been possible to consider hexamer frequencies (i.e. two adjacent codons), which seem to provide an even better signal than single codons.

Human gene mutations have also been observed more frequently at certain dinucleotides than others [9]. It is postulated that this dimer bias is related in part to the ability of certain amino acids to substitute for one another in a protein without significant loss of function and to their corresponding edit distance in the triplet genetic code.

The frequency of various amino acids in a protein sequence has consequences for the protein's behavior when dissolved in water (as in the cell's cytoplasm) or in the lipid environment of the cell membrane. Methods of distinguishing hydrophobic and hydrophilic regions of proteins (e.g. ref. [10]) have been used to predict the gross three-dimensional structure of proteins, their likely home within the cell, and their broad functional characteristics.

The frequencies of short oligomers can also be used to detect a variety of important biological features in a sequence. As an example of a short-range effect, the region just before the splice site at the end of an intron typically has high $C+T$ content. As with many such sequence patterns, this is detectable with statistical methods, but the pattern is not clear enough to be detected without also detecting many false positives. An even shorter pattern is the near-100% occurrence of the dinucleotide GT as the first two bases of an intron, and AG as the last two bases.

Another example of local sequence correlation is the ribosome binding site, which occurs just before the start codon that initiates protein translation. The sequence of this site, which varies in different organisms, shows clear preferences for some bases over others in particular positions. Recent evidence indicates that in addition to these positional preferences there are strong correlations between adjacent positions in the site. This makes sense from a biological standpoint, since the ribosome is a large molecular complex that binds to a region of DNA, not to individual bases. Sequence modeling techniques such as Markov chains are especially effective at capturing these local dependencies.

More subtle patterns can be detected by looking for long-range correlations between sites that, although distant in the sequence, are in close proximity in the three-dimensional environment of the cell. For example, transcriptional regulation occurs via proteins that bind to patterns in the DNA sequence at a considerable distance from the transcribed region. Regulatory sequences known as "promoters" and "enhancers" can occur anywhere from a few bases upstream of the start of transcription to thousands of bases away. They can even occur within the transcribed region itself. Sequence analysis software is just beginning to reliably identify these signals (e.g. ref [11]).

4. Measuring homology by pairwise alignment

Two proteins are homologous if they have similar three-dimensional structures; similar shape usually implies similar function. Because protein sequence and three-dimensional structure are closely related, we can take this a step further and say that if two protein sequences are sufficiently similar, then they are likely to have similar functions. This observation leads to a very useful method for determining the function of an unknown gene: by matching the predicted amino acid sequence of an uncharacterized but sequenced gene to the sequence of a known gene, we can infer the function of the new gene from the known one. The underlying assumption is that two sufficiently similar genes probably evolved from a common ancestor, and the degree of similarity visible today is a function of how long ago the sequences (and the organisms from which they were taken) diverged. Nucleic acid sequences are also compared, although the redundancy in the genetic code (whereby multiple codons map to the same amino acid) means that the corresponding amino acid sequences might be much more similar than the DNA sequences for any given pair of genes.

The process of comparing two sequences involves aligning them and inserting gaps in one or both sequences in order to produce an optimal score. An alignment is scored by adding up the match scores and subtracting penalties for the gaps. Match scores are contained in a matrix that assigns a score for each of the 400 possible amino acid pairs. For DNA sequence alignment, match scores usually just assign a fixed score for an identical nucleotide match and a different score for a mismatch. Given a scoring matrix and a gap penalty, the optimal alignment for two sequences can be computed by dynamic programming in $O(n^2)$ time. The most general formulation of the algorithm allows one to align just a subsequence of one input sequence to a subsequence of the other [12]; the entire sequences do not have to be similar.

When comparing and aligning protein sequences, it is important to consider that certain amino acids can substitute for others in a protein without dramatically altering the molecule's three-dimensional structure or function. Thus, mismatch penalties in protein alignment algorithms usually employ some sort of scoring matrix, a table which indicates the penalties associated with each possible amino acid substitution. Since there is no good way to measure the substitutability of one amino acid for another *a priori*, these scoring matrices have been empirically derived in the past based on a bootstrapping process which postulates an alignment of sequences with a putative evolutionary "distance," and then generates a score based on the frequency of substitution of alternative residues at each position. The most popular scoring matrices for general protein searching are the BLOSUM set of Henikoff and Henikoff [13], and the PAM matrices of Dayhoff [14]. See Altschul [15] for a discussion of scoring matrices and sequence searching.

It is also important to keep in mind that sequence alignment does not always imply functional similarity. This becomes an especially important caveat when the alignment score is relatively weak, indicating only distant similarity. The power and speed of sequence alignment programs makes it easier than ever to find weak homologies that might not represent true evolutionary relationships.

Once an alignment is calculated, the oldest but still much used method for displaying it is the dot plot, usually attributed to Gibbs and McIntyre [16]. The essential notion is that the two sequences to be compared are arrayed along each axis, and a point is plotted at every (x,y) position where the two sequences have an element in common. A perfect alignment between two identical sequences is thus immediately obvious as a continuous line along the main diagonal of the matrix. Similarly, common subsequences (e.g. possible translocations) show up as shorter off-diagonal lines, and inversions in the sequence appear as lines perpendicular to the main diagonal. Deletions are readily apparent as interruptions in these lines.

The spatial resolution of the display medium becomes a factor when trying to display long-range alignments of sequences several kilobases in length or longer. This is addressed by using a sliding window for the region of comparison, rather than a single sequence element. If the alignment within the window is better than some threshold value, a dot is displayed. In the course of their three decades of use, dot plots have been enhanced in numerous ways, including the introduction of a variety of filtering schemes to suppress statistically insignificant local alignments, the use of glyphs, color, dot size, etc., to indicate the nature or strength of the alignment, and so on. The program SIP, distributed in the Staden sequence analysis package [17], contains many of these features and is widely used.

5. Database searching

The incredible explosion of data resulting from the Human Genome Project has caused the rapid expansion of public repositories of nucleic acid and protein sequences. Databases such as GenBank [18], the EMBL Nucleotide Sequence Database [19], and SWISS-PROT [20] provide the wellspring for much of recent computational biology research.

It turns out that one of the most common sequence alignment applications is querying of sequence databases. While there is limited utility in more traditional sorts of database queries (e.g. "retrieve the sequences of all *E. coli* tRNA genes" or "retrieve all sequences greater than 800 kilobases in length"), the most common form of sequence database query is always "show me all sequences similar to this one" within some measure of statistical significance. Traditional database query languages and processors are not appropriate for this task. Smith–Waterman, while guaranteed to produce the optimal global alignment between a query and target sequence, is much too slow.

Fortunately, efficient sequence alignment algorithms have been specifically developed for this purpose. The most popular of these are FASTA [21] and BLAST [22]. Both methods rely at their core on the identification of brief subsequences (k-tuples) which serve as the core of an alignment. Multiple k-tuples can be combined and extended to serve as "seeds" for a more extended alignment, allowing for some number of insertions, deletions, or changes between the two sequences. Versions of these programs exist for comparing either a nucleic acid or protein query sequence to a database of one or the other kind of sequence. Pearson [23] reviewed some of the methods for comparing two sequences and comparing a single sequence to a database.

6. Multiple alignment

Much of our discussion has focused on the pairwise alignment of sequences, typically to explore the question of similarity between a query sequence and one or more targets. However, the task of simultaneously aligning a number of sequences to produce optimal global or local alignments (of similar subsequences) is also of great utility in biological research. The importance of these multiple sequence alignment algorithms is clearly evidenced by the large number of programs available for this task. They can be divided into two categories: those which attempt to find the best overall global alignment of the input sequence set, and those which focus on optimal local alignments of common subsequences across the input set. Algorithmic strategies vary widely, and include multiple pairwise alignment, clustering, hidden Markov models, and various iterative approximation methods. Examples of the global alignment programs include ClustalW [24], MAP [25], MSA [26], and PILEUP (derived from the work of Feng and Doolittle [27]). Programs which focus on producing local alignments (and in some cases global ones as well) include PIMA [28], Block Maker [29], MACAW [30], MEME [31], SAM [32], and the work of Myers et al. (1996) [33], to name just a few.

Complete multiple alignments are useful for studying protein families whose members have not diverged enough across evolution to show significant rearrangement of their individual sequences. In fact, these alignments can be analyzed as a measure of the evolutionary distance between species, a methodology that has almost completely replaced the analysis of large-scale morphological variation in bones or soft tissue. The widely used programs PHYLIP [34] and PAUP [35], along with many more recent systems infer evolutionary relationships (phylogenies) from multiple sequences. Local alignment methods are better for analyzing subdomains within sequences, and thus are useful for tasks such as identifying gene promoters, transcription factor binding sites, and critical protein regions.

7. Finding regions of interest in nucleic acid sequence

Sequence analysis methods are of fundamental importance to the process of DNA sequencing itself. The sequence of nucleotides along a DNA molecule cannot simply be read as one would envision reading a Turing machine tape. We cannot see DNA and generally speaking, we cannot manipulate individual DNA molecules. Also, while chromosomes are one million to several hundred million base pairs in length, our current fundamental sequencing technology can only read DNA sequence 500 to 1000 base pairs at a time!

Sequencing an entire genome thus demands a divide-and-conquer strategy. Multiple copies of the genome are randomly divided into smaller regions whose spatial relationships are then determined, usually by some method of estimating their overlap (a process known as "physical" mapping). These "large-insert clones," typically some 30 000 to 250 000 base pairs in length, are then once again individually copied and randomly divided into even smaller pieces 1000 to 5000 base pairs in length. These "sequencing clones" (or "subclones") are randomly sequenced for 500 to 1000 bases at one or both ends, and the resulting collection of sequence reads are then reassembled using one of a variety of "shotgun assembly" algorithms which compute potential pairwise alignments of the fragments and can take into account the paired relationships of sequence reads from either end of the same subclone. Examples of some currently popular sequence assembly programs include Alewife (ref. [36], in preparation), FAK II [37], CAP2 [38], GAP4 [39,17], and Phrap (Phil Green, personal communication).

It is now possible to sequence entire bacterial genomes, which range from approximately 0.5 to 5.0 million bases in length, by a single very large "shotgun" experiment. In this approach the entire genome is broken into pieces 1000 to 5000 base pairs in length, and the genome sequence is assembled from the tens of thousands of sequence reads from the ends of these small clones [40,41]. The TIGR Assembler [42] was specifically designed for this application.

Until recently, genomic DNA sequencing was sufficiently difficult that it was a major task to sequence even a 40 kilobase clone thought to contain a gene (or genes) of interest. Such projects were usually accompanied by many biological experiments aimed at identifying the coding regions in the sequence, and ultimately at understanding the regulation of the gene, the associated gene product, and its function. With the advent of large-scale genomic sequencing, we are now able to generate DNA sequence much faster than we can perform the bench work necessary to characterize it. As a result, there is growing interest in computational methods which can provide reliable characterizations of previously "anonymous" sequences, thus reducing the effort and cost involved in the confirming laboratory experimentation.

The detection of biological "regions of interest" in genomic sequence runs the gamut from the classification of lowly sequence repeats to the lofty identification of gene components, including exon/intron boundaries, promoters, enhancers, and transcription factor binding sites.

It turns out that the genomic DNA of higher organisms is full of repeated sequences of various sorts. These include many short repeated nucleotides (e.g. a dozen or more consecutive CA dinucleotides), longer tandem repeats, and various classes of short and long "interspersed" repeats scattered throughout the genome with varying frequency. Besides the obvious desire to label all biologically relevant portions of the sequence, identification and subsequent masking of repeat regions and other regions of low sequence complexity is critical to avoiding false homology reports in sequence database searches. For example, if one were to take a 10000 base pair stretch of new human genomic sequence data and BLAST it unprocessed against the contents of GenBank, you would be likely to bring back many spurious hits to your query because of the presence of perhaps *several* copies of the ubiquitous "ALU" short interspersed repeat in your query sequence. As this is a very important problem, there are a variety of programs for dealing with it. The programs DUST (Tatusov and Lipman, in preparation), SEG [43] and XNU [44] all detect and mask simple repeats and other regions of low sequence complexity. CENSOR [45] and RepeatMasker [46] both specialize in identifying short and long interspersed repeats based on specialized databases of sample and consensus sequences.

8. Gene finding

Computational gene finding is a crucial area of computational biology. By comparison with laboratory methods, finding genes by computer is extremely fast and cheap. The problem is, as always, how accurate can the computer be at this task? Several of the following chapters will address this issue in more detail and describe some of the current methods for solving this problem. Here we will outline some of the biological background and sketch the general approaches.

As mentioned above, only 1–3% of the human genome is translated into proteins. How can we focus a sequencing effort on just those widely scattered pieces that are translated? Since all genes are first transcribed into RNA, if we could just capture the RNA and sequence that, we would have the genes. Because sequencing technology prefers DNA, the RNA is extracted from the cytoplasm and then copied back into DNA using an enzyme called "reverse transcriptase". The resulting complementary DNA, or "cDNA," is then sample sequenced to generate an "expressed sequence tag" or EST [47]. What makes this even more useful is that EST sequencing methods capture the RNA after the introns have been spliced out, leaving just the translated parts plus a variable length region at the beginning and end of each RNA that is untranslated.

EST sequencing projects have already generated hundreds of thousands of ESTs from a variety of tissues and developmental stages in humans and other organisms. These databases represent direct biological evidence for many pieces of genes. A variety of projects have attempted to construct assemblies of these gene fragments to build up the complete coding region sequence. These include TIGR's tentative human consensus (THC) effort [48], the UniGene project [49], and the Merck Gene Index [50]. EST data are just beginning to be incorporated into computational techniques, which need to map the fragments and collected assemblies onto genomic sequence in order to pinpoint the locations of exons.

The difficulty with detecting genes in genomic DNA, as opposed to ESTs or other RNA sequences, is that introns (which on average are much longer than exons, at least in DNA from humans and other vertebrates) are spliced out before the sequence is translated into a protein. Therefore a program must identify the introns as part of the gene-finding process. The only clear pattern that indicates an intron are the dimers GT at the beginning and AG at the end. These dimers are necessary for splicing but certainly not sufficient; on average, each one occurs in 1/16 of the positions in a genome. Clearly the cellular processes involved in splicing recognize more than these two simple signals.

The direct approach to gene finding is to identify exactly where four specific signal types can be found. These are:
• the start codon;
• all donor sites (the beginning of each intron);
• all acceptor sites (the end of each intron);
• the stop codon.

If each of these can be identified correctly, then the protein product can be simply read off after removing the introns and concatenating all the sequence from the start to the stop codon. All gene-finding systems attempt to locate these signals, but thus far there is no completely accurate way to do so.

The start codon in eukaryotes is almost always ATG, which should occur on average in about 1 in 64 locations in random sequence. (In prokaryotes, the start codon is usually ATG, but is sometimes GTG or TTG.) This narrows down the search considerably, but it still leaves many candidates. As mentioned earlier, a Markov chain model of the sequence just before the start codon can narrow down the list considerably, but it is still usually impossible to find the correct start codon in every case from strictly local sequence patterns.

The splice sites can be recognized by the characteristic GT–AG patterns at the ends of the intron, combined with a model of consistencies in the local sequence surrounding the splice junction. Again, methods for doing this, such as Markov chains, are effective at narrowing down the list, but they still leave many choices. (Burge's chapter 8 of the present volume includes a detailed explanation of how to model these sites and build splice site detectors.)

Finally, the stop codon does not seem to be surrounded by any particular sequence pattern; apparently the translation process stops by recognizing this codon alone, and no other local sequence patterns are necessary to disengage the ribosome.

In addition to recognizing the signals that form important components of a gene, programs can analyze larger regions of sequence for other statistical patterns. These are known as content scoring methods. The basis of many of these methods is codon frequency, which is different in coding and noncoding regions as discussed previously. There are basically four different regions that one would like to distinguish:
• the coding portion of exons;
• the noncoding portion of exons (the upstream and downstream untranslated regions);
• introns;
• intergenic regions.

Another approach to distinguishing these regions is to compute their entropy (or, conversely, information content). This has been shown to be different among all these types of sequences, but the differences are not great enough to make effective gene finders by themselves. None of the four regions, incidentally, has the same information content as completely random DNA.

A slightly different approach to codon or dicodon frequency is to look for a 3-periodic pattern in the occurrences of A, C, G, or T [51]. Such a pattern would indicate a coding region, since only in coding regions are the bases grouped into triplets. This measure has been used effectively in several different gene-finding systems.

Gene-finding systems come in two major types: eukaryotic and prokaryotic. Eukaryotic gene finding is generally harder, because of the presence of introns and because coding density is low. There are many different integrated systems for gene finding in eukaryotes, and a smaller number for prokaryotes. (To be more precise, the prokaryotic systems are also used for the Archaea, a kingdom of single-celled organisms that is related to both prokaryotes and eukaryotes. Introns have not yet been observed in Archaea.)

There are many different systems available for finding genes in eukaryotes, and it would be impossible to summarize them all here. The underlying technologies include rules, hidden Markov models, decision trees, and neural networks. Four of the chapters in this book describe gene finders (or components of gene finders) in some detail: Xu and Uberbacher (chapter 7) describe GRAIL, a neural-net-based system. Krogh (chapter 4) and Burge (chapter 8) describe two different systems, HMMGene and GENSCAN, each based on a different type of hidden Markov model. Salzberg (chapter 10) describes MORGAN, a system based on decision trees. There are also systems that find genes entirely based on alignments to the sequence databases. Appendix A contains pointers to many of these systems, most of which are available either through Web servers or as complete systems for local installation.

Most integrated gene finders will accept as input a genomic DNA sequence, and will then parse the sequence to find all the coding portions of the exons. Many systems assume that the input contains just one gene, and will only output a single gene model. More recently, as larger contiguous pieces of genomic DNA sequence have become commonly available, several systems have been modified to remove this assumption.

Some of the active areas of research and development in eukaryotic gene finding include integration of database searches and *de novo* gene prediction, use of EST databases in gene finding, and specialization of gene finders for different organisms. The necessity for specialization derives from the fact that both content and signal detection methods work differently in different organisms. It pays to train a special gene finder (assuming sufficient data is available) on an organism such as the flowering plant *Arabidopsis*, whose genomic structure differs quite substantially from humans.

The problem of gene finding in prokaryotes seems much simpler, since prokaryotic genomes contain 85–90% coding sequence. One can be fairly successful simply by calling all the long open reading frames (ORFs) genes. However, there are six possible reading frames (three on one DNA strand and three on the reverse complement), and inevitably there are many ORFs that overlap. The most important task of a prokaryotic gene finder is to resolve these overlaps. Even though this task is easier in an absolute sense than eukaryotic gene finding, it is by no means completely solved. Rather, the requirements for a successful system are simply much more stringent.

There are currently two main systems for prokaryotic gene finding, both of which achieve well over 90% accuracy at finding genes. GeneMark [52] is based on 5th-order Markov chains, which are used to score open reading frames and thereby compare overlapping ones. Glimmer [53] uses a more general method known as an interpolated Markov model (IMM), which uses 1st-order through 8th-order Markov chains to score open reading frames. Glimmer computes a weighting scheme for interpolating between all these models, essentially allowing it to use some of the evidence from each of the different order Markov models. Both GeneMark and Glimmer use nonhomogeneous Markov chains; in particular, they use 3-periodic models that are specific to each codon position. The appendix of online resources contains pointers to both GeneMark and Glimmer.

9. Analyzing protein sequence

While sequence analysis focuses on the one-dimensional characteristics of nucleic acids and proteins, it is in fact their three-dimensional structure that underlies their structural and functional properties. Much computational biology research is devoted to the prediction of the precise three-dimensional structure of proteins given their amino acid sequence, and to the further elucidation of their resulting function.

A detailed description of protein biochemistry is beyond the scope of this chapter, but a quick sketch is in order. Structural biologists classify protein structure at four levels. A protein's primary structure is the sequence of amino acids in a polypeptide chain. As the protein folds in its native environment, local runs of amino acids often assume one of two secondary structures: a closely packed helical spiral ("alpha" helix), or a relatively flat structure where successive runs of peptides fold back on one another ("beta" or "pleated" sheet). Discussion of secondary structure also sometimes references a "coiled" region with a much less well-defined spiraling of the polypeptide chain, usually for only a short run of amino acids. The complete, detailed conformation of the molecule, describing how these helices, sheets, coils, and intervening sequences are precisely positioned in three dimensions, is referred to as the protein's tertiary structure. Computational methods have been developed for prediction of both secondary and tertiary structure, but both problems remain extremely difficult. Finally, some protein molecules actually can be assemblies of multiple polypeptide subunits which come together to form larger, even more complex structures. Thus, one can extend this classification to the quaternary structure of proteins.

The cell offers proteins several distinct environments. A substantial part of the cell contains water, and proteins in these aqueous compartments tend to fold up with their hydrophilic amino acids on the surface and their hydrophobic ones in the interior. The cell membrane is comprised of lipids, and proteins that are completely embedded in this environment must fold in the opposite manner, with their hydrophilic regions on the inside. Larger transmembrane proteins which are only partially embedded in the cell membrane but which extend into the cytoplasm or out into the extracellular space often pass back and forth through the membrane multiple times, with alternating subchains of hydrophobic or hydrophilic amino acids. A simple but sometimes effective method of identifying such proteins is to produce a hydrophobicity plot (e.g. by the method of Kyte and Doolittle [10]) and look for a series of alternating regions.

Sequence analysis of proteins is often aimed at solving one or more of the following problems.
• Finding conserved patterns of sequence in related proteins which may correspond to important functional domains in the three-dimensional molecule.
• Prediction of local secondary structure from a sequence of peptides.
• Prediction of tertiary structure (i.e. detailed three-dimensional form) from the peptide sequence, either based on homology with sequences whose tertiary structure is known, or from first principles based on fundamental atomic interactions.
• Prediction of function, for subdomains and whole molecules, again using methods relying on both homology with characterized proteins or from first principles.

The second major section of this book contains a number of chapters devoted to the description of computational methods for modeling protein structure.

10. Whole genome analysis

Now that a number of complete genomes have been sequenced, and larger contiguous pieces of vertebrate genomes are becoming available, we have entered the era of whole genome analysis. In addition to looking at genomes in their entirety, it is now possible to compare multiple genomes. Some of the first attempts at this kind of analysis include: characterization of those genes that can be assigned to functional roles by homology; examination of the G + C content and specific oligomer content across the whole genome; attempts to identify laterally transferred genes that have been incorporated from other organisms; and prediction of the chromosomal events which led to the present arrangements of genes in two genomes that have evolved from a common ancestor.

All of these analyses are built upon the computational tools developed for basic sequence analysis. For example, laterally transferred genes can be tentatively identified by comparing their codon content to that of other genes in the genome. Significant differences provide strong evidence for a foreign origin [54]. Similarly, understanding the evolutionary events that led to the divergence of two genomes with a common origin involves the prediction of duplication, deletion, insertion, and translocation of sequences from one portion of the genome to another (e.g. ref. [55]). Because this area of research is so new, the set of problems and discoveries that will become part of whole genome analysis are still just beginning to be explored.

Acknowledgments

KHF is supported by NIH Grant P50-HG00098 and DOE Award DE-FC02-91ER61230. SLS is supported in part by NIH Grant K01-HG00022-1, and by NSF Grant IRI-9530462.

References

[1] Gribskov, M. and Devereux, J. (1991) Sequence Analysis Primer. Oxford University Press, Oxford.

[2] Doolittle, R.F., ed. (1996) Computer Methods for Macromolecular Sequence Analysis, Vol. 266 of Methods in Enzymology, Series editors J.N. Abelson and M.I. Simon. Academic Press, San Diego, CA.

[3] Berg, P. and Singer, M. (1992) Dealing with Genes, The Language of Heredity. University Science Books, Mill Valley, CA.

[4] Casey, D. (1996) Primer on Molecular Genetics. U.S. Department of Energy Human Genome Management Information System.
URL: http://www.ornl.gov/hgmis/publicat/primer/intro.html.

[5] Lewin, B. (1997) Genes VI. Oxford University Press, Oxford.

[6] Watson, J.D., Gilman, M., Witkowski, J.A. and Zoller, M.J. (1992) Recombinant DNA, 2nd edition. Scientific American Books, New York.

[7] Staden, R. and McLachlan, A.D. (1982) Codon preference and its use in identifying protein coding regions in long DNA sequences. Nucleic Acids Res. 10, 141–156.

[8] Fickett, J.W. and Tung, C.-S. (1992) Assessment of protein coding measures. Nucleic Acids Res. 20, 6441–6450.

[9] Krawczak, M. and Cooper, D.N. (1996) Single base-pair substitutions in pathology and evolution, two sides of the same coin. Hum. Mutation 8, 23–31.

[10] Kyte, J. and Doolittle, R.F. (1982) A simple method for displaying the hydropathic character of a protein. J. Mol. Biol. 157, 105–132.

[11] Matis, S., Xu, Y., Shah, M., Guan, X., Einstein, J.R., Mural, R. and Uberbacher, E. (1996) Detection of RNA polymerase II promoters and polyadenylation sites in human DNA. Computer & Chemistry 20, 135–140.

[12] Smith, T.F. and Waterman, M.S. (1981) Comparison of biosequences. Adv. Appl. Math. 2, 482–489.

[13] Henikoff, S. and Henikoff, J.G. (1992) Amino acid substitution matrices from protein blocks. Proc. Natl. Acad. Sci. USA 89, 10915–10919.

[14] Dayhoff, M.O. (1978) Atlas of Protein Sequence and Structure. National Biomedical Research Foundation, Washington D.C.

[15] Altschul, S.F. (1991) Amino acid substitution matrices from an information theoretic perspective. J. Mol. Biol. 219, 555–565.

[16] Gibbs, A.J. and McIntyre, G.A. (1970) The diagram, a method for comparing sequences, its use with amino acid and nucleotide sequences. Eur. J. Biochem. 16, 1–11.

[17] Staden, R. (1996) The Staden sequence analysis package. Mol. Biotechnol. 5, 233–241. URL: http://www.mrc-lmb.cam.ac.uk/pubseq.

[18] Benson, D.A., Boguski, M.S., Lipman, D.J. and Ostell, J. (1997) GenBank. Nucleic Acids Res. 25, 1–6.

[19] Stoesser, G., Sterk, P., Tuli, M.A., Stoehr, P.J. and Cameron, G.N. (1997) The EMBL Nucleotide Sequence Database. Nucleic Acids Res. 25, 7–14.

[20] Bairoch, A. and Apweiler, R. (1997) The SWISS-PROT protein sequence data bank and its supplement TrEMBL. Nucleic Acids Res. 25, 31–36.

[21] Pearson, W.R. and Lipman, D.J. (1988) Improved tools for biological sequence comparison. Proc. Natl. Acad. Sci. USA 85, 2444–2448.

[22] Altschul, S.F., Gish, W., Miller, W., Myers, E.W. and Lipman, D.J. (1990) Basic local alignment search tool. J. Mol. Biol 215, 403–410.

[23] Pearson, W.R. (1995) Comparison of methods for searching protein sequence databases. Prot. Sci. 4, 1145–1160.

[24] Thompson, J.D., Higgins, D.G. and Gibson, T.J. (1994) CLUSTALW, Improving the sensitivity of progressive multiple sequence alignment through sequence weighting, position-specific gap penalties and weight matrix choice. Nucleic Acids Res. 22, 4673–4680.

[25] Huang, X. (1994) On global sequence alignment. Computer Applications in the Biosciences 10, 227–235.

[26] Lipman, D.J., Altschul, S.F. and Kececioglu, J.D. (1989) A tool for multiple sequence alignment. Proc. Natl. Acad. Sci. USA 86, 4412–4415.

[27] Feng, D.F. and Doolittle, R.F. (1987) Progressive sequence alignment as a prerequisite to correct phylogenetic trees. J. Mol. Evol. 25, 351–360.

[28] Smith, R.F. and Smith, T.F. (1992) Pattern-Induced Multi-sequence Alignment (PIMA) algorithm employing secondary structure-dependent gap penalties for comparative protein modeling. Protein Eng. 5, 35–41.

[29] Henikoff, S., Henikoff, J.G., Alford, W.J. and Pietrokovski, S. (1995) Automated construction and graphical presentation of protein blocks from unaligned sequences. Gene 163, GC17–26.

[30] Schuler, G.D., Altschul, S.F. and Lipman, D.J. (1991) A workbench for multiple alignment construction and analysis. Proteins Struct. Funct. Genet. 9, 180–190.

[31] Grundy, W., Bailey, T., Elkan, C. and M. Baker (1997) Meta-MEME, motif-based hidden Markov models of protein families. Computer Applications in the Biosciences 13, 397–403.

[32] Krogh, A., Brown, M., Mian, I.S., Sjolander, K. and Haussler, D. (1994) Hidden Markov models in computational biology, applications to protein modeling. J. Mol. Biol. 235, 1501–1531.

[33] Myers, E., Selznick, S., Zhang, Z. and Miller, W. (1996) Progressive multiple alignment with constraints. J. Comput. Biol. 3, 563–572.

[34] Felsenstein, J. (1989) PHYLIP – Phylogeny Inference Package (Version 3.2). Cladistics 5, 164–166. URL: http://evolution.genetics.washington.edu/phylip.html.

[35] Swofford, D.L. (1990) PAUP, Phylogenetic Analysis Using Parsimony, Users Manual, Version 3.5. Illinois Natural History Survey, Champaign, IL.

[36] Daly, M.J. and Lander, E.S. Alewife sequence assembler. Manuscript in preparation.

[37] Larson, S., Jain, M., Anson, E. and Myers, E. (1996) An interface for a fragment assembly kernel. Tech. Rep. TR96-04. Dept. of Computer Science, University of Arizona, Tucson, AZ.

[38] Huang, X. (1996) An improved sequence assembly program. Genomics 33, 21–31.

[39] Bonfield, J.K., Smith, K.F. and Staden, R. (1995) A new DNA sequence assembly program. Nucleic Acids Res. 24, 4992–4999.

[40] Fleischmann, R.D., Adams, M., White, O., Clayton, R., Kirkness, E., Kerlavage, A., Bult, C., Tomb, J.-F., Dougherty, B., Merrick, J., McKenney, K., Sutton, G., FitzHugh, W., Fields, C., Gocayne, J., Scott, J., Shirley, R., Liu, L.-I., Glodek, A., Kelley, J., Weidman, J., Phillips, C., Spriggs, T., Hedblom, E., Cotton, M., Utterback, T., Hanna, M., Nguyen, D., Saudek, D., Brandon, R., Fine, L., Fritchman, J., Fuhrmann, J., Geoghagen, N., Gnehm, C., McDonald, L., Small, K., Fraser, C., Smith, H.O. and Venter, J.C. (1995) Whole-genome random sequencing, assembly of *Haemophilus influenzae* Rd. Science 269, 496–512.

[41] Fraser, C., Gocayne, J., White, O., Adams, M., Clayton, R., Fleischmann, R., Bult, C., Kerlavage, A., Sutton, G., Kelley, J., Fritchman, J., Weidman, J., Small, K., Sandusky, M., Fuhrmann, J., Nguyen, D., Utterback, T., Saudek, R., Phillips, C., Merrick, J., Tomb, J.-F., Dougherty, B., Bott, K., Hu, P.-C.,

Lucier, T., Peterson, S., Smith, H.O., Hutchison, C. and Venter, J.C. (1995) The minimal gene complement of *Mycoplasma genitalium*. Science 270, 397–403.

[42] Sutton, G., White, O., Adams, M. and Kerlavage, A. (1995) TIGR Assembler, a new tool for assembling large shotgun sequencing projects. Genome Science Tech. 1, 9–19.

[43] Wootton, J.C. and Federhen, S. (1993) Statistics of local complexity in amino acid sequences and sequence databases. Comput. & Chem. 17, 149–163.

[44] Claverie, J.M. and States, D. (1993) Information enhancement methods for large scale sequence analysis. Comput. & Chem. 17, 191–201.

[45] Jurka, J., Klonowski, P., Dagman, V. and Pelton, P. (1996) CENSOR – a program for identification and elimination of repetitive elements from DNA sequences. Comput. & Chem. 20, 119–122.

[46] Smit, A.F.A. and Green, P. (1996) RepeatMasker.
URL: http://ftp.genome.washington.edu/RM/RepeatMasker.html.

[47] Adams, M.D., Kelley, J.M., Gocayne, J.D., Dubnick, M., Polymeropoulos, M.H., Xiao, H., Merril, C.R., Wu, A., Olde, B., Moreno, R.F., et al. (1991) Complementary DNA sequencing: expressed sequence tags and human genome project. Science 252, 1651–1656.

[48] Adams, M.D., Kerlavage, A.R., Fleischmann, R.D., et al. (1995) Initial assessment of human gene diversity and expression patterns based upon 83 million nucleotides of cDNA sequence. Nature 377 (Suppl.), 3–174.

[49] Boguski, M.S. and Schuler, G.D. (1995) ESTablishing a human transcript map. Nat. Genet. 10, 369–371.

[50] Aaronson, J.S., Eckman, B., Blevins, R.A., Borkowski, J.A., Myerson, J., Imran, S. and Elliston, K.O. (1996) Toward the development of a gene index to the human genome: an assessment of the nature of high-throughput EST sequence data. Genome Res. 6, 829–845.

[51] Fickett, J.W. (1982) Recognition of protein coding regions in DNA sequences. Nucleic Acids Res. 10, 5303–5318.

[52] Borodovsky, M. and Mcininch, J.D. (1993) GeneMark, parallel gene recognition for both DNA strands. Comput. & Chem. 17, 123–133.

[53] Salzberg, S., Delcher, A., Kasif, S. and White, O. (1998) Microbial gene identification using interpolated Markov models. Nucleic Acids Res. 26, 544–548.

[54] White, O., Dunning, T., Sutton, G., Adams, M., Venter, J.C. and Fields, C. (1993) A quality control algorithm for DNA sequencing projects. Nucleic Acids Res. 21, 3829–3838.

[55] Hannenhalli, S., Chappey, C., Koonin, E.V. and Pevzner, P.A. (1995) Genome sequence comparison and scenarios for gene rearrangements, a test case. Genomics 30, 299–311.

PART II

Learning and Pattern Discovery in Sequence Databases

S.L. Salzberg, D.B. Searls, S. Kasif (Eds.), *Computational Methods in Molecular Biology*

CHAPTER 4

An introduction to hidden Markov models for biological sequences

Anders Krogh

Center for Biological Sequence Analysis, Technical University of Denmark, Building 208,
2800 Lyngby, Denmark. Phone: +45 4525 2471; Fax: +45 4593 4808; E-mail: krogh@cbs.dtu.dk

1. Introduction

Very efficient programs for searching a text for a combination of words are available on many computers. The same methods can be used for searching for patterns in biological sequences, but often they fail. This is because biological "spelling" is much more sloppy than English spelling: proteins with the same function from two different organisms are almost certainly spelled differently, that is, the two amino acid sequences differ. It is not rare that two such homologous sequences have less than 30% identical amino acids. Similarly, in DNA many interesting signals vary greatly even within the same genome. Some well-known examples are ribosome binding sites and splice sites, but the list is long. Fortunately, there are usually still some subtle similarities between two such sequences and the question is how to detect these similarities.

The variation in a family of sequences can be described statistically, and this is the basis for most methods used in biological sequence analysis, see ref. [1] for a presentation of some of these statistical approaches. For pairwise alignments, for instance, the probability that a certain residue mutates to another residue is used in a substitution matrix, such as one of the PAM matrices. For finding patterns in DNA, e.g. splice sites, some sort of weight matrix is very often used, which is simply a position specific score calculated from the frequencies of the four nucleotides at all the positions in some known examples. Similarly, methods for finding genes use, almost without exception, the statistics of codons or dicodons in some form or other.

A hidden Markov model (HMM) is a statistical model, which is very well suited for many tasks in molecular biology, although they have been mostly developed for speech recognition since the early 1970s, see ref. [2] for historical details. The most popular use of the HMM in molecular biology is as a "probabilistic profile" of a protein family, which is called a profile HMM. From a family of proteins (or DNA) a profile HMM can be made for searching a database for other members of the family. These profile HMMs resemble the profile [3] and weight matrix methods [4,5], and probably the main contribution is that the profile HMM treats gaps in a systematic way.

The HMM can be applied to other types of problems. It is particularly well suited for problems with a simple "grammatical structure," such as gene finding. In gene finding several signals must be recognized and combined into a prediction of exons and introns, and the prediction must conform to various rules to make it a reasonable gene prediction.

An HMM can combine recognition of the signals, and it can be made such that the predictions always follow the rules of a gene.

Since much of the literature on HMMs is a little hard to read for many biologists, I will attempt in this chapter to give a non-mathematical introduction to HMMs. Whereas the biological background needed is taken for granted, I have tried to explain HMMs at a level that almost anyone can follow. Both profile HMMs and gene finding will be considered below. First HMMs are introduced by an example and then profile HMMs are described. Then an HMM for finding eukaryotic genes is sketched, and finally pointers to the literature are given.

2. From regular expressions to HMMs

Most readers have no doubt come across regular expressions at some point, and many probably use them quite a lot. Regular expressions are used in many programs, in particular on Unix computers. In programs like awk, grep, sed, and perl, regular expressions can be used for searching text files for a pattern. With grep for instance, you can search a file for all lines containing "*C. elegans*" or "*Caenorhabditis elegans*" with the regular expression "C[\.a-z]* elegans". This will match any line containing a C followed by any number of lower-case letters or ".", then a space and then elegans. Regular expressions can also be used to characterize protein families, which is the basis for the PROSITE database [6].

Using regular expressions is a very elegant and efficient way to search for some protein families, but difficult for other. As already mentioned in the introduction, the difficulties arise because protein spelling is much more free than English spelling. Therefore the regular expressions sometimes need to be very broad and complex. Imagine a DNA motif like this:

```
ACA---ATG
TCAACTATC
ACAC--AGC
AGA---ATC
ACCG--ATC
```

(I use DNA only because of the smaller number of letters than for amino acids). A regular expression for this is

```
[AT] [CG] [AC] [ACGT]*A [TG] [GC],
```

meaning that the first position is A or T, the second C or G, and so forth. The term "[ACGT]*" means that any of the four letters can occur any number of times.

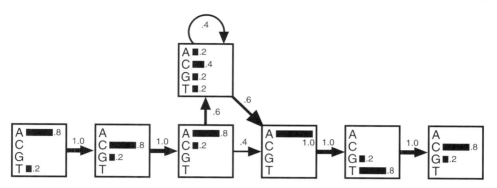

Fig. 1. A hidden Markov model derived from the alignment discussed in the text. The transitions are shown with arrows whose thickness indicate their probability. In each state the histogram shows the probabilities of the four nucleotides.

The problem with the above regular expression is that it does not in any way distinguish between the highly implausible sequence

```
TGCT--AGG
```

which has the exceptional character in each position, and the consensus sequence

```
ACAC--ATC
```

with the most plausible character in each position (the dashes are just for aligning these sequences with the previous ones). What is meant by a "plausible" sequence can of course be debated, although most would probably agree that the first sequence is not likely to be the same motif as the 5 sequences above. It is possible to make the regular expression more discriminative by splitting it into several different ones, but it easily becomes messy. The alternative is to score sequences by how well they fit the alignment.

To score a sequence, we say that there is a probability of $\frac{4}{5} = 0.8$ for an A in the first position and $\frac{1}{5} = 0.2$ for a T, because we observe that out of 5 letters 4 are an A and one is a T. Similarly in the second position the probability of C is $\frac{4}{5}$ and of G $\frac{1}{5}$, and so forth. After the third position in the alignment, 3 out of 5 sequences have "insertions" of varying lengths, so we say the probability of making an insertion is $\frac{3}{5}$ and thus $\frac{2}{5}$ for not making one. To keep track of these numbers a diagram can be drawn with probabilities as in Fig. 1.

This is a hidden Markov model. A box in the drawing is called a state, and there is a state for each term in the regular expression. All the probabilities are found simply by counting in the multiple alignment how many times each event occurs, just as described above. The only part that might seem tricky is the "insertion," which is represented by the state above the other states. The probability of each letter is found by counting all occurrences of the four nucleotides in this region of the alignment. The total counts are one A, two Cs, one G, and one T, yielding probabilities $\frac{1}{5}$, $\frac{2}{5}$, $\frac{1}{5}$ and $\frac{1}{5}$, respectively. After sequences 2, 3 and 5 have made one insertion each, there are two more insertions (from

Table 1

Probabilities and log-odds scores for the 5 sequences in the alignment and for the consensus sequence and the 'exceptional' sequence

	Sequence	Probability $\times 100$	Log odds
Consensus	A C A C - - A T C	4.7	6.7
Original sequences	A C A - - - A T G	3.3	4.9
	T C A A C T A T C	0.0075	3.0
	A C A C - - A G C	1.2	5.3
	A G A - - - A T C	3.3	4.9
	A C C G - - A T C	0.59	4.6
Exceptional	T G C T - - A G G	0.0023	−0.97

sequence 2) and the total number of transitions back to the main line of states is 3 (all three sequences with insertions have to finish). Therefore there are 5 transitions in total from the insert state, and the probability of making a transition to itself is $\frac{2}{5}$ and the probability of making one to the next state is $\frac{3}{5}$.

It is now easy to score the consensus sequence ACACATC. The probability of the first A is $\frac{4}{5}$. This is multiplied by the probability of the transition from the first state to the second, which is 1. Continuing this, the total probability of the consensus is

$$P(\text{ACACATC}) = 0.8 \times 1 \times 0.8 \times 1 \times 0.8 \times 0.6 \times 0.4 \times 0.6 \times 1 \times 1 \times 0.8 \times 1 \times 0.8$$
$$\simeq 4.7 \times 10^{-2}.$$

Making the same calculation for the exceptional sequence yields only 0.0023×10^{-2}, which is roughly 2000 times smaller than for the consensus. This way we achieved the goal of getting a score for each sequence, a measure of how well a sequence fits the motif.

The same probability can be calculated for the four original sequences in the alignment in exactly the same way, and the result is shown in Table 1. The probability depends very strongly on the length of the sequence. Therefore the probability itself is not the most convenient number to use as a score, and the log-odds score shown in the last column of the table is usually better. It is the logarithm of the probability of the sequence divided by the probability according to a null model. The null model is one that treats the sequences as random strings of nucleotides, so the probability of a sequence of length L is 0.25^L. Then the log-odds score is

$$\text{log-odds for sequence } S = \log \frac{P(S)}{0.25^L} = \log P(S) - L \log 0.25.$$

I used the natural logarithm in Table 1. Logarithms are proportional, so it does not really matter which one you use; it is quite common to use the logarithm base 2. One can of course use other null models instead. Often one would use the overall nucleotide frequencies in the organism studied instead of just 0.25.

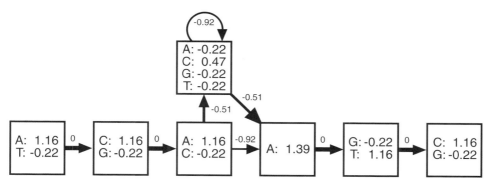

Fig. 2. The probabilities of the model in Fig. 1 have been turned into log-odds by taking the logarithm of each nucleotide probability and subtracting log(0.25). The transition probabilities have been converted to simple logs.

When a sequence fits the motif very well the log-odds is high. When it fits the null model better, the log-odds score is negative. Although the raw probability of the second sequence (the one with three inserts) is almost as low as that of the exceptional sequence, notice that the log-odds score is much higher than for the exceptional sequence, and the discrimination is very good. Unfortunately, one can not always assume that anything with a positive log-odds score is "a hit," because there are random hits if one is searching a large database. See section 5 for references.

Instead of working with probabilities one might convert everything to log-odds. If each nucleotide probability is divided by the probability according to the null model (0.25 in this case) and the logarithm is applied, we would get the numbers shown in Fig. 2. The transition probabilities are also turned into logarithms. Now the log-odds score can be calculated directly by adding up these numbers instead of multiplying the probabilities. For instance, the calculation of the log-odds of the consensus sequence is

$$\begin{aligned} \text{log-odds(ACACATC)} &= 1.16 + 0 + 1.16 + 0 + 1.16 - 0.51 + 0.47 - 0.51 \\ &\quad + 1.39 + 0 + 1.16 + 0 + 1.16 \\ &= 6.64. \end{aligned}$$

(The finite precision causes the little difference between this number and the one in Table 1.)

If the alignment had no gaps or insertions we would get rid of the insert state, and then all the probabilities associated with the arrows (the transition probabilities) would be 1 and might as well be ignored completely. Then the HMM works exactly as a weight matrix of log-odds scores, which is commonly used.

3. Profile HMMs

A profile HMM is a certain type of HMM with a structure that in a natural way allows position-dependent gap penalties. A profile HMM can be obtained from a multiple

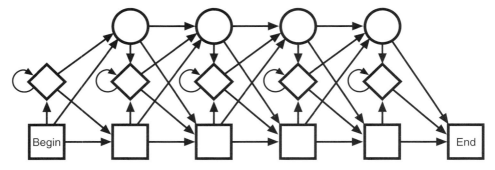

Fig. 3. The structure of the profile HMM.

alignment and can be used for searching a database for other members of the family in the alignment very much like standard profiles [3]. The structure of the model is shown in Fig. 3. The bottom line of states are called the main states, because they model the columns of the alignment. In these states the probability distribution is just the frequency of the amino acids or nucleotides as in the above model of the DNA motif. The second row of diamond shaped states are called insert states and are used to model highly variable regions in the alignment. They function exactly like the top state in Fig. 1, although one might choose to use a fixed distribution of residues, e.g. the overall distribution of amino acids, instead of calculating the distribution as in the example above. The top line of circular states are called delete states. These are a different type of state, called a silent or null state. They do not match any residues, and they are there merely to make it possible to jump over one or more columns in the alignment, i.e. to model the situation when just a few of the sequences have a "–" in the multiple alignment at a position. Let us turn to an example.

Suppose you have a multiple alignment as the one shown in Fig. 4. A region of this alignment has been chosen to be an "insertion," because an alignment of this region is highly uncertain. The rest of the alignment (shaded in the figure) are the columns that will correspond to main states in the model. For each non-insert column we make a main state and set the probabilities equal to the amino acid frequencies. To estimate the transition probabilities we count how many sequences use the various transitions, just like the transition probabilities were calculated in the first example. The model is shown in Fig. 5. There are two transitions from a main state to a delete state shown with dashed lines in the figure, that from "begin" to the first delete state and from main state 12 to delete state 13. Both of these correspond to dashes in the alignment. In both cases only one sequence has gaps, so the probability of these delete transitions is $\frac{1}{30}$. The fourth sequence continues deletion to the end, so the probability of going from delete 13 to 14 is 1 and from delete 14 to the end is also 1.

3.1. Pseudocounts

It is dangerous to estimate a probability distribution from just a few observed amino acids. If, for instance, you have just two sequences with leucine at a certain position, the probability for leucine would be 1 and the probability would be zero for all other amino

G	G	W	W	R	G	d	y	.	g	g	k	k	q	L	W	F	P	S	N	Y	V
I	G	W	L	N	G	y	n	e	t	t	g	e	r	G	D	F	P	G	T	Y	V
P	N	W	W	E	G	q	l	.	.	n	n	r	r	G	I	F	P	S	N	Y	V
D	E	W	W	Q	A	r	r	.	.	d	e	q	i	G	I	V	P	S	K	–	–
G	E	W	W	K	A	q	s	.	.	t	g	q	e	G	F	I	P	F	N	F	V
G	D	W	W	L	A	r	s	.	.	s	g	q	t	G	Y	I	P	S	N	Y	V
G	D	W	W	D	A	e	l	.	.	k	g	r	r	G	K	V	P	S	N	Y	L
–	D	W	W	E	A	r	s	l	s	s	g	h	r	G	Y	V	P	S	N	Y	V
G	D	W	W	Y	A	r	s	l	i	t	n	s	e	G	Y	I	P	S	T	Y	V
G	E	W	W	K	A	r	s	l	a	t	r	k	e	G	Y	I	P	S	N	Y	V
G	D	W	W	L	A	r	s	l	v	t	g	r	e	G	Y	V	P	S	N	F	V
G	E	W	W	K	A	k	s	l	s	s	k	r	e	G	F	I	P	S	N	Y	V
G	E	W	C	E	A	q	t	.	k	n	g	q	.	G	W	V	P	S	N	Y	I
S	D	W	W	R	V	v	n	l	t	t	r	q	e	G	L	I	P	L	N	F	V
L	P	W	W	R	A	r	d	.	k	n	g	q	e	G	Y	I	P	S	N	Y	I
R	D	W	W	E	F	r	s	k	t	v	y	t	p	G	Y	Y	E	S	G	Y	V
E	H	W	W	K	V	k	d	.	a	l	g	n	v	G	Y	I	P	S	N	Y	V
I	H	W	W	R	V	q	d	.	r	n	g	h	e	G	Y	V	P	S	S	Y	L
K	D	W	W	K	V	e	v	.	.	n	d	r	q	G	F	V	P	A	A	Y	V
V	G	W	M	P	G	l	n	e	r	t	r	q	r	G	D	F	P	G	T	Y	V
P	D	W	W	E	G	e	l	.	.	n	g	q	r	G	V	F	P	A	S	Y	V
E	N	W	W	N	G	e	i	.	.	g	n	r	k	G	I	F	P	A	T	Y	V
E	E	W	L	E	G	e	c	.	.	k	g	k	v	G	I	F	P	K	V	F	V
G	G	W	W	K	G	d	y	.	g	t	r	i	q	Q	Y	F	P	S	N	Y	V
D	G	W	W	R	G	s	y	.	.	n	g	q	v	G	W	F	P	S	N	Y	V
Q	G	W	W	R	G	e	l	.	.	y	g	r	v	G	W	F	P	A	N	Y	V
G	R	W	W	K	A	r	r	.	a	n	g	e	t	G	I	I	P	S	N	Y	V
G	G	W	T	Q	G	e	l	.	k	s	g	q	k	G	W	A	P	T	N	Y	L
G	D	W	W	E	A	r	s	n	.	t	g	e	n	G	Y	I	P	S	N	Y	V
N	D	W	W	T	G	r	t	.	.	n	g	k	e	G	I	F	P	A	N	Y	V

Fig. 4. An alignment of 30 short amino acid sequences chopped out of a alignment of the SH3 domain. The shaded areas are the most conserved and were chosen to be represented by the main states in the HMM. The unshaded area with lower-case letters was chosen to be represented by an insert state.

acids at this position, although it is well known that one often sees, for example, valine substituted for leucine. In such a case the probability of a whole sequence may easily become zero if a single leucine is substituted by a valine, or equivalently, the log-odds is minus infinity.

Therefore it is important to have some way of avoiding this sort of overfitting, where strong conclusions are drawn from very little evidence. The most common method is to use pseudocounts, which means that one pretends to have more counts of amino acids than those from the data. The simplest is to add 1 to all the counts. In the leucine example the probability of leucine would then be estimated as $\frac{3}{23}$ and for the 19 other amino acids it would become $\frac{1}{23}$. In Fig. 6 a model is shown, which was obtained from the alignment in Fig. 4 using a pseudocount of 1.

Adding one to all the counts can be interpreted as assuming *a priori* that all the amino acids are equally likely. However, there are significant differences in the occurrence of the 20 amino acids in known protein sequences. Therefore, the next

52

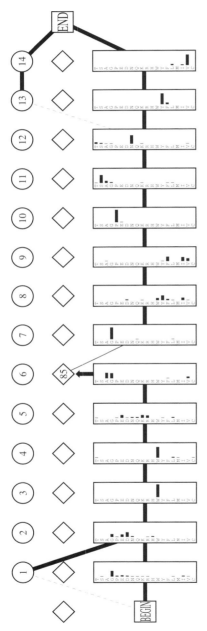

Fig. 5. A profile HMM made from the alignment shown in Fig. 4. Transition lines with no arrow head are transitions from left to right. Transitions with probability zero are not shown, and those with very small probability are shown as dashed lines. Transitions from an insert state to itself are not shown; instead the probability times 100 is shown in the diamond. The numbers in the circular delete states are just position numbers. (This figure and Fig. 6 were generated by a program in the SAM package of programs.)

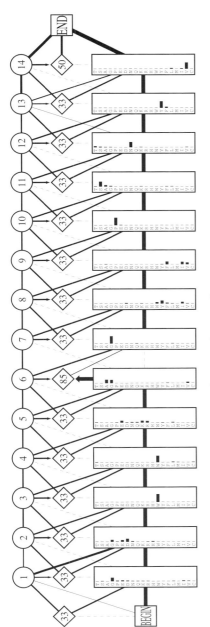

Fig. 6. Model obtained in the same way as Fig. 5, but using a pseudocount of one.

step is to use pseudocounts proportional to the observed frequencies of the amino acids instead. This is the minimum level of pseudocounts to be used in any real application of HMMs.

Because a column in the alignment may contain information about the preferred type of amino acids, it is also possible to use more sophisticated pseudocount strategies. If a column consists predominantly of leucine (as above), one would expect substitutions to other hydrophobic amino acids to be more probable than substitutions to hydrophilic amino acids. One can, for example, derive pseudocounts for a given column from substitution matrices. See section 5 for references.

3.2. Searching a database

Above we saw how to calculate the probability of a sequence in the alignment by multiplying all the probabilities (or adding the log-odds scores) in the model along the *path* followed by that particular sequence. However, this path is usually not known for other sequences which are not part of the original alignment, and the next problem is how to score such a sequence. Obviously, if we can find a path through the model where the new sequence fits well in some sense, then we can score the sequence as before. We need to "align" the sequence to the model. It resembles very much the pairwise alignment problem, where two sequences are aligned so that they are most similar, and indeed the same type of dynamic programming algorithm can be used.

For a particular sequence, an alignment to the model (or a path) is an assignment of states to each residue in the sequence. There are many such alignments for a given sequence. For instance, an alignment might be as follows. Let us label the amino acids in a protein as A_1, A_2, A_3, etc. Similarly we can label the HMM states as M_1, M_2, M_3, etc. for match states, I_1, I_2, I_3 for insert states, and so on. Then an alignment could have A_1 match state M_1, A_2 and A_3 match I_1, A_4 match M_2, A_5 match M_6 (after passing through three delete states), and so on. For each such path we can calculate the probability of the sequence or the log-odds score, and thus we can find the *best* alignment, i.e. the one with the largest probability. Although there are an enormous number of possible alignments it can be done efficiently by the above mentioned dynamic programming algorithm, which is called the Viterbi algorithm. The algorithm also gives the probability of the sequence for that alignment, and thus a score is obtained.

The log-odds score found in this manner can be used to search databases for members of the same family. A typical distribution of scores from such a search is shown in Fig. 7. As is also the case with other types of searches, there is no clear-cut separation of true and false positives, and one needs to investigate some of the sequences around a log-odds of zero, and possibly include some of them in the alignment and try searching again.

An alternative way of scoring sequences is to *sum* the probabilities of all possible alignments of the sequence to the model. This probability can be found by a similar algorithm called the forward algorithm. This type of scoring is not very common in biological sequence comparison, but it is more natural from a probabilistic point of view. However, it usually gives very similar results.

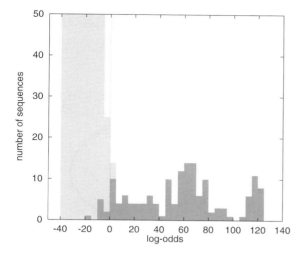

Fig. 7. The distribution of log-odds scores from a search of Swissprot with a profile HMM of the SH3 domain. The dark area of the histogram represents the sequences with an annotated SH3 domain, and the light those that are not annotated as having one. This is for illustrative purposes only, and the sequences with log-odds around zero were not investigated further.

3.3. Model estimation

As presented so far, one may view the profile HMMs as a generalization of weight matrices to incorporate insertions and deletions in a natural way. There is however one interesting feature of HMMs, which has not been addressed yet. It is possible to estimate the model, i.e. determine all the probability parameters of it, from *unaligned* sequences. Furthermore, a multiple alignment of the sequences is produced in the process. Like many other multiple alignment methods this is done in an iterative manner. One starts out with a model with more or less random probabilities, or if a reasonable alignment of some of the sequences are available, a model is constructed from this alignment. Then, when all the sequences are aligned to the model, we can use the alignment to improve the probabilities in the model. These new probabilities may then lead to a slightly different alignment. If they do, we then repeat the process and improve the probabilities again. The process is repeated until the alignment does not change. The alignment of the sequences to the final model yields a multiple alignment.[1]

Although this estimation process sounds easy, there are many problems to consider to actually make it work well. One problem is choosing the appropriate model length, which determines the number of inserts in the final alignment. Another severe problem is that the iterative procedure can converge to suboptimal solutions. It is not guaranteed that it finds the optimal multiple alignment, i.e. the most probable one. Methods for dealing with these issues are described in the literature pointed to in section 5.

---

[1] Another slightly different method for model estimation sums over all alignments instead of using the most probable alignment of a sequence to the model. This method uses the forward algorithm instead of Viterbi, and it is called the Baum–Welch algorithm or the forward–backward algorithm.

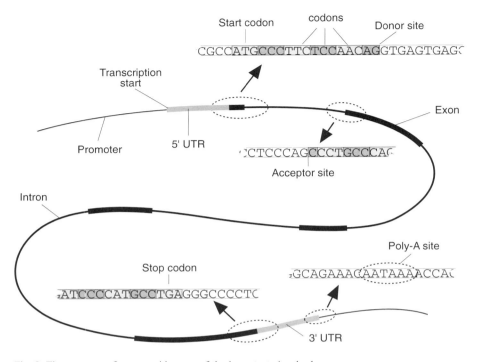

Fig. 8. The structure of a gene with some of the important signals shown.

4. HMMs for gene finding

One ability of HMMs, which is not really utilized in profile HMMs, is the ability to model *grammar*. Many problems in biological sequence analysis have a grammatical structure, and eukaryotic gene structure, which I will use as an example, is one of them. If you consider exons and introns as the "words" in a language, the sentences are of the form exon–intron–exon–intron· · ·intron–exon. The "sentences" can never end with an intron, at least if the genes are complete, and an exon can never follow an exon without an intron in between. Obviously, this grammar is greatly simplified because there are several other constraints on gene structure, such as the constraint that the exons have to fit together to give a valid coding region after splicing. In Fig. 8 the structure of a gene is shown with some of the known signals marked.

Formal language theory applied to biological problems is not a new invention. In particular, David Searls [7] has promoted this idea and used it for gene finding [8], but many other gene finders use it implicitly. Formally, the HMM can only represent the simplest of grammars, which is called a regular grammar [7,1], but that turns out to be good enough for the gene finding problem, and many other problems. One of the problems that has a more complicated grammar than the HMM can handle is the RNA folding problem, which is one step up the ladder of grammars, because base pairing introduces correlations between bases far from each other in the RNA sequence.

I will here briefly outline my own approach to gene finding with the emphasis on the principles rather than on the details.

4.1. Signal sensors

One may apply an HMM similar to the ones already described directly to many of the signals in a gene structure. In Fig. 9 an alignment is shown of some sequences around acceptor sites from human DNA. It has 19 columns and an HMM with 19 states (no insert or delete states) can be made directly from it. Since the alignment is gapless, the HMM is equivalent to a weight matrix.

There is one problem: in DNA there are fairly strong dinucleotide preferences. A model like the one described treats the nucleotides as independent, so dinucleotide preferences can not be captured. This is easily fixed by having 16 probability parameters in each state instead of 4. In column 2 we first count all occurrences of the four nucleotides given that there is an A in the first column and normalize these four counts, so they become probabilities. This is the *conditional probability* that a certain nucleotide appears in position two, given that the previous one was A. The same is done for all the instances of C in column 1 and similarly for G and T. This gives a total of 16 probabilities to be used in state 2 of the HMM. Similarly for all the other states. To calculate the probability of a sequence, say ACTGTC..., we just multiply the conditional probabilities

$$P(\text{ACTGTC}\cdots) = p_1(\text{A}) \times p_2(\text{C}|\text{A}) \times p_3(\text{T}|\text{C}) \times p_4(\text{G}|\text{T}) \times p_5(\text{T}|\text{G}) \times p_6(\text{C}|\text{T}) \times \cdots.$$

Here p_1 is the probability of the four nucleotides in state 1, $p_2(x|y)$ is the conditional probability in state 2 of nucleotide x given that the previous nucleotide was y, and so forth.

A state with conditional probabilities is called a first order state, because it captures the first order correlations between neighboring nucleotides. It is easy to expand to higher order. A second order state has probabilities conditioned on the two previous nucleotides in the sequence, i.e. probabilities of the form $p(x|y,z)$. We will return to such higher order states below.

Small HMMs like this are constructed in exactly the same way for other signals: donor splice sites, the regions around the start codons, and the regions around the stop codons.

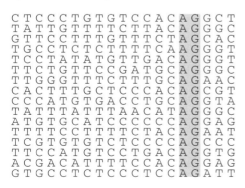

Fig. 9. Examples of human acceptor sites (the splice site 5′ to the exon). Except in rare cases, the intron ends with AG, which has been highlighted. Included in these sequences are 16 bases upstream of the splice site and 3 bases downstream into the exon.

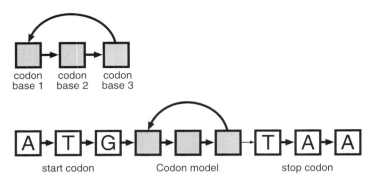

Fig. 10. Top: a model of coding regions, where state one, two and three match the first, second and third codon positions, respectively. A coding region of any length can match this model, because of the transition from state three back to state one. Bottom: a simple model for unspliced genes with the first three states matching a start codon, the next three of the form shown to the left, and the last three states matching a stop codon (only one of the three possible stop codons are shown).

4.2. Coding regions

The codon structure is the most important feature of coding regions. Bases in triplets can be modeled with three states as shown in Fig. 10. The figure also shows how this model of coding regions can be used in a simple model of an unspliced gene that starts with a start codon (ATG), then consists of some number of codons, and ends with a stop codon.

Since a codon is three bases long, the last state of the codon model must be at least of order two to correctly capture the codon statistics. The 64 probabilities in such a state are estimated by counting the number of each codon in a set of known coding regions. These numbers are then normalized properly. For example the probabilities derived from the counts of CAA, CAC, CAG and CAT are

$$p(A|CA) = c(CAA)/[c(CAA) + c(CAC) + c(CAG) + c(CAT)],$$
$$p(C|CA) = c(CAC)/[c(CAA) + c(CAC) + c(CAG) + c(CAT)],$$
$$p(G|CA) = c(CAG)/[c(CAA) + c(CAC) + c(CAG) + c(CAT)],$$
$$p(T|CA) = c(CAT)/[c(CAA) + c(CAC) + c(CAG) + c(CAT)],$$

where $c(xyz)$ is the count of codon xyz.

One of the characteristics of coding regions is the lack of stop codons. That is automatically taken care of, because $p(A|TA)$, $p(G|TA)$ and $p(A|TG)$, corresponding to the three stop codons TAA, TAG and TGA, will automatically become zero.

For modeling codon statistics it is natural to use an ordinary (zeroth order) state as the first state of the codon model and a first order state for the second. However, there are actually also dependencies between neighboring codons, and therefore one may want even higher order states. In my own gene finder, I currently use three fourth order states, which is inspired by GeneMark [9], in which such models were first introduced. Technically speaking, such a model is called an inhomogeneous Markov chain, which can be viewed as a subclass of HMMs.

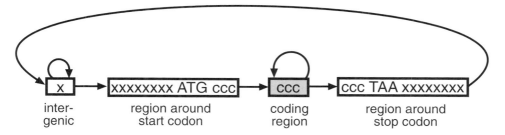

Fig. 11. A hidden Markov model for unspliced genes. In this drawing an "x" means a state for non-coding DNA, and a "c" a state for coding DNA. Only one of the three possible stop codons is shown in the model of the region around the stop codon.

4.3. Combining the models

To be able to discover genes, we need to combine the models in a way that satisfies the grammar of genes. I restrict myself to coding regions, i.e. the 5' and 3' untranslated regions of the genes are not modeled and also promoters are disregarded.

First, let us see how to do it for unspliced genes. If we ignore genes that are very closely spaced or overlap, a model could look like Fig. 11. It consists of a state for intergenic regions (of order at least 1), a model for the region around the start codon, the model for the coding region, and a model for the region around the stop codon. The model for the start codon region is made just like the acceptor model described earlier. It models eight bases upstream of the start codon[2], the ATG start codon itself, and the first codon after the start. Similarly for the stop codon region. The whole model is one big HMM, although it was put together from small independent HMMs.

Having such a model, how can we predict genes in a sequence of anonymous DNA? That is quite simple: use the Viterbi algorithm to find the most probable path through the model. When this path goes through the ATG states, a start codon is predicted, when it goes through the codon states a codon is predicted, and so on.

This model might not always predict correct genes, but at least it will only predict sensible genes that obey the grammatical rules. A gene will always start with a start codon and end with a stop codon, the length will always be divisible by 3, and it will never contain stop codons in the reading frame, which are the minimum requirements for unspliced gene candidates.

Making a model that conforms to the rules of splicing is a bit more difficult than it might seem at first. That is because splicing can happen in three different reading frames, and the reading frame in one exon has to fit the one in the next. It turns out that by using three different models of introns, one for each frame, this is possible. In Fig. 12 it is shown how these models are added to the model of coding regions.

The top line in the model is for introns appearing between two codons. It has three states (labeled ccc) before the intron starts to match the last codon of the exon. The

---

[2] A similar model could be used for prokaryotic genes. In that case, however, one should model the Shine–Dalgarno sequence, which is often more than 8 bases upstream from the start. Also, one would probably need to allow for other start codons than ATG that are used in the organism studied (in some eukaryotes other start codons can also be used).

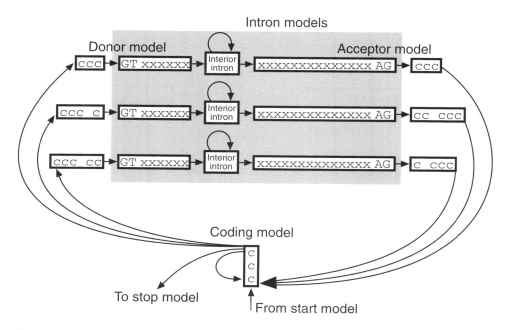

Fig. 12. To allow for splicing in three different frames three intron models are needed. To get the frame correct "spacer states" are added before and after the intron models.

first two states of the intron model match GT, which is the consensus sequence at donor sites (it is occasionally another sequence, but such cases are ignored here). The next six states matches the six bases immediately after GT. The states just described model the donor site, and the probabilities are found as it was described earlier for acceptor sites. Then follows a single state to model the interior of the intron. I actually use the same probability parameters in this state as in the state modeling intergenic regions. Now follows the acceptor model, which includes three states to match the first codon of the next exon.

The next line in the model is for introns appearing after the first base in a codon. The difference from the first is that there is one more state for a coding base before the intron and two more states after the intron. This ensures that the frames fit in two neighboring exons. Similarly in the third line from the top there are two extra coding states before the intron and one after, so that it can match introns appearing after the second base in a codon.

There are obviously many possible variations of the model. One can add more states to the signal sensors, include models of promoter elements and untranslated regions of the gene, and so forth.

5. Further reading

A general introduction can be found in ref. [2], and one aimed more at biological readers in ref. [1]. The first applications for sequence analysis are probably for modeling compositional differences between various DNA types [10] and for studying the compositional structure of genomes [11]. The initial work on using hidden Markov models as "probabilistic profiles" of protein families was presented at a conference in the spring of 1992 and in a technical report from the same year, and it was published in ref. [12,13]. The idea was quickly taken up by others [14,15]. Independently, some very similar ideas were also developed in ref. [16,17]. Also the generalized profiles [18] are very similar.

Estimation and multiple alignment is described in ref. [13] in detail, and in ref. [19] some of the practical methods are further discussed. Alternative methods for model estimation are presented in ref. [14,20]. Methods for scoring sequences against a profile HMM were given in ref. [13], but these issues have more recently been addressed in ref. [21]. The basic pseudocount method is also explained in ref. [13], and more advanced methods are discussed in ref. [22–26].

A review of profile HMMs can be found in ref. [27], and in ref. [1] profile HMMs are discussed in great detail. Also ref. [28] will undoubtedly contain good material on profile HMMs.

Some of the recent applications of profile HMMs to proteins are: detection of fibronectin type III domains in yeast [29], a database of protein domain families [30], protein topology recognition from secondary structure [31], and modeling of a protein splicing domain [32].

There are two program packages available free of charge to the academic community. One, developed by Sean Eddy, is called hmmer (pronounced "hammer"), and can be obtained from his web-site (http://genome.wustl.edu/eddy/hmm.html). The other one, called SAM (http://www.cse.ucsc.edu/research/compbio/sam.html), was developed by myself and the group at UC Santa Cruz, and it is now being maintained and further developed by Richard Hughey.

The gene finder sketched above is called HMMgene. There are many details omitted, such as special methods for estimation and prediction described in ref. [33]. It is still under development, and it is possible to follow the development and test the current version at the web-site http://www.cbs.dtu.dk/services/HMMgene/.

Methods for automated gene finding go back a long time, see [34] for a review. The first HMM-based gene finder is probably EcoParse developed for *E. coli* [35]. VEIL [36] is a recent HMM-based gene finder for human genes. The main difference from HMMgene is that it does not use high order states (neither does EcoParse), which makes good modeling of coding regions harder.

Two recent methods use so-called generalized HMMs. Genie [37–39] combines neural networks into an HMM-like model, whereas GENSCAN [40] is more similar to HMMgene, but uses a different model type for splice site. Also, the generalized HMM can explicitly use exon length distributions, which is not possible in a standard HMM. Web pointers to gene finding can be found at http://www.cbs.dtu.dk/krogh/genefinding.html.

Other applications of HMMs related to gene finding are: detection of short protein coding regions and analysis of translation initiation sites in *Cyanobacterium* [41,42], characterization of prokaryotic and eukaryotic promoters [43], and recognition of branch points [44].

Apart from the areas mentioned here, HMMs have been used for prediction of protein secondary structure [45], modeling an oscillatory pattern in nucleosomes [46], modeling site dependence of evolutionary rates [47], and for including evolutionary information in protein secondary structure prediction [48].

Acknowledgments

Henrik Nielsen, Steven Salzberg, and Steen Knudsen are gratefully acknowledged for comments on the manuscript. This work was supported by the Danish National Research Foundation.

References

[1] Durbin, R.M., Eddy, S.R., Krogh, A. and Mitchison, G. (1998) Biological Sequence Analysis. Cambridge University Press.
[2] Rabiner, L.R. (1989) Proc. IEEE 77, 257–286.
[3] Gribskov, M., McLachlan, A.D. and Eisenberg, D. (1987) Proc. Natl. Acad. Sci. USA 84, 4355–4358.
[4] Staden, R. (1984) Nucleic Acids Res. 12, 505–519.
[5] Staden, R. (1988) Comput. Appl. Biosci. 4, 53–60.
[6] Bairoch, A., Bucher, P. and Hofmann, K. (1997) Nucleic Acids Res. 25, 217–221.
[7] Searls, D.B. (1992) Am. Sci. 80, 579–591.
[8] Dong, S. and Searls, D.B. (1994) Genomics 23, 540–551.
[9] Borodovsky, M. and McIninch, J. (1993) Comput. & Chem. 17, 123–133.
[10] Churchill, G.A. (1989) Bull. Math. Biol. 51, 79–94.
[11] Churchill, G.A. (1992) Comput. & Chem. 16, 107–115.
[12] Haussler, D., Krogh, A., Mian, I.S. and Sjölander, K. (1993) Protein modeling using hidden Markov models: Analysis of globins. In: T.N. Mudge, V. Milutinovic and L. Hunter (Eds.), Proc. 26th Annu. Hawaii Int. Conf. on System Sciences. IEEE Computer Society Press, Los Alamitos, CA, Vol. 1, pp. 792–802.
[13] Krogh, A., Brown, M., Mian, I.S., Sjölander, K. and Haussler, D. (1994) J. Mol. Biol. 235, 1501–1531.
[14] Baldi, P., Chauvin, Y., Hunkapiller, T. and McClure, M.A. (1994) Proc. Natl. Acad. Sci. USA 91, 1059–1063.
[15] Eddy, S.R., Mitchison, G. and Durbin, R. (1995) J. Comput. Biol. 2, 9–23.
[16] Stultz, C.M., White, J.V. and Smith, T.F. (1993) Prot. Sci. 2, 305–314.
[17] White, J.V., Stultz, C.M. and Smith, T.F. (1994) Math. Biosci. 119, 35–75.
[18] Bucher, P., Karplus, K., Moeri, N. and Hofmann, K. (1996) Comput. & Chem. 20, 3–24.
[19] Hughey, R. and Krogh, A. (1996) CABIOS 12, 95–107.
[20] Eddy, S.R. (1995) Multiple alignment using hidden Markov models. In: C. Rawlings, D. Clark, R. Altman, L. Hunter, T. Lengauer and S. Wodak (Eds.), Proc. 3rd Int. Conf. on Intelligent Systems for Molecular Biology. AAAI Press, Menlo Park, CA, Vol. 3, pp. 114–120.
[21] Barrett, C., Hughey, R. and Karplus, K. (1997) Comput. Appl. Biosci. 13, 191–199.
[22] Brown, M., Hughey, R., Krogh, A., Mian, I.S., Sjölander, K. and Haussler, D. (1993) Using Dirichlet mixture priors to derive hidden Markov models for protein families. In: L. Hunter, D. Searls and J. Shavlik (Eds.), Proc. 1st Int. Conf. on Intelligent Systems for Molecular Biology. AAAI/MIT Press, Menlo Park, CA, pp. 47–55.
[23] Tatusov, R.L., Altschul, S.F. and Koonin, E.V. (1994) Proc. Natl. Acad. Sci. USA 91, 12091–12095.
[24] Karplus, K. (1995) Evaluating regularizers for estimating distributions of amino acids. In: C. Rawlings, D. Clark, R. Altman, L. Hunter, T. Lengauer and S. Wodak (Eds.), Proc. 3rd Int. Conf. on Intelligent Systems for Molecular Biology. AAAI Press, Menlo Park, CA, Vol. 3, pp. 188–196.
[25] Henikoff, J.G. and Henikoff, S. (1996) Comput. Appl. Biosci. 12, 135–143.

[26] Sjölander, K., Karplus, K., Brown, M., Hughey, R., Krogh, A., Mian, I.S. and Haussler, D. (1996) CABIOS 12, 327–345.
[27] Eddy, S.R. (1996) Curr. Opin. Struct. Biol. 6, 361–365.
[28] Baldi, P. and Brunak, S. (1998) Bioinformatics – The Machine Learning Approach. MIT Press, Cambridge, MA.
[29] Bateman, A. and Chothia, C. (1997) Curr. Biol. 6, 1544–1547.
[30] Sonnhammer, E.L.L., Eddy, S.R. and Durbin, R. (1997) Proteins 28, 405–420.
[31] Di Francesco, V., Garnier, J. and Munson, P.J. (1997) J. Mol. Biol. 267, 446–463.
[32] Dalgaard, J.Z., Moser, M.J., Hughey, R. and Mian, I.S. (1997) J. Comput. Biol. 4, 193–214.
[33] Krogh, A. (1997) Two methods for improving performance of a HMM and their application for gene finding. In: T. Gaasterland, P. Karp, K. Karplus, C. Ouzounis, C. Sander and A. Valencia (Eds.), Proc. 5th Int. Conf. on Intelligent Systems for Molecular Biology. AAAI Press, Menlo Park, CA, pp. 179–186.
[34] Fickett, J.W. (1996) Trends Genet. 12, 316–320.
[35] Krogh, A., Mian, I.S. and Haussler, D. (1994) Nucleic Acids Res. 22, 4768–4778.
[36] Henderson, J., Salzberg, S. and Fasman, K.H. (1997) J. Comput. Biol. 4, 127–141.
[37] Kulp, D., Haussler, D., Reese, M.G. and Eeckman, F.H. (1996) A generalized hidden Markov model for the recognition of human genes in DNA. In: D. States, P. Agarwal, T. Gaasterland, L. Hunter and R. Smith (Eds.), Proc. Conf. on Intelligent Systems in Molecular Biology. AAAI Press, Menlo Park, CA, pp. 134–142.
[38] Reese, M.G., Eeckman, F.H., Kulp, D. and Haussler, D. (1997) J. Comput. Biol. 4, 311–323.
[39] Kulp, D., Haussler, D., Reese, M.G. and Eeckman, F.H. (1997) Integrating database homology in a probabilistic gene structure model. In: R.B. Altman, A.K. Dunker, L. Hunter and T.E. Klein (Eds.), Proc. Pacific Symp. on Biocomputing. World Scientific, New York, pp. 232–244.
[40] Burge, C. and Karlin, S. (1997) J. Mol. Biol. 268, 78–94.
[41] Yada, T. and Hirosawa, M. (1996) DNA Res. 3, 355–361.
[42] Yada, T., Sazuka, T. and Hirosawa, M. (1997) DNA Res. 4, 1–7.
[43] Pedersen, A.G., Baldi, P., Brunak, S. and Chauvin, Y. (1996) Characterization of prokaryotic and eukaryotic promoters using hidden Markov models. In: Proc. 4th Int. Conf. on Intelligent Systems for Molecular Biology. AAAI Press, Menlo Park, CA, pp. 182–191.
[44] Tolstrup, N., Rouzé, P. and Brunak, S. (1997) Nucleic Acids Res. 25, 3159–3164.
[45] Asai, K., Hayamizu, S. and Handa, K. (1993) Comput. Appl. Biosci. 9, 141–146.
[46] Baldi, P., Brunak, S., Chauvin, Y. and Krogh, A. (1996) J. Mol. Biol. 263, 503–510.
[47] Felsenstein, J. and Churchill, G.A. (1996) Mol. Biol. Evol. 13, 93–104.
[48] Goldman, N., Thorne, J.L. and Jones, D.T. (1996) J. Mol. Biol. 263, 196–208.

S.L. Salzberg, D.B. Searls, S. Kasif (Eds.), *Computational Methods in Molecular Biology*

CHAPTER 5

Case-based reasoning driven gene annotation

G. Christian Overton and Juergen Haas

Center for Bioinformatics, University of Pennsylvania,
1312 Blockley Hall, 418 Boulevard, Philadelphia, PA 19104, USA

1. Introduction

The Human Genome Project and its follow-ons are generating an unprecedented flood of information on genetic and physical maps of chromosomes, biological structures and sequences, and increasingly, biological functions of genes and proteins in humans and other organisms. As was pointed out in a National Academy of Sciences report [1] prior to the start of the Human Genome Project, the sheer mass of data, and its explosive growth and complexity will overwhelm our capacity to understand and interpret it, if technologies are not developed to automate data management and analysis. Recent advances in computational methodologies have greatly increased the power to analyze information in nucleic acid and protein sequences, although even these techniques fall short in the many cases where too few biological examples are available to form robust statistical characterizations. Progress on automating data management has, however, lagged considerably behind with consequences that range, at a minimum, from the propagation of erroneous information to the more serious possibility of overlooking new scientific insights. Technologies under development for "data mining" and "knowledge discovery in databases" may address some of these problems.

Biological data have a curious quality whether it comes from the vast primary scientific literature or from genome-scale, high-throughput projects: it is at once extremely abundant and simultaneously relatively sparse. Much of our attention in recent years has focussed on the abundant sources of data, especially information in the nucleic acid and protein sequence databases. Statistical analysis of these data has led for example to an understanding of the bulk properties of genes and proteins, and in some cases the characterization of classes of functional regions in these sequences. However, elucidating the mechanisms of a biological system requires a deep analysis of the specific details of each particular system. For example, in the study of the control of gene expression during development, examination of actin expression during muscle differentiation would provide only a very rough, and perhaps inaccurate guide, to understanding hemoglobin gene expression during red blood cell development. Certainly many of the same general mechanisms are employed in each case, but biological systems have been wonderfully inventive in the specific strategies that generate a particular behavior. Because there are only one or few of each gene type in an organism, and few of these have been exhaustively analyzed in any organism, the information available on these systems is sparse and not generally suitable for analysis by statistical methods. Consider that when the more than

3 billion base pairs of the human genome are determined there will be only (depending on how one counts) two genes for the adult form of β-hemoglobin.

One of the challenges we face in managing biology data is capturing and maintaining an accurate representation of our current state of knowledge when faced with both abundant and sparse data. It is instructive to examine the principle nucleic acid sequence databases – GenBank, EMBL and DDBJ (which we will collectively refer to throughout as simply GenBank) – to understand some of the issues in information currency and maintenance. Information on gene features such as exons, introns, coding regions are captured in the feature table of GenBank. Unfortunately, many GenBank entries suffer from incomplete, noisy (ambiguous or contradictory), and erroneous feature information. To a degree, these types of errors arise because feature annotation is entered, along with the sequence itself, by the experimentalists who describe the sequence rather than by trained annotators. And this situation can not easily be remedied because there are not enough trained annotators to keep pace with the volume of data generated. One consequence of this situation is that once feature annotation is entered for a particular gene, it is rarely updated. Ironically then, the genes which have been in the database longest, and therefore about which the most information is known in the literature, often have the least annotation.

Here we examine case-based reasoning (CBR), a form of analogical reasoning, as a means of managing predictions of sparsely represented features in a biological database. CBR has the desirable property that it closely models the approach often taken by biologists when reasoning about structure and function in a new, partially characterized system from examples of well characterized systems. This method of reasoning is particularly powerful in biology because it reflects the underlying evolutionary history that ties all modern life-forms together. CBR can thus be viewed as a formalization of one form of reasoning about biological systems.

From a database perspective, CBR provides a convenient framework for exploring issues on maintaining automatically an accurate and consistent view of data. From a biology perspective, we consider CBR an appropriate framework for making heuristic predictions over sparse features and for building a composite view of our current state of knowledge. With respect to the biology, we are particularly interested in information on the control of gene expression during vertebrate development and differentiation. In the following sections, we will concentrate on CBR as a system for predicting gene regulatory elements in DNA sequence while only briefly touching on the database aspects of this work.

2. Case-based reasoning

Generalized CBR draws well-characterized instances (source cases) similar by some measure to a new situation (query case) from a case database (case-base) and then proceeds to reason about the query case by adapting information from the source cases to the new situation. The motivation for CBR originally came from cognitive models of human reasoning where it was observed by Shank [2] and others in the early 1980s that much of day-to-day reasoning was based on remembering previous experiences and adapting them to solve novel problems rather than reasoning from generalized rules or first principles. For example, planning the menu for a dinner party often begins by recalling the menus from previous dinner parties and modifying them for the upcoming dinner party. CBR attempts to formalize this process through the following broad steps:

(1) Build a case-base of known, characterized and indexed case instances.
(2) Retrieve a (set of) case(s) from the case-base which is similar to the query case.
(3) Adapt information from the known case(s) and apply to the query case.
(4) Evaluate the solution proposed in the adaptation step and if necessary repeat the adaptation step by, for example, selecting additional cases or repairing the solution proposed in the previous round of adaptation.
(5) Store the new solution in the case-base.

Of course, the details of building a case-base, identifying neighboring cases, and adapting the information from the source cases to the query case can be quite domain-specific. Rather than discuss these steps in the general case, we will examine and amplify the CBR process in its application to biological reasoning. The reader is directed to the comprehensive presentations on CBR found in Kolodner [3] and Leake [4] for more details on general methodologies.

3. CBR in biology

The strength of analogical reasoning in biology lies in the common evolutionary origin of homologous structures and systems, within and between organisms. For example, the many members of the hemoglobin family, which transport oxygen in vertebrate blood, are all descended from a common ancestor. Similarly, the vertebrate brain is generated by developmental processes common to all vertebrates. (As a cautionary note, the wings of insects, birds and bats arose by independent mechanisms and thus are not examples of similarity by descent.) It is not surprising then that reasoning by analogy plays a prominent role in biological research where model organisms and systems are valuable substitutes for human systems. Techniques developed by the artificial intelligence community for reasoning about analogies in the everyday world should be particularly well suited for reasoning about biological systems.

Aside from providing a formal representation of a type of reasoning used by biologists, analogical reasoning techniques, and specifically CBR, offer advantages over generalization-based methods in some situations. As previously mentioned, data are often too sparse to drive statistical analyses. Beyond that, frequently no sound biological model exists making the creation of a generalized model problematic. Finally, it may be more efficient and reliable to simply look up similar cases in a database, and adapt them to a novel situation rather than construct a statistical or general model and compute predictions from that. For example, CBR [5,6] and the closely related technique of memory-based reasoning [7,8] have been successfully applied to the problem of predicting secondary structure in globular proteins efficiently. Here, we examine CBR for the particularly difficult problem of predicting transcriptional regulatory DNA sequences involved in the control of vertebrate gene expression.

4. Annotating gene and promoter regions by CBR

The overall process of labeling sites and sub-sequences of DNA, RNA and protein with their features and functions is termed annotation. We have applied the CBR framework to the problem of annotating genes and the regulatory elements in their proximal promoter regions [9–11]. CBR was chosen for many of the reasons outlined above: the problem with sparse data is nowhere more evident than in the study of patterns in DNA, such as those in the proximal promoter region, necessary for the regulation of gene expression; no satisfactory models exist to explain the mechanistic details of gene regulation; and it is the goal of such research to discover these mechanisms. One aid to that end is the construction of a composite view derived from information on multiple different instances. By composite view, we mean piecing together a consistent representation of gene or regulatory features from studies done on different regions of genes with very similar or identical functions in different organisms.

When annotating genomic sequence from a high-throughput sequencing project, the primary interest is identifying genes, gene models (i.e. the exon/intron structure and coding regions) and protein products, and secondarily other features such as repetitive elements and STSs (markers for genetic and physical maps). Systems such as GRAIL [12], GENSCAN [13], and others are already able to identify the coding regions of genes with considerable reliability. However, signals in DNA sequence such as those for the regulatory elements in the proximal promoter region, polyA signals, matrix attachment sites and so on are far more difficult to predict computationally. Our interest is primarily in this latter category.

In addition to specifying structural information, genes must be turned on and off at precise times in the correct tissue of the developing and mature organism. This process is termed *gene regulation* and elucidating the detailed mechanisms of this process is one of the major challenges of modern biology. The first step in gene regulation is *transcription*, where the information in a gene is amplified by synthesizing multiple RNA copies. Considerable progress has been made in characterizing the components of the cellular machinery that regulate gene transcription (see, for example, [14] for review): short DNA sequences termed *transcription elements*, typically on the order of 10 base pairs (BP) in length, are recognized and bound by sequence-specific binding proteins termed *transcription factors* to form *transcription complexes* through protein–protein as well as DNA–protein interactions. As might be expected, important transcription elements are located immediately preceding the start of genes in the proximal promoter region. More surprisingly, transcription elements are also found thousands of bases upstream, downstream and even within the boundaries of a gene.

For illustrative purposes, we will focus on the gene model and regulatory elements of the vertebrate α- and β-hemoglobin gene families (Fig. 1), members of which are expressed during red blood cell differentiation. There are three distinct red blood cell developmental stages in mammals: the embryonic stage where the cell progenitor originates in the yolk sac, the fetal stage where the progenitor is found in the liver, and the adult stage where the progenitor is in the bone marrow. Specific subsets of the hemoglobin genes are expressed at each of these stages as shown in Fig. 1. (Note that the temporal expression corresponds to the linear layout of the genes.) The figure also shows the layout

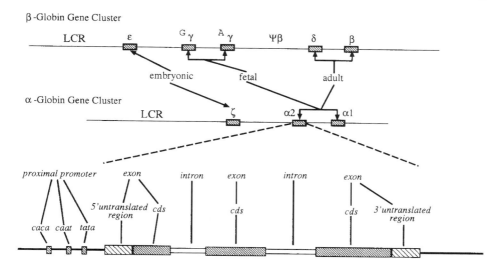

Fig. 1. Structure of the human α- and β-globin gene clusters. Schematics of the β-globin gene (approximately 70 KB on chromosome 11) and the α-globin gene (approximately 40 KB on chromosome 16) showing the physical layout of the five β-like and three α-like genes, and the global regulatory complex, the Locus Control Region (LCR). In the β-globin cluster, the ε-globin gene is expressed in the embryonic yolk sac during erythropoiesis, the Aγ and Gγ-globins in the liver during fetal development, and the δ- and β-globins in adult bone marrow. In the α-globin cluster, the ζ-globin is expressed at the embryonic stage, and α − 1- and α − 2-globins are expressed at both the fetal and adult stages. Detail of the α − 2-globin showing the gene model and the transcriptional regulatory elements in the proximal promoter region.

of a typical globin gene, in this case the α-2 globin, along with some of its transcription elements. For a comprehensive description of erythropoiesis and the globin genes see, e.g. [15].

The transcriptional state of a gene (i.e., its time, tissue and rate of expression) is determined by the process of transcription complex formation, the details of which are not well understood. In particular, transcription elements are necessary but not sufficient to define the transcriptional state of a gene. Other critical factors include the temporal presentation and quantitative level of transcription factors, the current, local chromosomal state, and DNA methylation patterns. Moreover, transcription factors bind to a family of transcription element sites only a few of which are generally well characterized. Computational models of these families are difficult to formulate and generally lead to both over and under prediction of sites when genomic sequence is analyzed (see for example, TESS [16]). In addition, information on the protein–protein interactions required for transcription complex formation is extremely limited.

In short, the transcriptional potential of a gene is extremely difficult, if not impossible, to predict from sequence information alone. On the other hand, a robust composite view of gene expression for a class of genes can be constructed by assembling the experimental information from related examples each of which contains part of the picture. CBR provides a framework in which these views can be built and maintained automatically.

5. Building the case-base

The actual construction of the case-base is by far the most time consuming and difficult step in the overall CBR process. Our case-base is part of a larger project – EpoDB – the goal of which is to create an informatics system for the study of gene expression in and functional analysis of red blood cell differentiation, a process termed erythropoiesis. Details on EpoDB can be found at the WWW site http://www.cbil.upenn.edu and in Salas et al. [17]. EpoDB integrates information from four types of resources by: data extraction from the online biology data banks and databases; manual entry from the published scientific literature; direct data capture from high-throughput data generation projects such as gene expression studies; and automated, computational analyses of the primary data.

In the first step, DNA, RNA and protein sequences for genes known to be expressed during erythropoiesis or which interact with members of the differentiating red blood cell lineage were extracted from GenBank and the protein data bank SwissProt. For the purposes of analyzing transcriptional regulatory elements, only genes for which sequence of several hundred base pairs in the proximal promoter region was available were included in the case-base. Annotation for each gene, such as exons, introns, and other features of the gene model and the corresponding protein, were accessed from the primary sequence data banks. Additional information on transcription factors, transcription elements, and gene expression patterns was obtained from the primary literature as part of an ongoing collaborative effort with Dr. N. Kolchanov, Dr. Alexander Kel and Dr. Olga Kel (Institute of Cytology and Genetics, Novosibirsk, Russia). Structural information on transcription elements and factors is stored in the TRRD and COMPEL databases from and coordinated with the TRANSFAC database [18]. Gene expression information is stored in GERD (unpublished results). Information relevant to EpoDB is extracted from each of these databases and then transformed into the internal EpoDB representation, an object-oriented extension to the CPL data exchange format [19].

5.1. Data cleansing with the sequence structure parser

In preparing the EpoDB case-base, the information extracted from the various source databases must be cleansed to reduce errors as a first step towards producing a resource of sufficiently high quality to drive CBR. For gene and protein features, information must then be transformed into an expanded, canonical form where all features are explicitly represented in a machine-understandable form so that they can be queried. It is worth examining the approach employed to cleanse and standardize information on gene structure extracted from GenBank because it introduces a grammar formalism for modeling biosequence structure conceptually similar to that used later in the adaptation step of the CBR process.

Each GenBank entry contains sequence data, literature references, keywords, a biotaxonomic classification of the organism from which the sequence was obtained, and in addition to other information, a feature table which lists the biologically significant features that, taken together, constitute GenBank's description of the structure of genes and other biologically meaningful sub-sequences. Unfortunately, many GenBank entries are idiosyncratic, and suffer from incomplete, noisy (ambiguous or contradictory) and erroneous listings in the feature table. Typical errors include mislabeled features (e.g. mRNA instead of prim_transcript or exon), incompatible boundary specifications among multiple features (e.g. between exon and CDS), and critical information such as gene names in multi-gene entries, buried in free-text fields. Some of this lost information can be recovered through analysis of the explicit information in the feature table to infer the implicit information, e.g. when only introns and the precursor mRNA boundaries are described and exons must be inferred (see Fig. 4). (Note also that it is the responsibility of the original submitter of the sequence data to perform updates, and consequently annotation is generally well out of date with respect to the current state of knowledge. One of the motivations for the EpoDB project is to remedy this situation for the subset of genes expressed during erythropoiesis.)

Obviously, the quality of prediction of new features depends critically on the quality of well characterized features. As a step towards improving the quality and consistency of information in GenBank, we have developed a system – the sequence structure parser (SSP) – which checks the logical constraints on gene structure implied in the feature table, deduces missing features and records a canonical form of the features in our database [20]. The logical constraints, which model the linear structures of genes and proteins, are expressed in a definite clause grammar (DCG) formalism [21,22], a type of logic grammar; however, because of the many and various exceptions and outright errors in the feature table, a large rule-base augments the otherwise clean and simple DCG formalism. Consequently, the SSP system should be regarded as hybrid between a grammar formalism and an expert system.

SSP is implemented in the logic programming language Prolog which has strong support for DCGs and expert systems. In operation, the SSP attempts to construct a parse tree expressing the structure of the gene or genes described in a GenBank entry repairing errors using the rule-base as it progresses (see [21–25] for a more complete discussion of grammatical representations of gene structure).

Standard DCG rules are of the form LHS ⇒ RHS, where the LHS (left-hand side) is a non-terminal and the RHS (right-hand side) any combination of terminals and non-terminals. *Terminals* correspond to words of the sentence being parsed (the leaf nodes of the parse tree), and *non-terminals* represent sets of phrases (sub-sequences of sentences) as defined by the grammar. Each interior node of a parse tree corresponds to a non-terminal in the grammar, and the sequence of terminals beneath a node is a phrase satisfying the grammar rule. The LHS non-terminal in the top level grammar rule is termed the start symbol (*trans_unit* in Fig. 2).

In grammars for biological sequences, the distinction between terminals and non-terminals is blurred because genes and proteins can be described and investigated at various levels of abstraction. As shown in the parse tree of Fig. 2, sub-sequences which may be considered as terminals in one context may become non-terminals in another. For example, exons as sub-sequences of a primary transcript may be considered terminals, whereas exons would be non-terminals when parsed into coding sequences and untranslated regions. Each feature of the feature table essentially describes one node (terminal or non-terminal) of the gene parse tree along with the DNA sub-sequence covered by it. These descriptions are often redundant, incomplete and even inconsistent

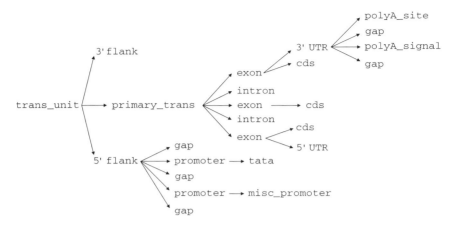

Fig. 2. Parse tree of a prototypical globin gene.

$$\begin{aligned}
trans\_unit &\Rightarrow 5' \text{ flank}, primary\_transcript, 3' \text{ flank}. & (1)\\
5' \text{ flank}, [primary\_transcript(S,E,I)] &\Rightarrow gap, [primary\_transcript(S,E,I)]. & (2)\\
5' \text{ flank} &\Rightarrow promoter, 5' \text{ flank}. & (3)\\
5' \text{ flank} &\Rightarrow gap, promoter, 5' \text{ flank}.\\
primary\_transcript &\Rightarrow exon, intron, primary\_transcript & (4)\\
primary\_transcript &\Rightarrow exon.\\
3' \text{ flank} &\Rightarrow gap. & (5)
\end{aligned}$$

Fig. 3. Simplified version of the grammar used by the Sequence Structure Parser.

in GenBank, and the task of the SSP is to assemble complete parse trees expressing the same information in a structured non-redundant, consistent fashion.

Since grammatical elements (terminals and non-terminals) correspond to sequence intervals [26], grammar rules can be naturally interpreted as interval relationships where "⇒" means the interval on the LHS contains the intervals on the RHS ("part–whole" relationship), and a " , " between intervals means that the end of the first interval is the beginning of the second interval ("order of parts" relationship). Techniques have been developed for reasoning about temporal intervals [27], and these techniques can be extended to cover nucleic acid intervals. Incorporating these techniques into grammar rule formalisms makes it possible to model other interval relationships such as overlaps, starts, and ends [28].

The SSP takes as input the feature table for a GenBank entry and returns a corrected, expanded feature table as shown in the example of Fig. 4. We have developed a complete, working grammar for typical eukaryotic protein coding genes [29], a much simplified version of which is shown in Fig. 3. Rule 1 is an example of a grammar rule expressing the fact that a transcription unit consists of a 5′ flanking region, followed by a primary transcript and then a 3′ flanking region. Alternative rule applications (disjunction) are expressed as shown in Rule 3 and recursion is illustrated in Rule 4. Practical grammars also need escapes to the underlying implementation language. Such escapes are also available in grammars to handle exceptional situations such as erroneous and missing input data.

In contrast to standard parsers that take as input a list of terminals, the input to the SSP may contain non-terminals as well. To facilitate efficient processing, the grammatical elements (features) on the input list are ordered by their start positions, lengths and ranks in the grammar hierarchy; for example, an exon occurs before a CDS fragment with the same boundaries. The square bracket notation, [ ], is used to remove and add elements to the input list. When used on the RHS of a rule, they remove grammatical elements, and when used on the LHS they add elements. Therefore, an element can be removed, examined and replaced on the input list as seen in Rule 2 which tests if the 5′ flank boundary has been reached. The logic variables S and E represent the start and end positions of the interval, and I provides context information about the features.

SSP is able to handle 60% of entries listed as having complete coding sequences in GenBank. In the remaining cases, the SSP is used to repair the entry in an interactive session where changes to the feature table are entered manually and then reprocessed by the SSP. It has been successfully applied to a large number of eukaryotic globin genes such as the human α-2-globin gene shown in Fig. 4. Note that the SSP inferred the missing exons, the mRNA boundaries, and the 5′ and 3′ UTRs. When a full CBR analysis is applied to this entry the start of transcription and the polyA addition site can also be determined precisely as discussed below.

This example illustrates only a few of the problems that complicate the development of grammars with broad coverage. Apart from mislabeling and omission of features, the SSP has to deal with feature tables describing multiple genes or alternative versions of genes including alternative splice sites, transcription start sites, and polyA additions sites. In these cases, manual intervention is almost always required to interpret information in the free-text fields of feature "qualifier" fields. Because qualifier field information is largely free-form, it is generally beyond the capability of SSP to interpret it.

5.2. Structure and content

EpoDB attempts to be exhaustive in its collection of genes and proteins expressed during or influencing vertebrate red blood cell differentiation. Because the members of the α- and β-hemoglobin gene families are among the most intensely studied of all genes, more than a third of EpoDB's entries are hemoglobin genes, mRNAs and proteins. And even among the hemoglobin genes, there is a strong skew towards entries on primates (especially human) and rodents (especially mouse). As always when performing a database search by sequence similarity, the problem of redundant information must be dealt with. EpoDB addresses this problem by coalescing information into a set of well characterized "reference" entries for each gene in each organism. While redundant entries are still available, most queries are against the reference gene set. There is still a skew towards primates and rodents, but other mammals, avians, amphibians, reptiles and even the occasional fish are now more visible to users.

Information is organized around genes or rather the transcripts (*transcript_unit*) for genes, not around sequence as in GenBank. In other words, the primary point of access to information is by querying reference transcripts rather than reference sequence (although

74

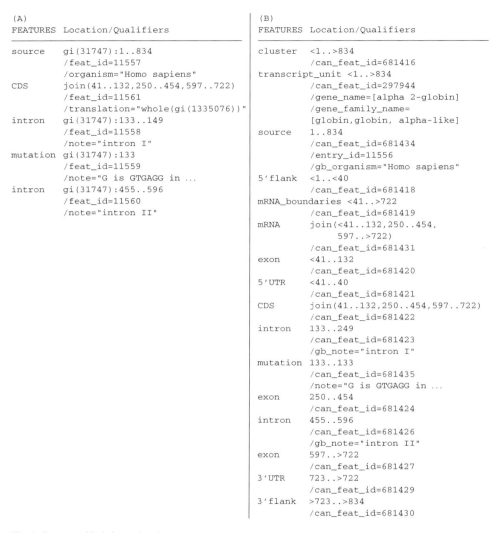

Fig. 4. Feature table information for the human α-2-globin gene (HSGL04) from (A) GenBank and (B) after data cleansing and augmentation by SSP. The qualifier fields for entry ID, change notes, justifications and the GenBank feature ID have been removed to make the output less verbose.

sequence queries are also available). This is of course essential when there is neither a one-to-one correspondence between a sequence and a gene (e.g. the multiple genes in the globin gene clusters) or between a gene and a sequence (e.g. human spectrin which has 50 separate exon sequence entries). Each reference transcript has a gene name and gene family name drawn from a controlled vocabulary, and associated information on its DNA, RNA and protein sequence, time of expression during development, transcription and promoter elements, bibliographic references, taxonomic classification and so on.

The latest version of EpoDB (2.1) contains cleansed and augmented information on more than 3700 nucleic acid sequence entries from GenBank and 1200 entries from SwissProt, as well as expression data on more than 70 genes "specific" for red blood cell differentiation. Of these entries, 166 have been fully qualified as reference sequences. Additional information on the database schema and information content can be found at the EpoDB WWW site.

6. Classification of new cases

Defining the set of source cases most similar to a query case depends on the particular problem being studied. The EpoDB case-base is sufficiently rich and varied to support a number of different CBR strategies tailored to the problem of interest. For example, one might be interested in studying how genes are coordinately expressed during red blood cell differentiation. In this case, the similarity measure would be based on the developmental stage and differentiated state of the cell when genes are expressed. The domain theory for this similarity measure would reflect the belief that coordinately expressed genes, regardless of the gene type, would be regulated by (some) common transcription factors and consequently have common transcription elements.

The problem we address here is that of building and maintaining a composite view of regulatory elements in functionally equivalent classes of vertebrate genes. In this case, our similarity measure is based on protein sequence comparisons which reflects the theory that genes whose proteins are very similar are more likely to be functionally equivalent and thus share transcriptional regulatory elements. The theory attempts to rank orthologous genes highest, then paralogous and finally genes which are merely homologous among organisms. Orthologous genes are those which play essentially identical functions in different species; paralogous genes perform similar but not identical functions; and homologous genes are ones which carry out functions which are even more distantly related. For example, the adult β-hemoglobins in mouse and human are orthologous in that they carry out identical functions in oxygen transport in adult red blood cells, whereas the human β- and γ-hemoglobin genes are paralogous in that they perform equivalent functions in oxygen transport but the former is expressed in the adult organism while the latter is expressed in the fetus. Finally, adult human α- and β-hemoglobins are homologs: both transport oxygen and show substantial sequence similarity, but have distinct roles in the overall mechanism of oxygen transport. (In fact, the functional hemoglobin complex is a tetramer of two *alpha*- and two *beta*-hemoglobins which are coordinately regulated during differentiation.)

For the studies described here, we focus entirely on globin genes because they provide the richest source of information at the DNA, RNA and protein levels. To identify source cases similar to the query case, we use a BLASTX [30] search of the complete nucleic acid sequence of the query case against the protein sequences of the source cases in EpoDB, rather than searching with just the coding sequence regions listed in GenBank. This choice was made to preserve generality of the method. As discussed below, the coding regions are discovered as part of the CBR process. BLASTX compares the query nucleotide sequence translated in all six reading frames against a protein sequence database. The query sequence is masked to filter out repetitive elements prior to the BLASTX search. The cases identified are ranked by the BLAST p-value score of their respective sequences.

If we were interested solely in gene regulatory elements, an alternative strategy, given the types of information available in EpoDB, would be to measure similarity based on some combination of gene name, gene family name, time of gene expression and biotaxonomic distance. We are not yet at a point in EpoDB where the information on gene and gene family name, and time of expression are unequivocally defined. In fact, one of our goals is to predict such information by CBR analysis. For the moment, protein sequence similarity remains the least biased method for predicting orthologous genes having similar patterns of gene expression.

7. Case adaptation

Our aim is to map well characterized features (which by-and-large means only those experimentally verified) annotated in the source cases onto the partially annotated or even completely unannotated sequence from the query case. While our interest here is primarily in mapping features involved in gene regulation, this is obviously a generic problem in which any type of feature could be mapped at the nucleic acid or protein sequence level. In the following examples, we will concentrate on regulatory features but even here other feature types, such as exons and introns, are mapped to build a framework necessary for annotating regulatory features. CBR analysis of regulatory elements in the proximal promoter regions of genes benefits from a cooperative comparison of the set of cases neighboring the query case. Again the reason for this is pragmatic: each gene in the set is expected to have incomplete but overlapping information about regulatory elements because no genes have been exhaustively experimentally analyzed. Consequently, our strategy is to retrieve all the cases similar to the query case ranked by their degree of similarity as described in the previous section.

Not surprisingly, we frame the adaptation process in terms of a grammar formalism where the associated parser can be specifically tailored to various analogical reasoning strategies. For the most part, these strategies rely on feature type dependent sequence alignment and pattern matching algorithms which range from simple heuristics to full-fledged sequence parsers along the lines of that described by Searls [24]. Conceptually, the structural information described in the feature table for each source and query case is converted into an "instance" grammar. The query case, including whatever structural information is known about it, is parsed with respect to each of the instance grammars identifying features above some reliability threshold in the source case annotation which are similar to features in the query case above some similarity threshold. The reliability threshold reflects our confidence in the feature. Generally, we only use experimentally verified features as input. The similarity threshold represents the acceptance level above which we consider the match significant. The acceptance level is, obviously, algorithm dependent. For example, if BLAST were to be used for pattern matching, then the BLAST p-value would be the similarity measure and some appropriate threshold would be chosen. The overall result of parsing the query case with the grammars from each of the source

```
ADAPT(set of SourceCase, QueryCase)
BEGIN
  QueryGrammar := GRAMMIFY(QueryCase);
  QuerySeq := SEQUENCE(QueryCase);
  All_Predictions := [];
  FOR_EACH SourceCase DO
    BEGIN
    SourceGrammar := GRAMMIFY(SourceCase);
    SourceSeq := SEQUENCE(SourceCase);
    Predicted_Parse := PARSE_STRATEGY(QueryGrammar,QuerySeq,
                                      SourceGrammar,SourceSeq);
    All_Predictions := All_Predictions + Predicted_Parse;
    END
  RETURN(RESOLVE(All_Predictions,QueryGrammar));
END
```

Fig. 5. Adaptation procedure: GRAMMIFY(Node) returns a grammar from the feature table of Node;
SEQUENCE(Node) returns the nucleotide sequence of Node;
PARSE_STRATEGY(QueryGrammar,QuerySequence,SourceGrammar,SourceSequence)applies
the source grammar and sequence to the query sequence and grammar (if known) and returns predicted features
in the query sequence;
RESOLVE(Predictions, QueryGrammar) builds a consensus prediction from the individual predictions
of each source case and returns a grammar (or feature table).

cases forms a composite view of the predicted features in the query case which are then
recorded back in EpoDB.

The logic for the generic adaptation procedure is shown in Fig. 5. For each source
case, the adaptation algorithm has four inputs: the source features (SF); the source
sequence (SS); the query case features (QF), which may be null; and the query case
sequence (QF). The basic approach is to generalize the set of source features into
a grammar describing the structure of the source sequence and then parse the query
sequence and whatever structure is known about the query case to identify corresponding
features. Variations in the adaptation strategies are specified by the execution model of
the parser.

As an example, we will describe a relatively naive strategy which has nonetheless
proven to be extremely reliable for the conservative prediction of regulatory elements
and the verification of gene models in EpoDB. In this two phase strategy, a gene model
for the query case is predicted by sequence alignment to the exons of the source cases.
An alignment algorithm such as FASTA is generally suitable for this step. If the query
case already has annotation for exon features, then this step serves to verify the exon
boundaries. Anomalies at the start of transcription and polyA sites in the GenBank
features, and even in internal exons, can often be detected and resolved at this step. If no
GenBank features are available the predicted exon/intron structure and their sub-structures
are accumulated to build a gene model. In the second phase, regulatory elements in the
proximal promoter region are predicted by "projection" from the regulatory elements
described in the source cases to the query case sequence. A simple pattern matching
algorithm with mismatches is employed where the location of the pattern is constrained by
the order and position information imposed by the source grammar rules. The following
sections describe the steps in more detail.

(A)

FEATURES	Location/Qualifiers	FEATURES (ct'd)	Location/Qualifiers
transcript_unit	<1..>2685	exon	770..919
source	1..2685	5' UTR	770..824
5' flank	1..769	CDS	join(825..919,1761..1965,
Sp1	522..533		2205..2333)
GATA-1	537..542	intron	920..1760
ZF1	549..576	repeat_region	979..992
ZF2	603..624	exon	1761..1965
GATA-1	663..669	intron	1966..2204
Sp1	673..696	repeat_region	1971..1975
CP1	705..709	exon	2205..2439
mRNA_boundaries	770..2439	3' UTR	2334..2439
mRNA	join(770..919,1761..1965,	polyA_signal	2419..2424
	2205..2439)	3' flank	2440..>2685

(B)

$$transcript\_unit \;\Rightarrow\; 5' \text{ flank}, mRNA\_boundaries, 3' \text{ flank}. \tag{6}$$

$$5' \text{ flank} \;\Rightarrow\; \text{gap, Sp1}(\_,[-247..-236]), \text{gap}, \tag{7}$$
$$\text{GATA-1}(\_,[-232..-227]), \text{gap},$$
$$\text{ZF1}(\_,[-220..-193]), \text{gap},$$
$$\text{ZF2}(\_,[-166..-145]), \text{gap},$$
$$\text{GATA-1}(\_,[-106..-100]), \text{gap},$$
$$\text{Sp1}(\_,[-96..-73]), \text{gap},$$
$$\text{CP1}(\_,[-64..-60]), \text{gap}.$$

$$mRNA\_boundaries \;\Rightarrow\; \text{exon-1}([1..150],\_,\_), \text{intron}([151..991],\_,\_), \tag{8}$$
$$\text{exon-2}(['992..1196'],[-678..-474],\_),$$
$$\text{intron}(\_,[-473..-235],\_),$$
$$\text{exon-3}(\_,[-234..0],\_).$$

$$\text{exon-1} \;\Rightarrow\; 5' \text{ UTR}([1..55],\_,\_), mRNA\_segment. \tag{9}$$

$$\text{exon-2} \;\Rightarrow\; mRNA\_segment. \tag{10}$$

$$\text{exon-3} \;\Rightarrow\; mRNA\_segment, 3' \text{ UTR}(\_,[-105..0],\_). \tag{11}$$

$$3' \text{ UTR} \;\Rightarrow\; \text{gap}, polyA\_signal(\_,[-20..-15],\_), \text{gap}. \tag{12}$$

Fig. 6. Converting a feature table to an instance grammar. (A) Feature table for the human ζ-globin gene (HUMHBA1) stripped of all qualifier fields. Features listed in bold are from the original GenBank feature table; the additional features have been deduced by SSP or manually added from information in the literature. (B) Instance grammar for HUMHBA1 derived from the feature table. Notation: feature_type([loc1],[loc2]) where loc1 is the location with respect to the start of the next higher level grammar symbol in which it is contained and loc2 is the location with respect to the end of the containing higher level grammar symbol.

7.1. Generating instance grammars

The set of feature table descriptions for each source case (and the query case if appropriate) is converted into an "instance grammar" suitable for input to the CBR parser. An example of an instance grammar for the human *zeta*-globin gene is shown in Fig. 6. The grammar specifies both the identity and the location of proposed features. Each grammar non-terminal is annotated with "positional references" defining the relative location of the corresponding feature. More precisely, the positional references of a feature are the distances of that feature from the start and end of the sequence covered by the next higher-level grammar non-terminal. Consider grammar rule 7 in Fig. 6. The Sp1 site in this rule is annotated with the distance of its sub-sequence interval from the start and end (_, [−247.. − 236]) of the 5′ *flank*; however, since the start of the 5′ flanking region is not precisely defined, only the reference to end of the 5′ flanking region can be given. The "_" indicates that the value is unknown. The specified sub-sequences are extracted from the source sequence as needed by the parser.

Once the instance grammars are formed, the ADAPT procedure parses the instance grammar and sequence for each source case along with the instance grammar and sequence for the query cases to the CBR parser. The instance grammar for the query case may be empty if no experimental features have been determined. In this particular strategy, the CBR parser splits the grammar rules into two parts: rules for the gene model (exon/intron structure) and rules for the proximal promoter or other regulatory regions given the gene model.

7.2. Predicting exon/intron structure

The goal of the first phase of the CBR parser is to predict, or confirm and correct boundaries of major features such as CDS segments and primary transcripts so that other specialized routines can be used effectively to identify regulatory features. This phase consists of essentially three steps: (1) alignment of source feature sequences to query sequence; (2) heuristic pruning of feature boundary predictions; and (3) unification of boundary predictions with a gene structure grammar. The output of UNIFY_SSD is a normalized feature table for the primary transcript structure. Each of these procedures is described in more detail below.

In the first step, a procedure ALIGN_SSD repeatedly aligns an annotated sequence with the query sequence in two stages. From information in the instance grammar, the sequences of the source case exons are extracted and aligned exon by exon with the complete genomic sequence of the query case using for example FASTA. This step predicts the exon boundaries. (More distant similarities could obviously be detected by searching in protein space but this would not help in identifying the CAP and polyA sites needed in the next phase.) These boundaries are then used as alignment starting points. Next the alignment is augmented with feature boundary marks as specified in the source grammar. Feature boundaries of the query sequence are then determined by projecting the boundary marks from the source. For each predicted feature boundary a score is computed reflecting how well the two sequences matched around this boundary. The list of potential feature boundaries of the second sequence along with their respective scores form the output of this procedure. In some cases the procedure is modified to include intronic sequence in the alignment phase.

Next, procedure UNIFY_SSD takes as input all the exon/intron features predicted from each of the source cases. First all predictions are removed for which there is a prediction for the same feature boundary with a higher score within a certain distance, generally set at about 10 BP. Next, predictions are eliminated based on how low their scores are relative to the average score, and based on whether their scores are below a fixed cut-off value. In the current implementation predictions are removed if their score is less than 25% of the average score. Finally, lowest score boundaries are removed until the remaining number of boundaries is less than the maximum number of expected distinct boundaries.

It then computes a parsable list of feature boundaries using the predicted boundary locations and the constraints implied in the grammar. That is, the output list of feature boundaries is consistent in that there is a feature end for every feature start and vice versa, and in that lower level features are placed appropriately within higher level features as specified by a universal grammar similar to that for SSP. The procedure non-deterministically removes (incorrect) boundary predictions from the input list. Then it executes the gene grammar using a left-to-right depth-first search strategy with iterative deepening taking as input the list of hypothesized feature boundaries. The parser executes partly in parsing mode and partly in generative mode to fill in features. Whenever the first feature boundary of the (ordered) list of feature boundaries matches with the grammar symbol being expanded by the parser/generator the first feature boundary from this list and its location information is incorporated into the feature table being constructed. If a feature boundary is a start of a feature, it is consumed before the matching grammar symbol is expanded. If it is an end, it is consumed after the matching grammar symbols are expanded. If the first feature on the list does not match the grammar symbol being expanded, a feature boundary is generated along with constraints for its location on the sequence as implied by the grammar/parse tree. These constraints are accumulated for all generated boundaries and evaluated at the end. The list of feature boundaries must be completely consumed at the end of the parse. In order to achieve acceptable performance, the procedure also requires that the list of feature boundaries is reduced to some degree during the parsing/generation process.

7.3. Predicting regulatory elements

There are no strong predictive techniques for the binding sites of particular transcription factors because the sites are short (usually around 10 BP or less) and can vary considerably in sequence. Despite sequence variations, transcription factors do not usually tolerate insertions and deletions in their transcription elements. Therefore, we can use a simple algorithm which aligns the regulatory elements defined in the source grammar to the proximal promoter region of the query sequence. The algorithm permits a preset percentage level of mismatches, but no insertions and deletions, and constrains the match to be within some fixed offset of the position specified in the source grammar. That is, prediction of transcription elements (or polyA signals) in the proximal promoter region (or introns and 3′ flanking region) relies on the boundaries determined in the predicted query case gene model. The procedure uses offsets from these boundaries to focus alignment of the regulatory elements by sequence, position and order as defined by the instance grammars. We refer to this strategy as offset CBR.

The algorithm must reconcile predictions made from different source grammars. The complication here is that the exact specification of a binding site depends on the experimental technique and conditions that defined it. Examination of the TRANSFAC database, for example, reveals that the experimentally characterized Sp1 sites range in size from 6 BP to more than 30 BP. We eliminate redundant predictions by taking the intersection of overlapping predictions for the same transcription factor as the output.

(One could argue for the union but this would result in an unreliable prediction.) If an empty intersection occurs, it is reported as an error.

Identifying redundant predictions is not straightforward, since in general, promoter sequences specified in GenBank are not precise; often a sub-sequence or super-sequence of the actual promoter is given. Thus, we eliminate redundant predictions by forming equivalence classes of predictions located within a short neighborhood of each other on gene sequence. Thus, all predictions of a given feature, in this case the same transcription factor, on a single gene instance that are within some small distance in base pairs from each other are considered to be the same prediction. Analysis of the available data (not shown) suggests 4 BP is a reasonable value for positional variability when predicting transcription elements in the proximal promoter region. This ensures that promoter sequences within an equivalence class overlap each other, giving credence to the belief that all sequences in an equivalence class are sub-sequences of a single transcriptional regulatory element.

7.4. Results

To get a measure of how reliably exon/intron structure can be predicted, a cross-validation test was performed in which each globin sequence in turn was stripped of all its feature table information and processed by CBR. A total of 398 exon boundaries were predicted. 370 (93%) of them could be confirmed, that is, they agree with the sequence annotations in GenBank, while 7% were determined to be inconsistent with the GenBank annotations. The system was able to complete an evaluation only for about one third of the test set because (1) the last step in the procedure (matching predicted boundaries onto a parse tree) involves some non-determinism which can lead to unacceptable time complexity for some cases in the current implementation; and 2) the system had to eliminate a large number of predictions during the evaluation since apparent false positives could not be positively identified as such due to incomplete data about the query cases ("fuzzy" boundary information). Note that in addition to indicating false predictions, "false positives" may also be the result of errors in GenBank. In fact, a number of such errors were confirmed. Also note that even if a prediction is off by only one base pair it is considered a false positive in this analysis.

Similarly, we applied offset CBR to a set of 38 globin genes for which promoter elements were available to test this predictive step. The maximum mismatch allowed was 20% and the offset from the known position was varied from 0 to 10 base pairs plus or minus. Again using a cross-validation procedure, up to 96% of the known promoters were correctly predicted depending on parameter settings. It is, if anything, more difficult to estimate the false positive rate here than in gene model prediction because many of the predictions for which there is no evidence may in fact be accurate but either have not been recorded in EpoDB or the experiment has not yet been performed.

The simple CBR strategy employed here appears to be extremely effective in building up a composite view of the regulatory elements of orthologous genes in EpoDB. The views of individual genes are mutually reinforcing and as described below, the methodology facilitates maintenance of the views in the database. Figure 7 is an example query displaying the features predicted when the mouse ζ-globin gene entry was

82

(A)

FEATURES	Location/Qualifiers
CAAT_signal	364..368
TATA_signal	401..407
exon	430..588
intron	589..1393
exon	1349..1598
intron	1599..1704
exon	1705..>1705
polyA_signal	1920..1925

(B)

FEATURES	Location/Qualifiers
transcript_unit	<1..>1925
5' flank	<1..429
GATA-1	323..328
Sp1	345..355
CP2	365..368
CAAT_signal	365..368
TATA_signal	401..406
mRNA_boundaries	430..>1925
exon	430..588
5' UTR	430..483
intron	589..1393
exon	1394..1598
intron	1599..1704
exon	1705..>1925
polyA_signal	1921..1925

(C)

FEATURES	Location/Qualifiers
c-Ets-2	70..75
NF-1-like	75..88
c-Myb	85..94
GCN4	89..94
GATA-1	95..105
Isl-1	99..104
Zeste	116..121
N-Oct-3	126..135
GR	129..134
GCN4	131..136
IRF-1	140..149
GR	148..153
TGT3	157..162
PPAR	175..180
RXR-alpha, COUP	175..188
SRF	181..190
Sp1	194..199
GATA-1	199..204
NIT2	200..210
GR	210..215
NF-1-like	219..233
Sp1	232..241
delta factor, YY1	240..250
SRF, MCM1	241..250
Sp1	249..254
UCRF-L	261..266
F-ACT1	262..267
GATA-1	263..270
TBF1	296..301
PTF1-beta	303..308
GATA-1	321..326
CACCC-binding factor	331..340
Ttk 88K	333..339
Sp1	344..354
MAZ	347..353
TGT3	381..386
muEBP-C2, TFE3-S	393..398
GAL4	400..405
TFIID, TBP	401..406
AP-2	411..420
AP-2	415..420
GR	417..422

Fig. 7. Predicted feature table for the mouse ζ-globin gene (MMEZGL). (A) The original GenBank feature table for MMEZGL with qualifier fields and the CDS feature removed. (B) The CBR predicted feature table. (C) Promoter elements predicted by TESS in a 430 BP region sequence including the the proximal promoter region of MMEZGL. The TESS mismatch stringency was set at 10%. Boxed features are verified predictions.

processed by CBR. The CBR predictions actually improve on the GenBank feature table listing in several ways. First, the location of the second exon is wrong; it should be 1394 to 1598 not 1349 to 1598 (the information is correct in the GenBank CDS description, however (not shown)). Second, the location of the third exon has been repaired. And finally, several additional promoter elements have been predicted.

Finally, it should be re-emphasized that the order, position and the specific binding site sequence of promoter elements are all critical factors in defining the subtle aspects of promoter sequences. Contrast the results determined by CBR for predicting the transcription elements in the mouse ζ-globin gene with those found by a search of the sequence using the TESS system [16] (Fig. 7). TESS searches the sequence with the complete set of transcription elements in TRANSFAC without regard for positional information. Even at a 10% mismatch level, as opposed to the 20% level used in CBR, the results from TESS vastly over-predict the real transcription factor binding sites (47 predicted sites). Furthermore, while TESS predicted three sites for which there is experimental justification (sites boxed in the figure), it missed the experimentally determined CAAT signal probably because the mismatch stringency was too high at 10%. In contrast, CBR predicted the two known promoter sequences recorded in GenBank plus three more highly likely sequences (GATA-1, Sp1 and Cp2).

8. Database issues

A major objective of the EpoDB project is to build and maintain a current and comprehensive view of information on gene expression during red blood cell development. The CBR methodology and its variants facilitate the automatic generation of corrected, composite views on the individual genes in the database, as well as generalizations over classes of genes as discussed elsewhere.

As a practical matter, incorporating the CBR methodology directly in the database system enables a process that tracks each prediction and correction. If the underlying primary data changes, for example, if the experimental data for a regulatory element is in error, then the database system can retract and if necessary re-compute the predictions over affected entries. This is a restricted form of truth maintenance [31] or view maintenance [32]. Similarly, if the CBR strategy changes then all predictions and corrections can be removed and re-computed for the whole database. This sort of wholesale bookkeeping and re-assessment of the information is simply not possible if it were to rely on manual intervention. Moreover, manual intervention introduces the type of idiosyncratic annotations found in GenBank that we are trying to avoid.

Furthermore, the results of CBR analysis can be viewed as a type of "cooperative database query" where a query on a particular instance triggers a search through related instances to satisfy the query. The information generated in building a composite view over each of the database instances can be considered the result of a cooperative query.

If the sequence of a predicted feature can actually be located near the predicted position, it would be a strong indication that the predicted feature has the biochemical properties attributed to that type of feature (such as promotion of gene expression). However, such properties can be reliably confirmed only by means of experimental analysis. Consequently, unconfirmed hypotheses may persist in the knowledge base for indeterminate periods of time, and may in turn contribute to further hypotheses. For this reason, because of the possibility of changes to the underlying databases (e.g. GenBank updates), and to support explanation of the derivation of hypotheses, some form of truth maintenance [31] is necessary. Truth maintenance in a database the size of GenBank is somewhat problematic but we are taking pains to preserve sufficient information during theory formation (grammar induction) and hypothesis generation to support limited forms of truth maintenance and explanation in EpoDB.

9. Conclusions

As should be expected, the CBR annotation framework performs extremely well for orthologous genes. The predictive capability is high, which allows the system to reinforce and extend predictions across organisms, i.e. discoveries on a gene in one organism provide support for belief in analogous features in orthologous genes. Conversely, discrepancies between orthologous genes suggest either fundamental differences in the orthologs or that the experimental evidence should be re-examined. Both reinforcement of belief and anomalies can be automatically maintained in the case-base.

For more distant relationships such as those between paralogous or even homologous genes, the performance of the system for analysis of promoter regions would likely degrade very quickly. Analysis of gene models could be substantially improved by enhancing the adaptation step but this would simply converge on what can be done with generalization-based gene recognition systems such as GRAIL and GENSCAN.

A more interesting problem would be to accommodate coordinately expressed genes, i.e. different types of genes that display the same temporal and spatial pattern of gene expression, by extending the grammar formalism and parser. For example, the embryonic α-globin (ζ-globin) gene and β-globin (ϵ-globin) show similar patterns of expression. One hypothesizes that coordinately expressed genes are controlled by some, but perhaps not all, of the same regulatory signals. It may well be that the signals to which the coordinately expressed genes respond are redundant or organized differently from gene to gene. A more sophisticated grammar formalism would be required to capture these variations. Fortunately, EpoDB already records information on gene expression patterns so that the first step in CBR, retrieval of analogous cases, would be straightforward.

As a final note, the CBR framework outlined here can be readily extended with a generalization component. As described in [9–11], we have built a hierarchy of gene classes representing levels of abstraction over vertebrate protein coding genes with the highest level class corresponding to the universal protein coding genes and lower classes corresponding to various families and sub-families of specific genes. At the leaves of the hierarchy are the gene instances in EpoDB. Associated with the highest level class is a universal grammar along the lines of that described in Fig. 3. Each of the classes representing gene families has an associated grammar induced from the instance grammars of the family members in EpoDB. The induced grammars can be used directly in exactly the same framework as CBR to reason about the structure of the individual members of the gene family. Perhaps more interesting, the induced grammars can be interpreted as a general description of the structure of the gene family thus distilling both a gene model for a gene class and a summary of its transcriptional regulatory features.

Acknowledgments

This work was supported by a grant from the National Center for Research Resources, National Institutes of Health (No. R01-RR04026).

References

[1] Morowitz, H.J., et al. (1985) Models for Biomedical Research: A New Perspective. National Academy Press, Washington, DC.

[2] Schank, R.C. (1982) Dynamic Memory: A Theory of Learning in Computers and People. Cambridge University Press, New York.

[3] Kolodner, J. (Ed.) (1993). Case-Based Reasoning. Morgan Kaufmann Publishers, San Mateo, CA.

[4] Leake, D.B. (Ed.) (1996) Case-Based Reasoning: Experiences, Lessons, and Future Directions. AAAI Press/MIT Press, Menlo Park, CA.

[5] Salzberg, S. and Cost, S. (1992). Predicting protein secondary structure with a nearest-neighbor algorithm. J. Mol. Biol. 227, 371–374.

[6] Leng, B., Buchanan, B. and Nicholas, H. (1994) Protein secondary structure prediction using two-level case-based reasoning. J. Comput. Biol. 1(1), 25–38.

[7] Zhang, X., Mesirov, J. and Waltz, D. (1992) Hybrid system for protein secondary structure prediction. J. Mol. Biol. 225(4), 1049–1063.

[8] Zhang, X., Fetrow, J., Rennie, W., Waltz, D. and Berg, G. (1993) Automatic derivation of substructures yields novel structural building blocks in globular proteins. In: ISMB, Vol. 1. AAAI Press, Menlo Park, CA, USA, pp. 438–446.

[9] Overton, G.C. and Pastor, J.A. (1991) A platform for applying multiple machine learning strategies to the task of understanding gene structure. In: Proc. 7th Conf. on Artificial Intelligence Applications, Vol. I. IEEE Computer Society Press, Los Alamitos, CA, pp. 450–457.

[10] Aaronson, J., Haas, J. and Overton, G.C. (1993) Knowledge discovery in GenBank. In: Proc. 1st Int. Conf. on Intelligent Systems for Molecular Biology. AAAI Press, Menlo Park, CA, pp. 3–11.

[11] Haas, J., Aaronson, J. and Overton, G.C. (1993) Using analogical reasoning for knowledge discovery in a molecular biology database. In: B. Bhargava, T. Finin and Y. Yesha (Eds.), Proc. 2nd Int. Conf. on Information and Knowledge Management. Association for Computing Machinery, New York, pp. 554–564.

[12] Uberbacher, E.C. and Mural, R.J. (1991) Locating protein-coding regions in human DNA sequences by a multiple sensor-neural network approach. Proc. Natl. Acad. Sci. USA 88, 11261–11265.

[13] Burge, C. and Karlin, S. (1997). Prediction of complete gene structures in human genomic DNA. J. Mol. Biol. 268(1), 78–94.

[14] Latchman, D.S. (1991) Eukaryotic Transcription Factors. Academic Press Inc, San Diego, CA.

[15] Stamatoyannopoulos, G., Nienhuis, A., Majerus, P. and Varmus, H. (1994) The Molecular Basis of Blood Diseases. Saunders Company, Philadelphia, PA.

[16] Schug, J. and Overton, G.C. (1997) Transcription element search software on the WWW. Technical report, CBIL, University of Pennsylvania. http://agave.humgen.upenn.edu/tess/index.html.

[17] Salas, F., Haas, J., Stoeckert, C.J. and Overton, G.C. (1997) Epodb: An erythropoiesis gene expression database/knowledge base. In: R. Hofest, T. Lengauer, M. Loffler and D. Schomburg (Eds.), Progress in Bioinformatics: German Conference on Bioinformatics, Vol. 1278 of Springer Lecture Notes in Computer Science. Springer, Berlin, pp. 52–61.

86

[18] Wingender, E., Kel, A., Kel, O., Karas, H., Heinemeyer, T., Dietze, P., Knuppel, R., Romaschenko, A. and Kolchanov, N. (1997) TRANSFAC, TRRD and COMPEL: towards a federated database system on transcriptional regulation. Nucleic Acids Res. 25(1), 265–268.

[19] Davidson, S., Overton, C., Tannen, V. and Wong, L. (1996) Biokleisli: A digital library for biomedical researchers. Int. J. Digital Libraries 1(1), 36–53.

[20] Overton, G.C., Aaronson, J., Haas, J. and Adams, J. (1994) QGB: A system for querying sequence database fields and features. J. Comput. Biol. 1(1), 3–13.

[21] Pereira, F.C. and Warren, D.H. (1980) Definite clause grammars for language analysis – a survey of the formalism and a comparison with augmented transition networks. Artificial Intelligence 13(3), 231–278.

[22] Searls, D.B. (1989) Investigating the linguistics of DNA with Definite Clause Grammars. In: E. Lusk and R. Overbeek (Eds.), Logic Programming: Proc. North American Conf., Vol. 1. MIT Press, pp. 189–208.

[23] Searls, D.B. (1992) The linguistics of DNA. Am. Sci. 80(6), 579–591.

[24] Searls, D.B. (1992) GenLang: A logic-based syntactic pattern recognition system for biological sequences. Computational Biology and Informatics Laboratory, University of Pennsylvania School of Medicine, User's Guide, Version 2.3.1 edition.

[25] Searls, D.B. and Dong, S. (1993) A syntactic pattern recognition system for DNA sequences. In: H.A. Lim, J. Fickett, C.R. Cantor and R.J. Robbins (Eds.), Proc. 2nd Int. Conf. on Bioinformatics, Supercomputing and Complex Genome Analysis. World Scientific, pp. 89–101.

[26] Overton, G.C., Koile, K. and Pastor, J.A. (1989) GeneSys: A knowledge management system for molecular biology. In: G. Bell and T. Marr (Eds.), Computers and DNA. Addison-Wesley, Reading, MA, pp. 213–240.

[27] Allen, J.F. (1983) Maintaining knowledge about temporal intervals. Commun. ACM 26(11), 832–843.

[28] Pastor, J.A., Koile, K. and Overton, G.C. (1991) Using analogy to predict functional regions on genes. In: J.F. Nunamaker Jr and R.H. Sprague Jr (Eds.), Proc. 24th Hawaii Int. Conf. on System Science, Vol. I. IEEE Computer Society Press, Los Alamitos, CA, pp. 615–625.

[29] Overton, G.C., Aaronson, J. and Haas, J. A grammar-based expert system for correcting errors in nucleic acid sequence databases. In preparation.

[30] Altschul, S.F., Gish, W., Miller, W., Myers, E.W. and Lipman, D.J. (1990) Basic local alignment search tool. J. Mol. Biol. 215, 403–410.

[31] McAllister, D. and McDermott, D. (1988) Truth maintenance systems. AAAI88 Tutorial Program.

[32] Zhuge, Y., Garcia-Molina, H., Hammer, J. and Widom, J. (1995) View maintenance in a warehousing environment. Proc. 1995 ACM, SIGMOD 24(2), 316–327.

S.L. Salzberg, D.B. Searls, S. Kasif (Eds.), *Computational Methods in Molecular Biology*
© 1998 Elsevier Science B.V. All rights reserved

CHAPTER 6

Classification-based molecular sequence analysis

David J. States and William C. Reisdorf, Jr.

Institute for Biomedical Computing, Washington University. States@ibc.wustl.edu, Reisdorf@ibc.wustl.edu

1. Introduction

Molecular sequence is an information-rich source of data that has been at the core of the revolution in molecular biology. Sequence data are universal, inexpensive to obtain, and are being generated at a prodigious rate. In and of itself, a molecular sequence is relatively uninformative; it is only in the context of other biological sequences that sequence data become truly illuminating. Sequence similarity analysis, primarily based on evolutionary homology, is the single most powerful tool for functional and structural inference available to computational biology. The process of identifying similar sequences and grouping related sequences into classes is a complex process that will be the subject of this chapter. There is an extensive literature on the identification and analysis of tertiary-structure-based classes (SCOP, CATH), but in most cases arising in the molecular biology laboratory crystallographic or NMR data are not available. For this reason, the current discussion will be limited to primary-sequence-based methods.

The space of molecular sequences is very large. For a typical 300 amino protein, there are 20^{300}, or more than 10^{390} possible sequences, much larger than the number of atoms in the universe. The genes found in nature are not a random sampling of the space of all possible combinations of sequences. This implies that evolution has explored only a tiny fraction of the space of possible proteins, and that contemporary proteins evolved from a much smaller set of ancestors by processes such as mutation, gene duplication and chromosomal rearrangements [1,2]. Similarity over an extended region of sequence almost certainly implies homology, descent from a common ancestral gene [3], but homology does not necessarily imply that recognizable and statistically significant sequence similarity persists between a pair of genes. A key value of classification-based analysis is the extension of similarity-based analysis to infer homology through the use of transitive relationships and the definition of class models that are statistically more powerful than pairwise sequence similarity measures.

There are three major classes of macromolecules with which we are concerned, genomic DNA, RNA transcripts, and protein products coded for by genes. In genomic DNA, transposable genetic elements and their residual products form a major class of repetitive sequence elements. Biological systems include many examples of RNA transcripts that are functional gene products. Their recognition and characterization will not be treated as a special case, although the techniques described here will identify and appropriately classify many of these genes. Most gene products are polypeptide chains derived by translation of mRNA transcripts. Well-developed rules for scoring both nucleic acid [4] and amino acid sequences (PAM [5]), BLOSUM [6] and many others) are available.

Sequence data are being generated at a rate of over 50 MB per year. Even with the comparatively modest rates of sequence generation, the system of manual sequence annotation is being severely stressed, and a great deal of data is entering the public repositories with only minimal annotation. The rate of sequence production is doubling approximately annually and will accelerate as the Human Genome Project begins full scale sequencing of the human genome. To complete the sequence of 3000 megabases in 6 years, sequencing must proceed at an average rate in excess of 500 megabases per year and peak throughput will likely exceed 2 megabases per day. For these data to be useful to the biological community, they must be annotated. To cope with the anticipated burden of sequence annotation, automated classification tools are essential, and it is likely that manually reviewed data will form a smaller and smaller part of the sequence annotation database.

1.1. Molecular evolution

The process of automated sequence classification forces us to be precise in our definition of a gene. A class or group is recognizable using automated classification technology only if it is present in multiple copies in the sequence database. While there are a small number of multiply represented sequences in GenBank, the vast majority of sequence similarities arise between biologically distinct molecules. In many cases, however, the region of similarity between biological sequences does not cover their entire extent. A commonly held explanation is that evolution has made use of a set of protein building blocks: collections of amino acids that form a functional unit and are combined or inherited as a group, which we will refer to as "domains". A variety of functionally related building blocks have been identified by biologists; for example, the "zinc finger" domain is a constituent of many DNA-binding proteins [7]. One hypothesis is that these building blocks correspond to exons, the protein coding regions in genes [8], and in some cases exon/intron junctions are observed near protein domain boundaries. In many other cases exon/intron junctions occur within domains or are not correlated with domain boundaries [9]. Regardless of whether the process of domain rearrangement occurred by exon shuffling, recombination, or another process, domain sequences represent independent units of heritable information. Viewed across species, a domain fits the definition of a gene, and a modular protein is a multigene complex.

If the distinction between domains and proteins is not made, proteins containing multiple domains result in the aggregation of unrelated sequences into groups. For example, the protein kinase domain is a component of many different receptors, some containing immunoglobulin like extracellular domains (e.g. platelet-derived growth factor receptor) and others a cysteine-rich domain (e.g. insulin receptor), but there is no evidence that these two extracellular domains are evolutionarily related.

Defining domains as independently heritable units of evolutionary information differs from a definition based on tertiary structure domain. The latter is likely to imply the former, but converse is not true. In particular, a set of folding domains may only be observed in a biological context where all are present and thus appear evolutionarily to behave as a single unit. Further, the limited experimental sampling of biological sequences may fail to identify cases where the individual folding domains occur independently. In either case, sets of folding domains may appear to behave evolutionarily as a single unit or evolutionary domain. At the level of primary sequence analysis, there is no way to appreciate that this evolutionary domain represents a composite structure.

Sequence similarities between a pair of proteins may be observed across the entirety of the protein, or regions of similarity may be limited subsequences. We refer to a set of proteins that are similar over their full extent as being in the same family. Similar subsequences of proteins found in proteins that are not in the same family are termed domains. Our terminology differs from that of Dayhoff[1] who defines a family as a group of proteins which share greater than 50% sequence identify among them. The term superfamily is commonly used to denote a group of proteins whose sequence similarity is sufficient that there is less than one chance in a million of its occurring at random. We discourage use of the Dayhoff nomenclature because it is based on an erroneous test for statistical significance (Z-scores) and fails to take modular domain structure into account.

Examples of protein families include the insulins, immunoglobulins, bacterial heat shock proteins and cytochromes. Protein domains are more specialized regions that appear in many different proteins. For example, the kinase domain, which is found in many unrelated proteins, functions to add a phosphate group to a substrate. Kinase domains are often coupled to a specialized receptor domain (which recognizes, e.g. insulin or a growth factor) in cell membrane proteins. In some cases, the distinction between protein families and domains is not entirely clear; a sequence that is in one case an entire protein may also appear as a domain of a larger protein, examples include the kinase domain and serine proteases.

Many residue positions in proteins are not constrained by functional or structural considerations and are free to mutate. This implies that many sequences will share a common gene function. Observationally, function and structure evolve more slowly than does primary sequence [10], so statistically significant sequence similarity almost certainly implies conservation of structure and function. The fact that genes and gene products that are functionally important and are conserved in evolution can be recognized as recurrent patterns or motifs in sequence databases is the basis for classification-based annotation. It is important to appreciate that the rates of sequence and functional evolution are strongly dependent on the specific case at hand. For example, globins divergent to the point that primary sequence similarity is no longer statistically recognizable nevertheless share a common fold and function as an oxygen transporter [11]. In contrast, the red and green pigment proteins responsible for human color vision are more than 95% identical at a sequence level and yet have diverged in function [12].

1.2. Sequence annotation

The traditional standard for sequence curation has been manual annotation, but there are many difficulties in scaling manual annotation to the capacities needed for whole genome analysis. Individual curators are inconsistent, biased by personal backgrounds and interests, and error prone. If failure to note a feature is considered an error, the absence of annotated features is by far the most prevalent class of error in the manually annotated portions of contemporary sequence databases. Lack of scalability is another problem with the manual annotation. Molecular sequence databases and the rate of new sequence data generation are growing exponentially. Analysis and annotation tasks grow more rapidly that the linear size of the database because it is necessary to compare sequences with each other and to propagate new annotation back to older database entries. The costs of manual data curation are already substantial and will continue to grow independent of whether the analysis is performed at a single site, the NCBI model, or distributed as in the community annotation model.

Sequence annotation consists of several phases. Repetitive elements such as LINE and ALU and regions of simple sequence repeat [13,14] are identified and masked. Similar sequences are identified by database search, and the sequence at hand is compared with a multiple alignment for the family. Domains are labeled either through pattern search or through the recognition of sequence similarity matches. Exon/intron junctions are identified by comparison of mRNA and genomic sequences. Gene models may be predicted (Uberbacher and Mural 1991 [15], and many others). Performing each of these steps manually and collating the data is a time-consuming and painstaking process. The point is that most of the steps in sequence annotation now involve computational analysis and in particular sequence similarity-based analysis. The process appears well suited to automation with the aim being to develop a system that requires little or no manual supervision.

To fill the gap and make the results of high throughput genome sequencing available to the biological research community, an alternative, which might be referred to as minimal automated analysis is being used at many sequencing centers. This approach consists of running a repetitive sequence filter, running BLAST, and running the gene identification tool of local choice on the data, recording the results with little or no manual review, and depositing the data in the public repositories. While this is economically efficient, it leaves a great deal to be desired. Sequence similarity search is a powerful method for identifying protein coding regions in anonymous nucleotide sequence [16], but it does a poor job of identifying promoters, regulatory elements, origins of replication, chromatin binding sites, or other functional elements of the genome. A sequence similarity hit may associate a sequence with a known gene family but does little to place the sequence in context, and with increasing frequency, the highest scoring hits are to other sequences which are also only minimally annotated. The inference of function through sequence similarity depends on longer and longer chains of association to reach an experimentally established reference, and the process becomes less and less reliable. Minimal automated analysis may also be inconsistent between centers because there is considerable variation in the choice of tools, the settings of adjustable parameters, and the reference databases used for comparison.

The groups defined in sequence classifications must be composed of sequences long enough to achieve statistical significance in database searches. This makes it extremely unlikely that they arose as a result of convergent evolution [3]. Since each of our groups almost certainly corresponds to a family of truly homologous sequences (i.e. derived from a common ancestor by a process or biological evolution), the use of hierarchical tree model is appropriate for characterization of member relationships within a group. Numerous tools for hierarchical data analysis are available [17–19]. Although hidden Markov models (HMMs) have been useful in extending phylogenetic analysis [20], attempts to derive detailed phylogenies using hierarchic HMMs have failed [21].

1.3. Artifacts due to sequence and database errors

Gene prediction is improving in both correctly detecting the presence of exons and in assembling full gene models [22], but gene prediction software tools remain error prone. Exon boundaries, small exons, and non-coding exons are frequently mispredicted [23], and none of the current software considers the possibility of alternative transcript splicing. Unfortunately, many unverified gene predictions are conceptually translated into peptide sequences and entered into the protein sequence databases. While gene prediction software may be successful in recognizing many of the exons in a gene, even a single error in locating and exon/intron junction may catastrophically corrupt the conceptually translated protein sequence.

Errors are present in nucleotide sequence databases at levels that prevent reliable translation of coding regions [24,4]. Many genes and protein coding regions are not annotated, and many artifacts such as the presence of vector sequence or polylinker tails are present and unannotated in the database [25]. Among the coding regions, many are untranslatable [16,26,27]. The presence of partial and fragmentary sequences in the databases is of particular importance to automated sequence classification. The original authors may not recognize the partial nature of a sequence, and the annotation of partial sequence data is quite unreliable in current databases. Partial sequences may make a single protein family appear as a set of multidomain families in classification analysis.

1.4. Sequence classification

Molecular sequence classification is a practical application of machine learning: the induction of a meaningful classification from anonymous character strings. There are many challenges in applying machine learning (ML) methods to large, real-world problems. The classification of sequence data is a clustering problem that is several orders of magnitude larger than any other application of unsupervised learning technology of which we are aware. There are several features of protein building blocks that are important in trying to design machine learning methods for identifying them. First, the building blocks are not of uniform size. Of the hundreds of putative domains identified by biologists, see e.g. [28,29], the shortest is six nucleotides long, and the longest contains nearly 400 amino acids. Second, domains evolve over time; insertions, deletions and substitutions in the amino acid sequence of each domain occur through evolution, leading to significant variation in the composition of instances of any particular domain. Despite the variation, each domain has an underlying characteristic pattern, or motif. Third, the databases that must be searched to find these motifs are quite large. Efforts related to the Human Genome Project and other biological research have produced large numbers of protein sequences. The GenBank [30] sequence database includes more than 100 000 sequences containing a total of more than 100 000 000 nucleotides and associated databases such as the dbEST [31] collection of expressed sequence tags (ESTs) now contain over 1 300 000 entries. These databases continue to grow exponentially. The goal of classification is to use as much as possible of these data to identify domains that have been used repeatedly throughout evolution.

The underlying assumption of classification-based analysis is that biologically important features in sequence data can be identified as motifs, or repeating patterns, in molecular sequence data. Such motifs could, in principle arise either as a result of evolutionary homology, descent from a common ancestor, or by convergent evolution. For sequence similarities extensive enough to yield a statistically significant similarity hit in searching contemporary molecular sequence databases, the region of similarity must cover several dozens of residues with a fraction of identity greater than 25% at an amino acid level. The likelihood of such a similarity arising by convergent evolution is extremely low [3]. For this reason, we assume that evolutionary rather than structural factors form the basis for the similarities used for primary sequence classification. This is in contrast to the case for transcription factor binding sites and tertiary-structure-based classification. Many regulatory sites in the genome are quite short, 6–12 nucleotide residues in length. The odds that such site could occur at random in a freely mutating region of the genome are high so independent origin and convergent evolution must be considered as possibilities. Similarly, the limited number of folds available to a peptide chain has led to suggestions that some structures may have multiple independent evolutionary origins and may exhibit convergent evolution.

The sparse sampling of sequence space implies that similarity-based clustering can be a robust and reliable process. Although mutations result in stochastic noise, there are very few ambiguously classified sequences. Rather than discriminating between multiple possible classes for a sequence, the issue is primarily one of finding the closest sequence class for a given sequence.

1.5. Sequence similarity search

Sequence similarity search is an essential component of sequence analysis, but the increasing size and complexity of sequence databases makes it difficult for non-expert users to interpret the results of sequence similarity searches directly. For example, a query sequence with an immunoglobulin domain or a region of heavily biased composition may have a thousand or more similarity hits in a database search. To see if the query contains other domains of interest, the user must scan through this entire list and decide which of the various target sequence names are just additional immunoglobulins and which might be new domains worth following up. Large-scale classification techniques address the issue of annotation and generate a higher level view of the database that facilitates sequence analysis.

Whole database classification is made feasible by the availability of high performance search tools such as BLAST [32], WU2BLAST (Gish, unpublished) and SENSEI [33]. These programs rapidly identify statistically significant partial matches between pairs of protein sequences. providing binary (yes/no) similarity judgments, and identifying variable length stretches of a pair of protein sequences that have greater than chance similarity. These similarity judgments form the basis for essentially all classification methods. As will be described in more detail below, these assessments are noisy, sometimes finding fragmented regions of similarity when there is only one, and sometimes misidentifying the extent of the similar regions. BLASTing the entire Genbank database against itself, looking for all statistically significant pairs of similar regions results in the identification of 1.5×10^7 similarity relationships.

An issue that arises in the practical use of the similarity analysis is its susceptibility to overaggregation due to false positive similarity judgments. A single false positive merges two unrelated groups of sequences. We take a variety of steps in the clustering algorithm to avoid this problem, including the use of sequence filters that eliminate repetitive and low-entropy sequences, such as XNU [14] and SEG [34]. The use of these filters dramatically reduces the number of high scoring false positive alignments generated in the course of a sequence database self-comparison. However, these filters do not completely eliminate false positives. The problem is compounded by the fact that false positives often occur in sets. If a high scoring alignment is seen between two members of biologically unrelated sequence classes, sequence correlations within the classes often imply that many high scoring alignments will be observed between closely related members of the two classes. That means that increasing the strictness of the similarity measure (e.g. increasing the number of similar sequences required for two groups to be merged) does not solve the problem. Although testing of the method on synthetic data shows that this problem occurs in fewer than 1% of groups [35], current databases produce many thousand groups, and overaggregation does occur.

Because the number of overaggregated groups can be expected to be relatively low (a few dozen out of thousands), it is possible to identify incorrectly merged groups manually after a classification has been performed. This has proven to be a difficult task because of the size and complexity of the individual classes. The overaggregated groups are going to be the largest ones, and these can include several thousand sequences and millions of similarity pairs. The transitive closure representation of large groups clarifies the sequence relationships within them and greatly facilitates manual review to identify and eliminate false positive hits and falsely merged sequence classes.

2. A taxonomy of existing classifications

Several classifications of molecular sequence data are now available. These methods differ in fundamental ways that impact on their characteristics and application.

2.1. Domains versus molecules

At the core of any classification is a set of atomic elements upon which the classification is built. The classic Dayhoff "Atlas of Protein Structure", assumed that these elements were complete proteins and that similarity relationships defined on the protein sequence would imply evolutionary relationships amongst the corresponding genes. In practice, this has proven to be only partially correct. Composite proteins consisting of two or more mobile domains are present in all kingdoms of biology. These mobile domains are the elements upon which evolutionary homology can be defined. Classification that operate on entire

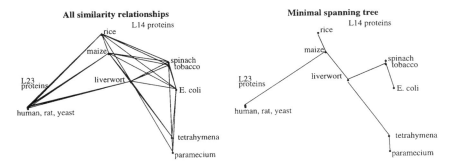

Fig. 1. Comparison of full similarity relationship with a minimal spanning tree for a group of ribosomal proteins defined in the HHS classification. Note that sequence similarity analysis fails to identify many of the similarity relationships in the group. The spanning tree contains 12 edges, fewer than 1/3 the number of edges in the former representation.

protein sequences may be misleading because homology between one pair of domains need not imply homology over the full extent of the molecule.

2.2. Sequence versus structure

A second fundamental distinction that must be made is that of evolutionary as opposed to structural similarity. Sequence similarity-based approaches are the major focus of this chapter. Because structure evolves more slowly than does sequence [10], proteins classified together on the basis of sequence similarity almost certainly share the same fold. Whether the converse holds as well is a subject of active debate, and it is difficult to exclude the possibility of multiple independent evolutionary origins for small domains such as the beta sandwich. When the structural classification is generalized to groups of folds, almost certainly there will be several independent evolutionary origins within each group.

2.3. Clustering methods

Having agreed on what is to be classified, the next step is to establish a classification algorithm. There are several options, including transitive closure, fully connected sets, and fit to a model. In transitive closure, a class is defined as the set of all examples similar to at least one other member of the class. In a fully connected set, all members must be similar to all other members. In a model-based approach, a pattern, signature, motif or other model is defined a priori to define each class and members are assigned to classes based on their fit to the model.

The choice of algorithm affects both the computational complexity of the classification problem and the sensitivity of the classification to errors in similarity judgements. In transitive closure, the algorithmic complexity scales linearly with the number of similarity relationships in the data set. Transitive closure is relatively robust to false negative similarity judgements, particularly for large sequence classes, but the resulting classification is very sensitive to false positive similarity judgements.

Defining an optimal set of fully connected groups from an error prone set of similarity judgements is a graph partition problem and is NP complete in computational complexity. Such a classification would be more robust than transitive closure to false positive errors in similarity assessment, but for the data sets encountered in molecular biology, rigorous application of the approach is not feasible.

Pattern or signature-based classifications, in which each class is described by a class model, are formally equivalent in computational complexity to graph partition. In practice these methods are implemented using local optimization methods in which the class model is iteratively refined based on current class membership assignments and the class membership are redefined based on the updated model. This heuristic is guaranteed to converge, but it is not guaranteed to find a globally optimal solution.

2.4. Hierarchic versus atomic classifications

Classifications vary in the structure that they assign within a class. Classification such as HHS built on transitive closure may retain these transitive relations as a first order structuring of the class. While computationally efficient [36], the direct use of the transitive similarity relationships used to build a class are suboptimal in several respects. Spanning trees are frequently deep but not broad and bear little resemblance to the phylogenetic trees with which biologists are familiar. Spanning trees can, however, be very useful in identifying false positive similarity hits that artifactually merge sets of unrelated sequences. By making explicit all of the relationships in the path between members, the path connecting two unrelated sequences can be traced and the false hits located.

Classes themselves can be viewed as independent entities or as members of a tree or graph. The PIMA [37] classification built a hierarchy of progressively smaller and less informative fingerprints. At the limit, these signatures merge into a single pattern that encompassed all proteins in the database. One cannot exclude the possibility that all proteins, at some point near the origins of life, evolved from a single gene and are therefore all technically homologous. But such a meta-class by definition contains all functions and is therefore completely uninformative from the perspective of functional inference.

2.5. Manual, supervised and fully automated

Sequence annotation began as a manual process with only very limited computational support and has evolved to include a range of approaches varying in their level of automation. HHS is a fully automated classification that runs independent of any user input or supervision. Pfam builds class models in an automated manner but requires the manual selection of seed sequences for each class. Several classifications (Blocks, ProClass, ProDom, PROF_PAT) are built-in using automated procedures based on the PROSITE database of sequence signatures, but the PROSITE signatures are manually derived.

3. Databases of protein sequence/structure classification

Many classifications of molecular sequence have been developed. These are summarized below. They are listed in alphabetic order.

Blocks http://blocks.fhcrc.org/
   Henikoff and Henikoff (1991) [38]. Nucleic Acids Res. 19, 6565–6572;
   Henikoff and Henikoff (1996) [39]. Methods Enzymol. 266, 88–105;
   Pietrokovski, Henikoff and Henikoff (1996) [40]. Nucleic Acids Res. 24, 197–200;
   Henikoff, Pietrokovski and Henikoff (1997) [41]. Nucleic Acids Res. 25, 222–225.
   Sequence-based collection of short, ungapped, aligned sequence regions that can
   be used to characterize the most highly conserved portions of a sequence family.
   Once the set of families are chosen (based on those defined in PROSITE), the
   MOTIF algorithm (ref. [42], PNAS 87, 826–830) is used to generate a number of
   candidate blocks for each family set. The blocks are merged, extended and refined
   into a non-overlapping best set by the MOTOMAT program. Position-based sequence
   weights are then computed for each segment of a block, to compensate for possible
   overrepresentation of some included sequences. Although constructed differently, the
   PRINTS database is in a sense complementary to Blocks, and contains some families
   not represented in Blocks.
COG http://www.ncbi.nlm.nih.gov/COG
   Tatusov, Koonin and Lipman (1997) [43]. Science 278, 631–637.
   Sequence-based Clusters of Orthologous Groups (COGs) were delineated by com-
   paring protein sequences encoded in 7 complete genomes, representing 5 major
   phylogenetic lineages. Each COG consists of individual proteins or groups of paralogs
   from at least 3 lineages and thus corresponds to an ancient conserved domain.
HHS http://www.ibc.wustl.edu/hhs
   States, Harris and Hunter (1993) [36]. Proc. Intelligent Systems Mol. Biol. 1, 387–394.
   Sequence-based method for automated clustering and analysis of large databases.
   Sequence comparisons are generated by the WUBLAST algorithm, combined with
   filtering of repetitive regions. The similarity data are represented by a minimal spanning
   tree (MST) to reduce memory storage requirements, and similar clusters are generated
   by an approximate transitive closure criterion. For each group of similar sequences, the
   CLUSTALW program (ref. [44], Nucleic Acids Res. 22, 4673–4680) is used to perform
   a multiple alignment from which a consensus sequence is derived. The relationships
   between members of groups is visualized graphically as a tree derived from the
   CLUSTAL dendrogram. At present, the databases available are: PDB (2 versions) and
   SwissProt. In the near future dbEST (and others) will also be made available.
Pfam (Washington University) http://genome.wustl.edu/Pfam/
   Pfam (Sanger Center) http://www.sanger.ac.uk/Software/Pfam/
   Sonnhammer, Eddy and Durbin (1997) [45]. Prot. Struct. Funct. Genet. 28, 405–420.
   Sequence-based method using hidden Markov models (HMM). For each domain family
   a seed alignment of representative sequences is assembled (usually by hand), which is
   then used for training an HMM. The HMM is then used to search a sequence database
   (e.g. SwissProt) for all occurrences of the domain. If some known occurrences of the
   domain are missed, the seed alignment is revised and the process repeated until the
   HMM is able to find all known examples of the domain on a single pass. The current
   release (Pfam2.0) contains 527 families. The Pfam database is maintained at both the
   Sanger Center (UK) and Washington University (US). The underlying data are the
   same, but the organization and www interfaces are site-specific.

PIMA (no www-site)

Smith and Smith (1990) [46]. Proc. Natl. Acad. Sci. USA 87, 118–122;

Smith and Smith (1992) [37]. Prot. Eng. 5, 35–41.

Sequence-based automated classification based on the inference of fingerprints in a reduced alphabet to represent a set of proteins – Pattern-Induced Multi-sequence Alignment (PIMA). The fingerprints are defined logically as the most informative signature present in all member of a set and are applied absolutely rather than probabilistically. The algorithm builds a hierarchy of classes with lower levels defining small sets with highly informative signatures and progressively large classes defined by progressively shorter and less specific signatures. The algorithm can be run to a limit in which all proteins are grouped into a single tree. This is *not* an actively maintained data set.

PIR http://www-nbrf.georgetown.edu/pir/

George, Hunt and Barker (1996) [47]. Methods Enzymol. 266, 41–59;

Barker, Pfeiffer and George (1996) [48]. Methods Enzymol. 266, 59–71;

George et al. (1997) [49]. Nucleic Acids Res. 25, 24–27.

Sequence-based classification of sequences into FAMILY and SUPERFAMILY categories is an ongoing project. A FASTA version of PIR has been generated, which can be used to identify possible family relationships for new additions to the sequence database. Stringent requirements for similarity threshold and length congruence must be met for automatic inclusion into a family; borderline cases are examined manually by the scientific staff. Multiple alignments of families containing more than two members are generated by the PILEUP program from the GCG package. Families are then clustered into superfamilies. Homology domains are also annotated as such, based on sequence similarity measures. The domain boundaries are only approximate.

PROBE (no www-site available yet)

Neuwald et al. (1997) [50]. Nucleic Acids Res. 25, 1665–1677.

Sequence-based fully automated method for constructing a multiple alignment model of a protein family, starting from a single sequence. First, a pairwise transitive search using BLAST is run iteratively to gather sequences from the database which are related to the query. PROBE then aligns the sequences and generates an alignment model to use in searching for additional family members. This is done via a combination of Gibbs sampling and a genetic algorithm. Applying this method to a randomly chosen protein sequence generates an average of four times as many relationships as a pairwise comparison alone.

ProClass http://diana.uthct.edu/proclass.html

Wu, Zhao and Chen (1996) [51]. J. Comput. Biol. 3, 547–561.

Sequence-based "second-generation" database organized according to family relationships. A non-redundant set of entries from SwissProt and PIR are grouped into families, as defined by PROSITE groups and PIR superfamily annotations. False negative family members missed by both PROSITE and PIR are detected by a neural network family identification method. The current release of ProClass contains 85 165 sequence entries, approximately half of which are grouped into 3072 ProClass families. ProClass also includes 10 431 newly established SwissProt/PIR links.

ProDom http://protein.toulouse.inra.fr/prodom.html

Sonnhammer and Kahn (1994) [52]. Prot. Sci. 3, 482–492.

Sequence-based method of collecting and defining domains and domain boundaries in an automated fashion. Sequence comparisons are generated by BLAST with a very strict cutoff (P-value = 10^{-6}). Ungapped alignments are clustered, and an overlap parameter was optimized to allow detection of small domains. The clustered domains are then trimmed at the domain boundaries. Multiple alignments and consensus sequences are generated for each domain. The SwissProt database was used for input to ProDom. The methodology used to construct ProDom was similar in several aspects to that used in HHS, although the latter does not attempt to explicitly define domain boundaries.

PROSITE http://expasy.hcuge.ch/sprot/prosite.html

Bairoch and Bucher (1994) [53]. Nucleic Acids Res. 22, 3583–3589;

Bairoch, Bucher and Hofmann (1996) [28]. Nucleic Acids Res. 24, 189–196;

Bairoch, Bucher and Hofmann (1997) [54]. Nucleic Acids Res. 25, 217–221.

Sequence-based collection of profiles – tables of position-specific amino acid weights and gap costs – designed for use as a flexible tool for identifying protein families with high sequence diversity. Specific profiles may occur more then once in a given sequence, provided that they are non-overlapping (e.g. tandem repeats of Ig or FN3 domains). PROSITE is designed to emphasize: (1) completeness; (2) high specificity; (3) documentation; (4) periodic review/updating; and (5) links with SwissProt.

PRINTS http://www.biochem.ucl.ac.uk/bsm/dbbrowser/PRINTS/PRINTS.html

Attwood et al. (1996) [55]. Nucleic Acids Res. 24, 182–188;

Attwood et al. (1997) [56]. Nucleic Acids Res. 25, 212–216.

Sequence-based method for generating fingerprints for a sequence. These fingerprints may be composed of several motifs within the sequence, allowing for more flexible matching than with simple patterns or profiles. For example, a sequence can be found to match a target even if only a subset of the motifs are present, so long as their ordering and spacings are consistent with the target. The source database is OWL, a non-redundant composite of SwissProt, PIR, Genbank, and NRL_3D. Sequence alignments and motif definition are performed with SOMAP (ref. [57], CABIOS 7, 233–235) and correlation of hits is done using the ADSP sequence analysis package (ref. [58], Prot. Eng. 7, 195–203). Regular updates of PRINTS – iterative rescanning of the OWL database for new additional hits, along with extensive annotation of families to include links to other databases – are released.

PROF_PAT → e-mail protein sequences to: bachin@vector.nsk.su

DOS versions of the programs and database (~20 MB) are also available on diskette or tape from the authors.

Bachinsky et al. (1997) [59]. CABIOS 13, 115–122.

Sequence-based method to construct a "best set" of elements (pattern) for a given group of related proteins, which can be used for database searching and discrimination between different families. This approach combines some of the advantages of other databases such as Blocks (automated generation of multiple elements for protein groups), PROSITE (simple presentation of these elements), and profile methods. Families were generated by combining SwissProt entries with similar DE and KW fields. Elements are based on physical/chemical properties of amino acids, as defined by Kidera et al. (1985) [60], J. Prot. Chem. 4, 23–55, which form a short, gap-free interval of aligned protein positions. Mean and standard deviation of the element properties are also computed as a function of position.

SBASE http://base.icgeb.trieste.it/sbase/

Pongor et al. (1993) [61]. Prot. Eng. 6, 391–395;

Pongor et al. (1994) [62]. Nucleic Acids Res. 22, 3610–3615;

Murvai et al. (1996) [63]. Nucleic Acids Res. 24, 210–213;

Fábián et al. (1997) [64]. Nucleic Acids Res. 25, 240–243.

Sequence-based collection of annotated protein domains, designed to facilitate detection of distant similarities between modules of multidomain proteins. Domain definitions are based on protein sequence segments of known structure and/or function, and are classified as STRUCTURAL (e.g. Ig-like, ser/thr-rich), HOMOLGY (e.g. SH2, SH3), LIGAND-BINDING (e.g. DNA, metal), or CELLULAR LOCATION (e.g. signal peptide, transmembrane). Entries are clustered on the basis of BLAST similarity, and redundancy is kept to a minimum. Extensive cross-references to other databases are supported.

3.1. Nucleic acid sequence classifications

By comparison with protein sequence classifications, there have been relatively few attempts to classify nucleic acid sequences directly. Undoubtedly this reflects the difficulty in defining the atomic elements upon which any such classification is built. One widely used resource is the REPBASE collection of repetitive sequence elements.

REPBASE ftp://ncbi.nlm.nih.gov/repository/repbase

Jurka and Pethiyagoda (1995) [13]. J. Mol. Evol. 40, 120–126.

REPeats DataBASE (REPBASE) created by manual curation and sequence similarity search using known seed members as queries. The database includes a manually maintained taxonomy of subclasses for common sequence elements such as ALU.

RPT http://www.ibc.wustl.edu/archive/rpt

Agarwal and States (1994) [65]. Proc. 2nd Int. Conf. on Intelligent Systems for Molecular Biology. AAAI Press, Menlo Park, CA, pp. 1–9.

RePeat Toolkit (RPT) is a collection of sequence similarity search and classification tools based on the transitive closure approach to similarity-based classification. Groups are induced in a fully automated approach that collects elements into families of repeated sequences based on similarity to any other member of the family. A collection of repetitive sequence families present in the *C. elegans* genome is included. In RPT, specific endpoints are not enforced so that fragmented elements and partial reduplications can be collected into a single family. A consequence of this lack of endpoints is lack of clearly defined model or prototypic sequences for each family. In some cases, the transitive closure families include members with no overlap between each other.

3.2. Tertiary-structure-based classifications

For completeness, we mention several of the major tertiary-structure-based classifications that have been developed, although this is not the major subject of this chapter. In a recent survey [66], only 29 531 of 120 068 protein sequences could be assigned to a tertiary structure class, even when homology and similarity relationships were used. In more than three fourths of cases, tertiary structure classification can not be applied.

CATH http://www.biochem.ucl.ac.uk/bsm/cath/

Orengo et al. (1993) [67]. Prot. Eng. 6, 485–500;

Orengo, Jones and Thornton (1994) [68]. Nature 372, 631–634;

Orengo et al. (1996) [69]. PDB Q. Newslett. 78, 8–9;

Orengo et al. (1997) [70]. Structure 5, 1093–1108.

Structure-based classification by clustering at various hierarchical levels – Class, Architecture, Topology, Homologous (CATH) superfamily. Each domain is assigned to one of three classes (α, β, $\alpha\beta$) by an automated procedure. Assignment to an architecture group within each class (3 types for α, 17 for β, 10 for $\alpha\beta$) is done manually. The topological family and homologous superfamily groups are identified by sequence alignment and structure comparisons using the SSAP algorithm. The CATH numbering system is used to assign a unique number to each domain, describing its classification.

DDBASE $\rightarrow$ anonymous ftp to: ftp.cryst.bbk.ac.uk

tar file is located in the directory /pub/ddbase.

Sowdhamini, Rufino and Blundell (1996) [71]. Folding & Design 1, 209–220.

Structure-based clustering of representative family members into similar domain folds – DIAL-derived domain DataBASE (DDBASE). A non-redundant subset of PDB structures is used for input. Secondary structural elements are assigned through main chain hydrogen bonding patterns, and a proximity index is calculated between each pair of secondary structural elements. Clustering of domains is based on these average Cα distances, using the DIAL program (ref. [72], Prot. Sci. 4, 506–20) to identify domain boundaries. Clustering and dendrograms are generated by KITSCH and DRAWTREE from the PHYLIP package. A "disjoint factor" is also computed, as a measure of how strongly a domain interacts with other domains in a multidomain protein.

DEF

Reczko and Bohr (1994) [73]. Nucleic Acids Res. 22, 3616–3619;

Reczko, Karras and Bohr (1997) [74]. Nucleic Acids Res. 25, 235.

Sequence/structure-based method for predicting the fold-class of a protein sequence – Database of Expected Fold (DEF) classes. Implementation is based on neural networks, the training of which optimizes both weights and the number of hidden units in a feed-forward mechanism. Sequences are assigned to one of (currently) 49 fold-classes, where each class is required to contain at least three member structures of low pairwise sequence identity. Networks are trained on half the members of known structure in each class and tested on the remaining members to assess the prediction quality.

FSSP/Dali http://croma.ebi.ac.uk/dali/fssp/fssp.html

Holm and Sander (1996a) [75]. Methods Enzymol. 266, 653–662;

Holm and Sander (1996b) [76]. Nucleic Acids Res. 24, 206–209;

Holm and Sander (1997) [77]. Nucleic Acids Res. 25, 231–234.

Structure-based classification based on an all-against-all fold comparison of PDB entries using the Dali program (ref. [78], J. Mol. Biol. 233, 123–138). The Dali method computes structural similarity based on a weighted sum of interresidue $C\alpha$ coordinates for residues in the common core. A representative set of proteins, all with less than 25% sequence identity to other PDB entries is chosen for further classification into fold families, whose members have very similar structures despite very low sequence similarity. The current release of FSSP contains 813 members in the representative set, which cluster into 253 distinct fold classes.

HSSP http://www.sander.embl-heidelberg.de/
Sander and Schneider (1991) [79]. Proteins 9, 56–68;
Schneider and Sander (1996) [80]. Nucleic Acids Res. 24, 201–205;
Schneider, de Daruvar and Sander (1997) [81]. Nucleic Acids Res. 25, 226–230.
Structure/sequence-based database created by merging 3D and 1D information – Homology-derived StructureS of Proteins (HSSP). For each entry in the DB, a list of homologs is generated by searching the SwissProt database with a modified Smith–Waterman dynamic programming algorithm which allows for gaps. Likely candidates for homology were screened with the MaxHom algorithm before being accepted. At each position in the multiple alignment, the residue variability is assessed by two different criteria, with secondary structure and solvent accessibility data derived from the PDB entry. Thus HSSP is also a database of implied secondary and tertiary structure, and can be used for homology-based structural modeling.

LPFC http://www-camis.stanford.edu/projects/helix/LPFC/
Altman and Gerstein (1994) [82]. Proc. 2nd Int. Conf. on Intelligent Systems for Molecular Biology 2, 19–27;
Gerstein and Altman (1995a) [83]. J. Mol. Biol. 251, 161–175;
Gerstein and Altman (1995b) [84]. CABIOS 11, 633–644;
Schmidt et al. (1997) [85]. Prot. Sci. 6, 246–248.
Structure-based method for automated estimation of the residues which comprise the conserved core of a protein fold family – Library of Protein Family Cores (LPFC). Input is a multiple structural alignment, and the output consists of the core residues (whose atoms occupy the same relative positions in all family members) and non-core residues (whose atoms do not). The core elements are displayed as ellipsoids, the volumes of which are proportional to the positional variability of the constituent atoms across all family members. These positional variabilities are not related to the crystallographic B-factors in any simple way. Images are viewable over the Internet in VRML format.

SCOP http://scop.mrc-lmb.cam.ac.uk/scop/
Murzin et al. (1995) [86]. J. Mol. Biol. 247, 536–540;
Brenner et al. (1996) [87]. Methods Enzymol. 266, 635–643;
Hubbard et al. (1997) [88]. Nucleic Acids Res. 25, 236–239.
Structure-based database which also provides information on evolutionary relationships – Structural Classification Of Proteins (SCOP). Entries are also included for proteins of known structure whose coordinates have not been deposited in the Brookhaven PDB. The primary unit of protein classification is the domain. Domains from multidomain proteins are classified individually. Proteins are grouped into family (sequence identities of 30% or more, or known similarity of structure/function with lower sequence identity); superfamily (families with low sequence identity but probable common evolutionary origin); common fold (proteins containing the

102

same major secondary structural elements in the same arrangement with identical topological connections); class (all alpha, all beta, alpha and beta, alpha plus beta, multidomain) [89].

3.3. Assessing classifications

Figure 2 illustrates the distribution of group sizes in the HHS classification applied to the SwissProt 34 database. While the various classifications differ in detail, the overall trends are typical. A small number of classes contain a very large number of members while a very large number of classes contain only one or two members.

Several of the classification engines attempt to assign domain identifications. We applied each of these to a test sequence, the human CSF-1 receptor, to illustrate the performance of these tools. CSF-1R is a membrane spanning receptor tyrosine kinase that mediates all of the effects of the hematologic growth factor CSF-1. The extracellular domain consists of 5 immunoglobulin-like domains and is responsible for ligand binding. The intracellular domain is a tyrosine kinase.

As is shown in Table 1, most of the classification failed to identify all of these elements, and in some cases failed altogether in their domain assignments. Pfam, a database of hidden Markov models, provided a highly informative view of this query in a format that is close to traditional biomedical nomenclature.

In summary, molecular sequence classification has evolved into a diverse field with a variety of data sets now available on the Internet spanning a range of algorithms and assumptions. While many aspects of these classifications are similar, their detailed application to specific test sequences reveals a wide variety of results.

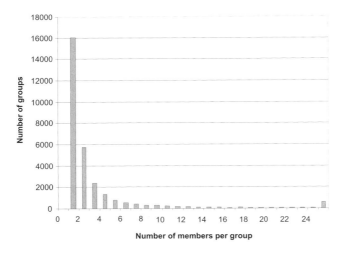

Fig. 2. Histogram showing the distribution of the number of groups in the classification as a function of the number of members in the group. More than 16 000 groups have only a single member while fewer than 400 groups have more than 25 members.

Table 1
Classification of CSF-1 receptor domains[a]

Database	Igs	Kinase	Related proteins
HHS	yes (group 1065)	yes (groups 170 and 755)	fms, kit, flt, pdgfr, vegfr, fgfr
Pfam	yes (domains 1,3,5 only)	yes (finds split kinase domain)	1351 Igs / 928 kinases
ProDom	(finds many N-term domains, but does not identify them as Igs)	no (does not find ANY domains)	fms, kit, pdgfr, flt
PROSITE	(mentions Igs in ECD in annotation; has separate Ig class)	yes (identifies as type III receptor tyrosine kinase)	fms, kit, pdgfr, vegfr, flt
PRINTS	no	yes (identifies as TYRKINASE)	many
Blocks	no	yes	variety of kinases and others
PROF_PAT	no	yes	fms, pdgfr, fgfr, fes; identifies as CSF-1R family
PIR	yes (finds all 5 Igs)	yes	
SBASE	yes (entries for 5 individual Igs and full ECD)	yes	kit, ncam, vcam, pdgfr, icam, mag
ProClass	yes (immunoglobulin homology)	yes (protein kinase homology)	many

[a] Domain assignments are given for the CSF-1 receptor (PIR entry TVHUMD). Note that several classifications fail to identify well-known domains in both the extracellular and intracellular components of the receptor.

Acknowledgments

We wish to thank Dr. Angel Lee for her introduction to the structure and biology of the CSF-1 receptor and assistance in interpretation of analytical results.

References

[1] Dayhoff, M.O., Schwartz, R.M. and Orcutt, B.C. (1978) In: Atlas of Protein Sequence and Structure 5, suppl. 3. Nat. Biomed. Res. Found., Washington, DC, pp. 345–352.
[2] Doolittle, R. (1992) Reconstructing history with amino acid sequences. Prot. Sci. 1, 191–200.
[3] Patterson, C. (1988) Homology in classical and molecular biology. Mol. Biol. Evol. 5, 603–625.
[4] States, D.J., Gish, W. and Altschul, S.F. (1992) Improved sensitivity in nucleic acid database searches using task specific scoring systems. Methods, A Companion to Methods in Enzymology 3, 66–70.
[5] Schwartz, R.M. and Dayhoff, M.O. (1978) In: Atlas of Protein Sequence and Structure 5, suppl. 3. Matrices for detecting distant relationships. Nat. Biomed. Res. Found., Washington, DC, pp. 353–358.
[6] Henikoff, S. and Henikoff, J.G. (1992) Amino acid substitution matrices from protein blocks. Proc. Natl. Acad. Sci. USA 89, 10915–10919.
[7] Berg, J.M. (1990) Zinc finger domains: hypotheses and current knowledge. Annu. Rev. Biophys. & Biophys. Chem. 19, 405–421.
[8] Dorit, R.L., Schoenbach, L. and Gilbert, W. (1990) How big is the universe of exons? Science 250(7), 1377–1382.

[9] Stoltzfus, A., Logsdon Jr, J.M., Palmer, J.D. and Doolittle, W.F. (1997) Intron 'sliding' and the diversity of intron positions. Proc. Natl. Acad. Sci. USA 94(20), 10739–10744.

[10] Chothia, C. and Lesk, A. (1987) The evolution of protein structures. Cold Spring Harbor Symp. Quantum Biol. 52, 399–405.

[11] Suyama, M., Matsuo, Y. and Nishikawa, K. (1997) Comparison of protein structures using 3D profile alignment. J. Mol. Evol. 44(4), 163–173.

[12] Nathans, J., Merbs, S.L., Sung, C.H., Weitz, C.J. and Wang, Y. (1992) Molecular genetics of human visual pigments. Annu. Rev. Genet. 26, 403–424.

[13] Jurka, J. and Pethiyagoda, C. (1995) Simple repetitive DNA sequences from primates: compilation and analysis. J. Mol. Evol. 40, 120–126.

[14] Claverie, J.-M. and States, D.J. (1993) Information enhancement methods for large scale sequence analysis. Comput. Chem. 17, 191–201.

[15] Uberbacher, E.C. and Mural, R.J. (1991) Locating protein-coding regions in human DNA sequences by a multiple sensor-neural network approach. Proc. Natl. Acad. Sci. USA 88(24), 11261–11265.

[16] Gish, W. and States, D.J. (1993) Identification of protein coding regions by database similarity search. Nat. Genet. 3(3), 266–272.

[17] Felsenstein, J. (1988) Phylogenies from molecular sequences: inference and reliability. Annu. Rev. Genet. 22, 521–565.

[18] Cheeseman, P., Self, M., Kelly, J., Taylor, W., Freeman, D. and Stutz, J. (1988) Bayesian classification. In: Proc. AAAI 88, Saint Paul, MN.

[19] Feng, D.F. and Doolittle, R.F. (1988) A flexible method to align large numbers of biological sequences. J. Mol. Evol. 28, 161–169.

[20] Felsenstein, J. and Churchill, G.A. (1996) A Hidden Markov Model approach to variation among sites in rate of evolution. Mol. Biol. Evol. 13, 93–104.

[21] Mitchison, G. and Durbin, R. (1995) Tree-based maximal likelihood substitution matrices and Hidden Markov Models. J. Mol. Evol. 41, 1139–1151.

[22] Burge, C. and Karlin, S. (1997) Prediction of complete gene structures in human genomic DNA. J. Mol. Biol. 268(1), 78–94.

[23] Burset, M. and Guigo, R. (1996) Evaluation of gene structure prediction programs. Genomics 34(3), 353–367.

[24] Krawetz, S.A. (1989) Sequence errors described in GenBank: A means to determine the accuracy of DNA sequence interpretation. Nucleic Acids Res. 17, 3951–3957.

[25] Lamperti, E.D., Kittelberger, J.M., Smith, T.F. and Villa-Komaroff, L. (1992) Corruption of Genomic Databases with Anomalous Sequence. Nucleic Acids Res. 20, 2741–2747.

[26] Posfai, J. and Roberts, R.J. (1992) Finding errors in DNA sequences. Proc. Natl. Acad. Sci. USA 89, 4698–4702.

[27] States, D.J. and Botstein, D. (1991) Molecular sequence accuracy and the analysis of protein coding regions. Proc. Natl. Acad. Sci. USA 88(13), 5518–5522.

[28] Bairoch, A., Bucher, P. and Kofmann, K. (1996) The PROSITE database, its status in 1995. Nucleic Acids Res. 24, 189–196.

[29] Ghosh, D. (1992) TFD: the transcription factors database. Nucleic Acids Res. 20, 2091–2093.

[30] Benson, D.A., Boguski, M., Lipman, D.J. and Ostell, J. (1996) GenBank. Nucleic Acids Res. 24, 1–5.

[31] Boguski, M.S., Lowe, T.M. and Tolstoshev, C.M. (1993) dbEST – database for "Expressed Sequence Tags". Nat. Genet. 4, 332–333.

[32] Altschul, S.F., Gish, M., Miller, W., Myers, E.W. and Lipman, D.J. (1990) Basic local alignment tool. J. Mol. Biol. 215, 403–410.

[33] States, D.J. and Agarwal, P. (1996) Compact encoding strategies for DNA sequence similarity search. ISMB 4, 211–217.

[34] Wootton, J.C. and Federhen, S. (1993) Statistics of local complexity in amino acid sequences and sequence databases. Comput. & Chem. 17, 149–163.

[35] Hunter, L., Harris, N. and States, D.J. (1992) Efficient classification of massive, unsegmented datastreams. In: D. Sleeman and P. Edwards (Eds.), International Machine Learning Workshop. Morgan Kaufman, San Mateo, CA, pp. 224–232.

[36] States, D.J., Harris, N.L. and Hunter, L. (1993) Computationally efficient cluster representation in molecular sequence megaclassification. Proc. Intell. Systems Mol. Biol. 1, 387–394.

[37] Smith, R.F. and Smith, T.F. (1992) Pattern-induced multi-sequence alignment (PIMA) algorithm employing secondary structure-dependent gap penalties for use in comparative protein modelling. Prot. Eng. 5, 35–41.

[38] Henikoff, S. and Henikoff, J.G. (1991) Automated assembly of protein blocks for database searching. Nucleic Acids Res. 23, 6565–6572.

[39] Henikoff, J.G. and Henikoff, S. (1996) Blocks database and its applications. Methods Enzymol. 266, 88–105.

[40] Pietrokovski, S., Henikoff, J.G. and Henikoff, S. (1996) The Blocks database: a system for protein classification. Nucleic Acids Res. 24, 197–200.

[41] Henikoff, J.G., Pietrokovski, S. and Henikoff, S. (1997) Recent enhancements to the Blocks database servers. Nucleic Acids Res. 25, 222–225.

[42] Smith, H.O., Annau, T.M. and Chandrasegaran, S. (1990) Finding sequence motifs in groups of functionally related proteins. Proc. Natl. Acad. Sci. USA 87, 826–830.

[43] Tatusov, R.L., Koonin, E.V. and Lipman, D.J. (1997) A genomic perspective on protein families. Science 278, 631–637.

[44] Thompson, J.D., Higgins, D.G. and Gibson, T.J. (1994) ClustalW: improving the sensitivity of progressive multiple sequence alignment through sequence weighting, position-specific gap penalties and weight matrix choice. Nucleic Acids Res. 22, 4673–4650.

[45] Sonnhammer, E.L.L., Eddy, S.R. and Durbin, R. (1997) Pfam: a comprehensive database of protein domain families based on seed alignments. Prot. Struct. Funct. Genet. 28, 405–420.

[46] Smith, R. and Smith, T.F. (1990) Automatic generation of primary sequence pattterns from sets of related protein sequences. Proc. Natl. Acad. Sci. USA 87, 118–122.

[47] George, D.G., Hunt, L.T. and Barker, W.C. (1996) PIR-International protein sequence database. Methods Enzymol. 266, 41–59.

[48] Barker, W.C., Pfeiffer, F. and George, D.G. (1996) Superfamily classification in PIR-International protein sequence database. Methods Enzymol. 266, 59–71.

[49] George, D.G., Dodson, R.J., Garavelli, J.S., Haft, D.H., Hunt, L.T., Marzec, C.R., Orcutt, B.C., Sidman, K.E., Srinivasarao, G.Y., Yeh, L.-S.L., Arminski, L.M., Ledley, R.S., Tsugita, A. and Barker, W.C. (1997) The protein information resource (PIR) and the PIR-International protein sequence database. Nucleic Acids Res. 25, 24–27.

[50] Neuwald, A.F., Liu, J.S., Lipman, D.J. and Lawrence, C.E. (1997) Extracting protein alignment models from the sequence database. Nucleic Acids Res. 25, 1665–1677.

[51] Wu, C., Zhao, S. and Chen, H.L. (1996) A protein class database organized with PROSITE protein groups and PIR superfamilies. J. Comput. Biol. 3, 547–561.

[52] Sonnhammer, E.L.L. and Kahn, D. (1994) Modular arrangement of proteins as inferred from analysis of homology. Prot. Sci. 3, 482–492.

[53] Bairoch, A. and Bucher, P. (1994) PROSITE: recent developments. Nucleic Acids Res. 22, 3583–3485.

[54] Bairoch, A., Bucher, P. and Hofmann, K. (1997) The PROSITE database, its status in 1997. Nucleic Acids Res. 25, 217–221.

[55] Attwood, T.K., Beck, M.E., Bleasby, A.J., Degtyarenko, K. and Parry-Smith, D.J. (1996) Progress with the PRINTS protein fingerprint database. Nucleic Acids Res. 24, 182–188.

[56] Attwood, T.K., Beck, M.E., Bleasby, A.J., Degtyarenko, K., Michie, A.D. and Parry-Smith, D.J. (1997) Novel developments with the PRINTS protein fingerprint database. Nucleic Acids Res. 25, 212–216.

[57] Parry-Smith, D.J. and Attwood, T.K. (1991) SOMAP: a novel interactive approach to multiple protein sequences alignment. CABIOS 7, 233–235.

[58] Attwood, T.K. and Findlay, J.B.C. (1994) Fingerprinting G-protein-coupled receptors. Prot. Eng. 7, 195–203.

[59] Bachinsky, A.G., Yarigin, A.A., Guseva, E.H., Kulichkov, V.A. and Nizolenko, L.Ph. (1997) A bank of protein family patterns for rapid identification of possible functions of amino acid sequences. CABIOS 13, 115–122.

[60] Kidera, A., Konishi, Y., Oka, M., Ooi, T. and Scheraga, H.A. (1985) Statistical analysis of physical properties of the 20 naturally occurring amino acids. J. Prot. Chem. 4, 23–55.

[61] Pongor, S., Skerl, V., Cserzo, M., Hátsági, Z., Simon, G. and Bevilacqua, V. (1993) The SBASE domain library: a collection of annotated protein segments. Prot. Eng. 6, 391–395.

[62] Pongor, S., Hátsági, Z., Degtyarenko, K., Fábián, P., Skerl, V., Hegyi, H., Murval, J. and Bevilacqua, V. (1994) The SBASE protein domain library, release 3.0: a collection of annotated protein sequence segments. Nucleic Acids Res. 22, 3610–3615.

[63] Murvai, J., Gabrielian, A., Fábián, P., Hátsági, Z., Degtyarenko, K., Hegyi, H. and Pongor, S. (1996) The SBASE protein domain library, release 4.0: a collection of annotated protein sequence segments. Nucleic Acids Res. 24, 210–213.

[64] Fábián, P., Murvai, J., Hátsági, Z., Vlahovicek, K., Hegyi, H. and Pongor, S. (1997), The SBASE protein domain library, release 5.0: a collection of annotated protein sequence segments. Nucleic Acids Res. 25, 240–243.

[65] Agarwal, P. and States, D.J. (1994) The repeat pattern toolkit (RPT): analyzing the structure and evolution of the *C. elegans* genome. In: Proc. 2nd Int. Conf. on Intelligent Systems for Molecular Biology. AAAI Press, Menlo Park, CA, pp. 1–9.

[66] Gerstein, M. and Levitt, M. (1997) A structural census of the current population of protein sequences. Proc. Natl. Acad. Sci. USA 94, 11911–11916.

[67] Orengo, C.A., Flores, T.P., Taylor, W.R. and Thornton, J.M. (1993) Identification and classification of protein fold families. Prot. Eng. 6, 485–500.

[68] Orengo, C.A., Jones, D.T. and Thornton, J.M. (1994) Protein superfamilies and domain superfolds. Nature 372, 631–634.

[69] Orengo, C., Michie, A., Jones, S., Jones, D., Swindells, M. and Thornton, J. (1996) The CATH classification scheme of protein domain structural families. PDB Q. Newslett. 78, 8–9.

[70] Orengo, C.A., Michie, A.D., Jones, S., Jones, D.T., Swindells, M.B. and Thornton, J.M. (1997) CATH: a hierarchic classification of protein domain structures. Structure 5, 1093–1108.

[71] Sowdhamini, R., Rufino, S.D. and Blundell, T.L. (1996) A database of globular protein structural domains: clustering of representative family members into similar folds. Folding & Design 1, 209–220.

[72] Sowdhamini, R. and Blundell, T.L. (1995) An automatic method involving cluster analysis of secondary structures for the identification of domains in proteins. Prot. Sci. 4, 506–520.

[73] Reczko, M. and Bohr, H. (1994) The DEF database of sequence based protein fold class predictions. Nucleic Acids Res. 22, 3616–3619.

[74] Reczko, M., Karras, D. and Bohr, H. (1997) An update of the DEF database of protein fold class predictions. Nucleic Acids Res. 25, 235.

[75] Holm, L. and Sander, C. (1996) Alignment of three-dimensional protein structures: network server for database searching. Methods Enzymol. 266, 653–662.

[76] Holm, L. and Sander, C. (1996) The FSSP database: fold-based on structure–structure alignment of proteins. Nucleic Acids Res. 24, 206–209.

[77] Holm, L. and Sander, C. (1997) Dali/FSSP classification of three-dimensional protein folds. Nucleic Acids Res. 25, 231–234.

[78] Holm, L. and Sander, C. (1993) Protein structure comparison by alignment of distance matrices. J. Mol. Biol. 233, 123–138.

[79] Sander, C. and Schneider, R. (1991) Database of homology-derived protein structures and the structural meaning of sequence alignment. Prot. Struct. Funct. Genet. 9, 56–68.

[80] Schneider, R. and Sander, C. (1996) The HSSP database of protein structure–sequence alignments. Nucleic Acids Res. 24, 201–205.

[81] Schneider, R., de Daruvar, A. and Sander, C. (1997) The HSSP database of protein structure–sequence alignments. Nucleic Acids Res. 25, 226–230.

[82] Altman, R.B. and Gerstein, M. (1994) Finding an average core structure: application to the globins. ISMB 2, 19–27.

[83] Gerstein, M. and Altman, R.B. (1995) Average core structures and variability measures for protein families: application to the immunoglobulins. J. Mol. Biol. 25, 161–175.

[84] Gerstein, M. and Altman, R.B. (1995) Using a measure of structural variation to define a core for the globins. CABIOS 11, 633–644.

[85] Schmidt, R., Gerstein, M. and Altman, R.B. (1997) LPFC: an Internet library of protein family core structures. Prot. Sci. 6, 246–248.

[86] Murzin, A.G., Brenner, S.E., Hubbard, T. and Chothia, C. (1995) SCOP: a structural classification of proteins database for the investigation of sequence and structures. J. Mol. Biol. 247, 536–540.

[87] Brenner, S.E., Chothia, C., Hubbard, T.J. and Murzin, A.G. (1996) Understanding protein structure: using SCOP for fold interpretation. Methods Enzymol. 266, 635–643.

[88] Hubbard, T.J.P., Murzin, A.G., Brenner, S.E. and Chothia, C. (1997) SCOP: a structural classification of proteins database. Nucleic Acids Res. 25, 235–239.

[89] Brenner, S.E., Chothia, C. and Hubbard, T.J.P. (1997) Population statistics of protein structures: lessons from structural classifications. Curr. Opin. Struct. Biol. 7, 369–376.

S.L. Salzberg, D.B. Searls, S. Kasif (Eds.), *Computational Methods in Molecular Biology*
© 1998 Elsevier Science B.V.   All rights reserved

CHAPTER 7

Computational gene prediction using neural networks and similarity search

Ying Xu and Edward C. Uberbacher

Computational Biosciences Section, Life Sciences Division,
Oak Ridge National Laboratory, Oak Ridge, TN 37831-6480, USA

1. Introduction

One of the most fundamental questions that can be asked about a DNA sequence is whether or not it encodes protein. Localization of protein-coding regions (*exons*) in anonymous DNA sequence by pure biological means is both time-consuming and costly. A number of computational methods have been proposed and used to predict protein-coding regions and gene structures in the past few years [1–13]. Though the performance of these computational methods is currently imperfect, the computer-based approach may soon be the only one capable of providing analysis and annotation at a rate compatible with worldwide DNA sequencing throughput.

Computer-based gene prediction methods range from database searches for homology with known proteins to the more general pattern recognition approaches. In this chapter, we describe a pattern recognition-based method for exon predictions, and a homology information-based approach for parsing the predicted exons into gene structures.

The basis for most pattern recognition-based exon prediction methods is the positional and compositional biases imposed on the DNA sequence in coding regions by the genetic code and by the distribution of amino acids in proteins. Though recognition of each of these biases provides a useful indication of exons it is unrealistic to expect a single "perfect" indicator, given the incomplete state of our understanding of the underlying biological processes around genes. We previously proposed an approach to combine information from several exon-prediction algorithms, each designed to recognize a particular sequence property, using a neural network to provide more powerful exon recognition capabilities. The algorithm has been implemented as a computer system, called GRAIL [4,14].

To determine the likelihood that a DNA segment is an exon involves measuring coding potentials of the region and evaluation of the strength of edge signals of the region, e.g. the strength of potential splice junctions or translation starts bounding the region. A number of coding measures have been proposed based on the frequency of nucleotide "words" of a fixed length. Different types of DNA sequence (exons, introns, etc.) have different distributions of word occurrence [15]. In GRAIL, we have used a frame-dependent 6-tuple preference model [4] and a 5th order non-homogeneous Markov chain model [2] to measure coding potentials. A number of measures including a 5-tuple preference model, long-distance correlations between single bases, etc. have been used to measure the strength of a potential splice junction or a translation start. These measures along with a few other features are fed into a neural network for the final exon candidate evaluation. This neural network is trained, based on empirical data, to effectively weigh the various features in scoring the possibility of each sequence segment (exon candidate) being an actual exon. The use of empirical data for training allows the system to optimally utilize each feature in the presence of the others, without *a priori* assumptions about the independence of the features or their relative strengths.

Parsing the predicted exons into gene structures represents the second phase of a computational gene prediction problem. The problem involves identification of the start and the end of a gene, and determination of whether a predicted exon should be part of a gene. This problem could be easily solved if there existed a complete cDNA or protein sequence database. Short of such a complete database, gene structures of the predicted exons can be determined in such a way that is most consistent with the known protein/cDNA sequences, and with the predicted exon scores and translation frames.

We have recently proposed a practical approach to apply expressed sequence tags (ESTs) as reference models when parsing the predicted exons into gene models [16]. *ESTs* are single-pass sequence fragments of cDNA clones. They have been used for gene discovery in genomic sequences for a number of years [17–21]. With the significant effort which has gone into EST sequencing in the past few years, it is now estimated that about 60% of human genes are partially represented by ESTs [22]. Though ESTs alone often are insufficient to make complete gene structure predictions due to their incomplete coverage of genes, they can provide useful information to help determine the 3′ end of a gene and the minimal extent of a gene, identify falsely predicted exons, identify and locate missed exons, and correct the false boundaries of the predicted exons.

Within each gene model, adjacent exons should be consistent in their translation frames. This simple condition on a gene model provides a very effective constraint in the computational gene modeling process, especially in the absence of EST reference models. Our gene modeling algorithm constructs gene models, from the predicted exons, that are most consistent with the available EST information, under the constraint that the translation frame consistency is maintained between adjacent exons within each gene model. The optimality of the predicted gene models is achieved through solving an optimization problem using a dynamic programming algorithm.

This chapter is organized as follows. Section 2 describes the neural network-based GRAIL exon prediction system. Section 3 addresses issues related to exon predictions in coding sequences in which translation frames may be disrupted. Section 4 presents an algorithm to construct (multiple) gene models based on the available EST information. Section 5 summarizes this chapter.

2. Exon prediction by neural networks

GRAIL (GRAIL II version 1.3) predicts an exon by recognizing a coding region bounded by exon-edge signals [14]. Three types of exons, initial, internal and terminal exons, are recognized in GRAIL. An *initial* exon is an exon that starts with a translation start; an *internal* exon is bounded on the left by an acceptor splice junction and on the right by a donor splice junction; A *terminal* exon ends with a translation stop codon. We use the internal exon as an example to illustrate how exons are recognized in GRAIL. GRAIL uses different algorithms to recognize coding signals and edge signals (related to acceptor/donor splice junctions) separately, and combines the recognized signals using a neural network to make an exon prediction.

2.1. Coding region recognition

A coding DNA sequence encodes protein by encoding each amino acid of the protein into a triplet of nucleotides, also called a *codon*. Recognition of a coding region essentially involves a determination of whether the DNA sequence can be partitioned into segments of three and this sequence of nucleotide triplets may possibly correspond to a "valid" protein, a sequence of amino acids, and also reflect the typical codon usage in a particular organism. A number of models have been proposed to measure the coding potential of a DNA sequence, based on the distribution of consecutive amino acids in a protein. GRAIL uses two of those models, a frame dependent 6-tuple preference model [4] and a 5th order non-homogeneous Markov chain model [2], as basic coding measures. The coding of amino acids in nucleotide triplets means that there are three possible ways to translate a DNA to protein, i.e. the three possible translation frames (two of which are incorrect).

The *frame dependent 6-tuple preference model* consists of three preference values, $pf_0(X)$, $pf_1(X)$, $pf_2(X)$, for each of the 4096 possible 6-tuples X, which are defined as follows:

$$pf_r(X) = \log \frac{f_r(X)}{f_n(X)}, \quad \text{for } r = 0, 1, 2, \tag{1}$$

where $f_r(X)$ is the frequency of 6-tuple X appearing in a coding region and in the actual translation frame $+r$, for $r = 0, 1, 2$, and $f_n(X)$ is the frequency of X appearing in a non-coding region. In GRAIL, all the 6-tuple frequencies were calculated from a set of 450 DNA sequences containing 462 608 coding bases and 2 003 642 non-coding bases.

Let $a_1 \cdots a_n$ be a DNA sequence of n bases long. The preference model calculates the coding potential of a segment $a_k \cdots a_m$ in each of the three possible translation frames, $r = 0, 1, 2$, as follows:

$$\begin{aligned} pf_r(a_k \cdots a_m) = \big[& pf_{(k+5-r) \bmod 3}(a_k \cdots a_{k+5}) + pf_{(k+6-r) \bmod 3}(a_{k+1} \cdots a_{k+6}) \\ & + pf_{(k+7-r) \bmod 3}(a_{k+2} \cdots a_{k+7}) + \cdots + pf_{(m-r) \bmod 3}(a_{m-5} \cdots a_m) \big] \\ & \times \frac{1}{m - k + 1}, \end{aligned} \tag{2}$$

where mod is the modulo function. We call the translation frame with the highest preference value (or score) among the three possible ones the *preferred* translation frame. Figure 1 shows a plot of preference scores (in the preferred translation frames) for a large set of coding regions versus their flanking non-coding regions (a 60-base region to each side of an exon).

A DNA sequence can be considered as a stochastic process. Under the assumption that a DNA sequence forms a 5th order non-homogeneous Markov chain, GRAIL uses

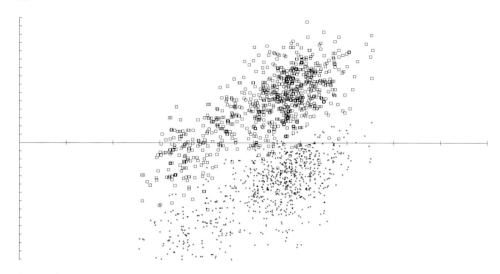

Fig. 1. The X-axis represents the $G+C$ composition of an exon candidate and the Y-axis represents the 6-tuple scores measured by the frame-dependent preference model. Each tick mark on the horizontal axis represents 10% in $G+C$ composition with 0% on the left and 100% on the right. The large squares represent the coding regions and the small dots represent the regions flanking coding regions.

Bayes' Law to measure the coding potential of a DNA segment $a_k \cdots a_m$ in each of the three possible translation frames, $r = 0, 1, 2$, as follows:

$$P_r(\text{coding}|a_k \cdots a_m) = \frac{P_r(a_k \cdots a_m|\text{coding})}{\sum_{f=0}^{2} P_f(a_k \cdots a_m|\text{coding}) + CP(a_k \cdots a_m|\text{non}-\text{coding})}, \quad (3)$$

where by the Markov chain assumption,

$$P_r(a_k \cdots a_m|\text{coding}) = P_{(k+5-r)\bmod 3}(a_k \cdots a_{k+4}|\text{coding})$$
$$\times P_{(k+5-r)\bmod 3}(a_{k+5}|a_k \cdots a_{k+4}, \text{coding})$$
$$\times P_{(k+6-r)\bmod 3}(a_{k+6}|a_{k+1} \cdots a_{k+5}, \text{coding}) \quad (4)$$
$$\cdots$$
$$\times P_{(m-r)\bmod 3}(a_m|a_{m-5} \cdots a_{m-1}, \text{coding}).$$

and

$$P_n(a_k \cdots a_m|\text{non}-\text{coding}) = P_n(a_k \cdots a_{k+4}|\text{non}-\text{coding})$$
$$\times P_n(a_{k+5}|a_k \cdots a_{k+4}, \text{non}-\text{coding})$$
$$\times P_n(a_{k+6}|a_{k+1} \cdots a_{k+5}, \text{non}-\text{coding}) \quad (5)$$
$$\cdots$$
$$\times P_n(a_m|a_{m-5} \cdots a_{m-1}, \text{non}-\text{coding}).$$

and C is the estimation of the ratio of coding versus non-coding bases in DNA sequences, $P_r(X|Y)$ and $P_n(X|Y)$ are the conditional probabilities of X in coding regions (in translation frame $+r$) in the presence of Y and in non-coding regions, respectively. These conditional probabilities can be estimated using the preference values pf_r and pf_n, defined above.

Though not totally independent measures, each of these two models has its own coding recognition strengths and weaknesses according to our test results. GRAIL uses both models as the basic coding feature extraction methods, and combines them along with other measures in the neural network exon recognition system.

2.2. Splice junction recognition

GRAIL recognizes acceptor junctions having the usual YAG (i.e. CAG or TAG) consensus, as well as the non-standard AAG consensus, and also recognizes donor junctions containing the GT consensus. Our recognition method is based on a number of relative frequency measures of nucleotide "words" appearing in the neighborhood of true splice sites versus false splice sites (containing minimal splice consensus) as each of those measures exhibits some discriminative power among true and false splice junctions. A large set of true and false splice sites are used to calculate these frequencies. As a result, a profile of frequencies is obtained for true and false splice sites, respectively. For the YAG acceptor recognition, seven relative frequencies are measured, which are given as follows:

Let $a_{-60} \cdots a_{35}$ represent the DNA segment containing a YAG consensus with $a_0 a_1 a_2 = YAG$.

(1) $\displaystyle\sum_{i=-23}^{-4} \log\left(\frac{F_t^i(a_i \cdots a_{i+4})}{F_f^i(a_i \cdots a_{i+4})}\right),$

where $F_t^i(\cdot)$ and $F_f^i(\cdot)$ represent the positionally dependent (position i) 5-tuple frequencies in true and false splice junction regions, respectively.

(2) $\displaystyle\sum_{i=-27}^{0} \log\left(\frac{F_t(a_i)}{F_f(a_i)}\right) + \sum_{i=3}^{4} \log\left(\frac{F_t(a_i)}{F_f(a_i)}\right),$

where $F_t(\cdot)$ and $F_f(\cdot)$ are defined similarly to (1) except that they are not positionally dependent.

(3) $\displaystyle\sum_{i=-27}^{0} PY(a_i)\sqrt{i+28},$

where $PY(a_i)$ is 1 if a_i is a pyrimidine (C or T) otherwise 0, and $\sqrt{i+28}$ is a weighting factor for position i.

(4) The (normalized) distance between a_0 and the nearest upstream YAG.

(5) $\displaystyle\sum_{i=-27}^{4} \sum_{j \geqslant i}^{4} \log\left(\frac{F_t^i(a_i a_j)}{F_f^i(a_i a_j)}\right),$

where $F_t^i(\cdot)$ and $F_f^i(\cdot)$ are defined similarly to (1).

(6,7) Coding potentials in regions of $a_{-60} \cdots a_{-1}$ and $a_3 \cdots a_{35}$ measured using a 6-tuple preference model. This is to give an indication of a transition between non-coding and coding sequences.

A simple feed-forward neural network is trained, using the standard back-propagation learning algorithm [23], to score the true and false YAG acceptors (more discussion on

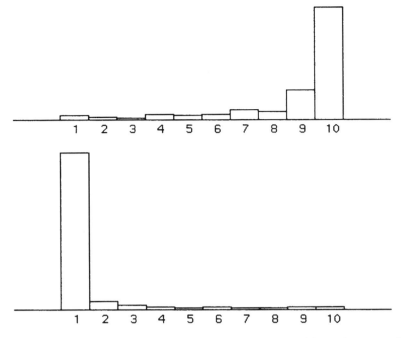

Fig. 2. YAG acceptor prediction. A total of 227 true (top) and 5127 false YAG acceptors (bottom) were tested. The X-axis represents the YAG scores by the neural network (ranging from 0 to 10, divided into 10 intervals), and the Y-axis represents the percentage of acceptor candidates which were scored in the interval.

the neural networks is given in section 2.3). The neural network is trained to assign a numerical value between 0 and 1 to each of the training cases, with 1 representing a true acceptor and 0 a false acceptor. Figure 2 shows the performance statistics of a YAG acceptor neural network recognizer on an independent test set.

Acceptors with non-standard AAG consensus are recognized using basically the same measures but with different frequency profiles. A separate neural network was trained for this type of acceptor. Similarly, donor splice junctions are recognized.

2.3. Information fusion

This section describes how information from different sources can be fused together to provide a powerful exon recognizer. In GRAIL, an (internal) exon candidate is located by finding an open translation frame bounded from the left by a potential acceptor site and from the right by a potential donor site. The possibility of such a region being an actual exon is evaluated as follows.

The goal of the exon recognition process in GRAIL is not just to discriminate exons from non-exonic regions but also to score the degree of correctness of an exon candidate that overlaps actual exons. For example, we consider a candidate which extends past one edge of an exon, but otherwise overlaps it, to be partially correct. To achieve this scoring, we use coding measures in the flanking areas in addition to the coding measures of a candidate region. The rationale is that a strong coding indication from the neighboring areas indicates that the candidate may be just a portion of an exon. As the candidate more closely approximates an actual exon, more non-coding elements will be included in its surrounding areas and hence the surroundings will exhibit a weaker coding score. GRAIL uses 60 bases on each side of an exon candidate as the flanking regions.

The recognition of coding regions using the 6-tuple (or in general k-tuple, for any fixed k) method is known to have strong dependence on the G+C (bases G and C) composition, and is more difficult in G+C poor domains. Our observation on the relationship of 6-tuple coding measures and G+C composition supports this belief. If we estimate the frequencies of coding 6-tuples and non-coding 6-tuples in the high G+C domain, and use these frequencies to calculate coding measures for a set of coding regions and their 60-base flanking regions in all ranges of G+C composition, the result is shown in Fig. 1. The coding measures for both the coding regions and their flanks are much lower in the G+C poor domain compared to the G+C rich domain. A very similar behavior is observed if the 6-tuple frequencies are collected from low G+C DNA sequences. Interestingly, though the relative separation between coding regions and their flanking regions is similar at both ends of the G+C composition range; that is, many of the non-exonic regions in a high G+C isochore have a higher coding measure than many coding regions in G+C poor regions. This certainly highlights the necessity to include G+C composition as one piece of information in the information fusion process for exon recognition.

Splice junction scores help determine the correct exon boundaries. Though false splice junction prediction may occur, in general true splice junctions have higher scores than nearby false splice junctions. One interesting observation we made indicates that shorter exons tend to have stronger splice junction sites and hence higher splice scores. Hence, it is necessary to provide information on exon lengths to the information fuser.

We have used a layered feed-forward neural network as the information fuser, using the measures described above, to recognize exons. A schematic of this neural network is shown in Fig. 3. This neural network has 13 inputs, two hidden layers with seven and three nodes, respectively, and one output, which can be represented mathematically by the following.

$$\text{output} = g\left(\sum_{k=1}^{3} W_k^3 \, g\left(\sum_{j=1}^{7} W_{kj}^2 \, g\left(\sum_{i=1}^{13} W_{ji}^1 (\text{input}_i) \right) \right) \right), \tag{6}$$

where

$$g(x) = \frac{1}{1 + exp(-x)},$$

and W's are edge weights to be "learned" using the standard back-propagation learning algorithm [23].

In training the network (determining the weights), our goal is to develop a network which can score the "partial correctness" of each potential exon candidate. A simple

Exon Candidate
Parameters

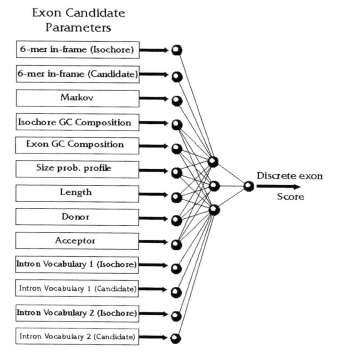

Fig. 3. A schematic of the neural network for evaluating internal protein-coding exons in GRAIL.

matching function $M(\cdot)$ is used to represent the correspondence of a given candidate with the actual exon(s):

$$M(\text{candidate}) = \frac{\sum_i m_i}{\text{length(candidate)}} \frac{\sum_i m_i}{\sum_j \text{length(exon}_j)}, \tag{7}$$

where $\sum_i m_i$ is the total number of bases of the candidate that overlap the actual exons (in the same translation frame), and $\sum_j$ length(exon$_j$) is the total length of all the exons that overlap the candidate. Using such a function helps "teach" the neural network to discriminate between candidates with different degrees of overlap with actual exon(s).

The network was trained on a training set containing about 2000 true, partially true and false exon candidates. Each training example contains a vector of feature values, defined above, along with its corresponding $M(\cdot)$ value for the exon candidate. The back-propagation learning algorithm determines all the weights W to optimally "fit" the training data, which is done in an iterative fashion.

Figure 4 shows a typical example of GRAIL neural network exon predictions. There may be more than one prediction for each actual exon. As can be seen, predictions for the same exon form a natural cluster, and in general the candidate that matches the actual exon exactly has the highest neural network score in the cluster.

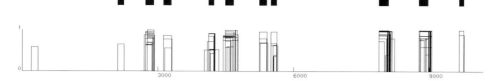

Fig. 4. Exon clusters. The X-axis is the sequence axis and the Y-axis is the neural network score axis. The solid bars on the top represent the positions of the actual exons, and the hollows rectangles represent the predicted exon candidates with different edge assumptions.

3. Predicting coding regions in erroneous DNA

Performance of the GRAIL exon recognition algorithm depends on the correctness of the input DNA sequence. Insertion or deletion of DNA bases (or simply *indels*) can change the performance significantly as indels may disrupt the translation frames, and GRAIL coding recognition methods are highly translation frame-dependent. To deal with sequences with high indel rates, we have developed an algorithm to detect and "correct" indels appearing in DNA coding regions [24], and implemented it as a front-end subsystem of GRAIL. While indel detection can be achieved using the information present in the DNA sequence, the basic goal in indel correction is to recover a consistent translation frame within a (presumed) coding region.

Figure 5 shows how indels affect the preferred translation frames. Note that in Fig. 5, the transition points of the preferred translation frames quite accurately correspond to the positions of the indels, where a *transition point* is a position in the sequence where the preferred translation frame with the highest coding score switches to a new frame. Our algorithm predicts indel positions by identifying the transition points in the presumed coding regions [24]. The algorithm first locates all transition points in a given DNA sequence, and then measures the coding potential around each located transition point to determine if the transition point is in a (presumed) coding region. The identified transition points in a (presumed) coding region are predicted as indels.

3.1. Localization of transition points

We formulate the transition point localization problem as follows. For a given DNA sequence, we want to find a set of locations at which a base can be inserted or deleted to obtain one consistent preferred translation frame for the whole sequence in such a way that the total coding potential is maximized along this frame. To prevent short-range fluctuations, we require that two such locations should be at least a certain number of bases apart.

More specifically, for a given DNA $S = a_1 \cdots a_n$, we want to find a set of locations on S, with the distance between two such points larger than $\mathcal{K}$ ($\mathcal{K} = 30$ in our implementation), so that the base at each such location can be deleted or a new base C (we always insert a letter C to prevent potentially forming stops) can be inserted right in front of it; this

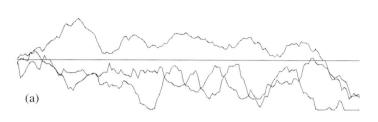

(a)

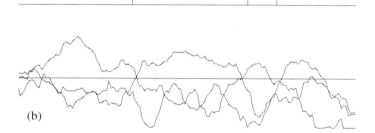

(b)

Fig. 5. Transitions of preferred translation frames. The X-axis is the sequence axis and the Y-axis is the coding score axis. The solid rectangles on the top in both (a) and (b) represent the positions of actual exons of HUMBAR. The three curves represent the 6-tuple preference model scores in three translation frames. The hashmarks represent the positions of indels.

produces a "corrected" sequence S^c with one consistent preferred translation frame, and the following objective function is maximized:

$$\max_{S^c \in \mathcal{F}(C),\ r \in \{0,1,2\}} pf_r(S^c). \tag{8}$$

where $\mathcal{F}(C)$ denotes the set of all possible "corrected" sequences by deleting bases from or inserting bases "C" to S under the constraint indels have to be at least $\mathcal{K}$ bases apart.

We now present a fast dynamic programming algorithm to solve this optimization problem. The core of the algorithm is a set of recurrence relations as follows. Based on these recurrences, the algorithm computes a maximum solution for each subsequence $a_1 \cdots a_i$, based on maximum solutions for its subsequences $a_1 \cdots a_j$, with $j < i$. The algorithm stops when i reaches n.

First we introduce some notation. Let S_i represent the subsequence $a_1 \cdots a_i$ of S. For $i \in [6, n]$ and $r \in \{0, 1, 2\}$, we define $C_0(i, r)$ to be the highest $pf_r(\cdot)$ score among all the "corrected" sequences S_i^c (of S_i), and $C_1(i, r)$ to be the highest $pf_r(\cdot)$ score among all those "corrected" sequences S_i^c, whose right-most transition positions are at least $\mathcal{K}$ bases apart from position i. Formally,

$$C_0(i, r) = \max_{S_i^c \in \mathcal{F}(S_i)} \{pf_r(S_i^c)\},$$

$$C_1(i, r) = \max_{S_i^c \in \mathcal{F}(S_i)} \left\{ pf_r(S_i^c) \,\middle|\, \begin{array}{l} \text{distance between the last transition} \\ \text{point and position } i \text{ is larger than } \mathcal{K} \end{array} \right\}. \tag{9}$$

We use $X_{i,r}$ to represent the right-most 6mer of the optimally "corrected" subsequence corresponding to $C_0(i, r)$ (note that $X_{i,r}$ is not necessarily the right-most 6mer of S_i due to "corrections"), and $X_{i,r}(a)$ to represent a 6mer formed by appending the letter a to the (right) end of $X_{i,r}$ after removing the first base of $X_{i,r}$.

Recurrences (10)–(13) give the relationship between $C_0(i, r)$ (and also $C_1(i, r)$) with $C_0(j, r)$ and $C_1(j, r)$, for $j < i$. They can be proved using an inductive argument on i, which we omit here. There are three possible cases in calculating $C_0(i, r)$: (1) i is not a transition point; (2) i is a transition point, and a base should be inserted in front of i; and (3) i is a transition point, and the base in front of it should be deleted. For $r \in \{0, 1, 2\}$ and $i \geqslant 7$, we have the following:

Case 1. When there is no change in the translation frame at position i, we have:

$$C_0(i, r) = C_0(i - 1, r) + pf_{(i-r) \bmod 3}(X_{i-1,r}(a_i)); \tag{10}$$

Case 2. When there is a translation frame change at position i, and a letter C is inserted in front a_i, we have:

$$C_0(i, r) = C_1(i - 1, (r + 1) \bmod 3) + pf_{(i-(r+1)) \bmod 3}(X_{i-1, r+1}(C))$$
$$+ pf_{(i-r) \bmod 3}(a_{i-4} \cdots a_{i-1} C a_i); \tag{11}$$

Case 3. When there is a translation frame change at position i, and the base a_{i-1} is deleted, we have:

$$C_0(i, r) = C_1(i - 2, (r + 2) \bmod 3) + pf_{(i-r) \bmod 3}(a_{i-6} \cdots a_{i-2} a_i); \tag{12}$$

In the general situation, $C_0(i, r)$ is equal to the highest value of the right hand sides of (10)–(12). And for $r \in \{0, 1, 2\}$ and $i > \mathcal{K}$, we have:

$$C_1(i, r) = C_0(i - \mathcal{K}, r) + \sum_{j=i-\mathcal{K}+1}^{i} pf_{(j-r) \bmod 3}(X_{j-1, r}(a_j)). \tag{13}$$

The initial values for $C_0(\cdot)$ and $C_1(\cdot)$ are defined as follows:

$$\begin{aligned} C_0(j, r) &= pf_{(6-r) \bmod 3}(a_1 \cdots a_6), \\ C_1(i, r) &= -\infty, \end{aligned} \qquad \text{for } 1 \leqslant j \leqslant 6 \text{ and } 1 \leqslant i \leqslant \mathcal{K}. \tag{14}$$

Note that by definition,

$$\max_{r \in \{0, 1, 2\}} C_0(n, r) \tag{15}$$

corresponds to the optimal solution to the optimization problem (8). Hence, to locate the transition points and to obtain the information on how the sequence S should be "corrected", we only need to calculate $C_0(n, r)$, for $r \in \{0, 1, 2\}$, using the recurrences (10)–(14) in the increasing order of i until i reaches n. Some bookkeeping is needed to record all the necessary information to recover the "corrected" sequence from the optimal solution, which can be done in linear time and space of the input sequence.

To consider the computational resources used by this algorithm, note that for each i, $C_0(\cdot)$ can be calculated in a constant time. To calculate $C_1(\cdot)$ fast, we can precalculate $\sum_{j=6}^{i} pf_{(j-r) \bmod 3}(a_{j-5} \cdots a_j)$ for each $i \geqslant 6$ and $r \in \{0, 1, 2\}$, which takes a total of $O(n)$ time and space. Hence it only takes a constant time to calculate each $C_1(\cdot)$ by doing a subtraction operation of two precalculated sums. Thus, the total time and space of the transition point localization algorithm is $O(n)$.

Determining the coding potential around a transition point can be done using two fixed size windows, one on each side of the transition point, to calculate the scores of the 6-tuple preference model. A more sophisticated method for this problem is given in [24].

The indel detection and correction algorithm has greatly improved the prediction results of the GRAIL gene recognition system in the presence of indel sequencing errors. On a test set containing 202 DNA sequences with 1% randomly implanted indels, this algorithm has helped improve GRAIL's coding recognition from a true positive rate of 60% to 81% with only 1% increase in false positive rate.

4. Reference-based gene structure prediction

The key issues in parsing the predicted exons into gene models lie in the determination of the boundaries of a gene, and the determination of how exons are spliced together within a gene. Because of the difficulty of both problems, most of the current gene parsing systems only deal with the single-gene modeling problem, and treat the issue of exon splicings in a simplistic manner, e.g. splicing exons together in such a way that the total scores of the predicted exons is maximized under the condition that adjacent exons are translation-frame consistent (which we will define in the following).

The availability of large numbers of EST sequences facilitates a practical solution to the multiple gene parsing problem. Exons can be parsed in such a way that is most consistent with the EST sequences that they "match", where a match between two sequences is determined by the BLAST search program (version 1.4.9) [25]. Our gene parsing system first constructs reference models using the matched EST sequences, and then parses the predicted exons into genes based on these reference models.

4.1. EST-based reference model construction

A public effort to sequence both 3' and 5' EST sequences from human cDNA clones started in 1990. These sequences have been placed into the dbEST database. Currently the dbEST database [26] (release January 1997) contains 798 908 entries, of which 435 351 (55%) entries are labeled as 5' ESTs, and the rest are labeled either as 3', both 3' and 5', or unlabeled. Among all the labeled EST sequences, 187 484 pairs of 3' and 5' ESTs share the same clone names, indicating that each of such pairs are sequenced from the same cDNA clone. Typically, an EST sequence has 300–500 bases. For each

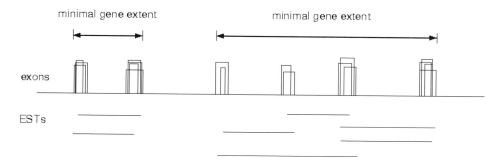

Fig. 6. Minimal gene extent. The X-axis is the sequence axis. The hollow rectangles represent GRAIL-predicted exons with their height representing the prediction confidence level. The starting and ending positions of each EST-representing line segment represent its leftmost and rightmost matched positions in the genomic sequence.

predicted exon, the gene-parsing system finds all the matched ESTs (with a match score above a specified threshold) using BLAST. The search results are organized as a list of alignments between exon candidates and EST segments. For each alignment, the name of the matched EST, the positions of the matched portions in both the genomic and EST sequences, the alignment score, and the $3'/5'$-end label of the EST are recorded. From this list, the following information can be extracted and used as reference models in the gene modeling algorithm.

GRAIL exons are rescored based on the matched EST information. We have applied three practical rules when rescoring GRAIL exons: (1) scores for exons with EST matches are increased; (2) exons that are inconsistent with the EST information are labeled as false exons; (3) scores for the rest of the exons are unchanged.

On average an exon is about 150 bases long. Hence, a typical EST sequence can match three consecutive (possibly partial) exons. Matching a number of consecutive exons by a single EST sequence provides the "adjacency" information, based on which falsely predicted/missed exons are identified and the minimal extent of a gene is determined. More generally, two EST sequences are said to be *overlapping* if their matched portions (with exons) in the genomic sequence overlap (in the same DNA strand). A set of all overlapping ESTs determines *the minimal extent of a gene*. Two minimal gene extents A and B are merged into one if A and B contain the $5'$ and $3'$ ESTs of the same cDNA clone, respectively, and A precedes B in the DNA sequence. This structural information is used as a reference in the gene parsing process given in section 4.2. Figure 6 gives a schematic of determining minimal extent of a gene.

4.2. Reference-based gene modeling

This section describes the second step of the gene modeling algorithm, reference-based gene structure prediction. The algorithm builds a (multiple) gene model that is most consistent with the given EST information and GRAIL predictions. More specifically, the predicted gene structure is consistent with the matched $3'$ ESTs, the identified gene segments (minimal gene extents possibly merged through the same cDNA clone names),

122

and the adjacency relationship of the matched EST segments of the adjacent exons. To maximally use the adjacency information and hence to facilitate the identification of falsely predicted exons and missed exons, we reward using long stretches of ESTs as reference models in the gene modeling process. To achieve all these, we have formulated the reference-based gene modeling problem as an optimization problem and solved the problem using a dynamic programming algorithm.

We first define some terminology. Exons are predicted by GRAIL with a fixed type $\in \{\text{initial, internal, terminal}\}$, and a fixed translation frame $\in \{0, 1, 2\}$. In a single-gene modeling process, an exon E is said to be *spliceable* to exon E' if
(1) E is a non-terminal exon, and E' is a non-initial exon,
(2) $l(E') - r(E) \geqslant \mathcal{I}$,
(3) $f(E') = (l(E') - r(E) - 1 + f(E)) \bmod 3$,
(4) no in-frame stop is formed at the joint point when appending E and E',
where $f(E)$ represents E's translation frame, and $\mathcal{I}$ represents the minimal intron size and $\mathcal{I} = 60$ in GRAIL. We can extend this spliceability condition to a multiple gene modeling process as follows. Exons are said to be *related* if they belong to the same minimal gene extent. Two exons are *spliceable* if they are spliceable in the single-gene modeling process, and they are either related, or if E belongs to a $3'$ minimal gene extent then E' belongs to the same minimal gene extent. A list of non-overlapping exons $\{E_1, \cdots, E_k\}$ form a *gene* model if (a) E_i is spliceable to E_{i+1} for all $i \in [1, k-1]$, and (b) E_1 is an initial exon, and E_k is a terminal exon. More generally, $\{E_1, \ldots, E_k\}$ form a *partial* gene model if condition (a) holds.

A *reference-based multiple gene modeling problem* is defined as follows. Given are a set of N predicted exons and a list of M EST reference models $\{R_1, \ldots, R_M\}$. Each exon E has a score score(E, R) with respect to each EST reference model R. For simplicity of discussion, we define score$(E, \emptyset) = \text{score}_G(E)$ and always use R_0 to represent $\emptyset$ as a special reference model. The goal is to select a list $\{E_1, \cdots, E_n\}$ of non-overlapping exon candidates from the given exon set, a mapping $\mathcal{M}$ from $\{E_1, \cdots, E_n\}$ to the (extended) EST reference model list $\{R_0, R_1, \ldots, R_M\}$, and a partition of $\{E_1, \ldots, E_n\}$ into D (not predetermined) sublists $\{E_1^1, \ldots, E_{n_1}^1\}, \ldots, \{E_1^D, \ldots, E_{n_D}^D\}$ (corresponding to D (partial) gene models) in such a way that the following function[1] is maximized,

$$\text{maximize} \sum_{g=1}^{D} \left(\sum_{i=1}^{n_g} \text{score}(E_i^g, \mathcal{M}(E_i^g)) + \sum_{i=2}^{n_g} \text{link}(\mathcal{M}(E_{i-1}^g), \mathcal{M}(E_i^g)) \right.$$
$$\left. + \mathcal{P}_f(E_1^g) + \mathcal{P}_t(E_{n_g}^g) \right) \tag{16}$$

subject to: (1) $l(E_1^{g+1}) - r(E_{n_g}^g) \geqslant \mathcal{L}$, for $g < D$,
(2) $E_{n_g}^g$ and E_1^{g+1} are not related, for $g < D$,
(3) E_i^g is spliceable to E_{i+1}^g, for all $i \in [1, n_g - 1]$ and $g \leqslant D$,

[1] One possible variation of this optimization problem is to relax the hard constraint that adjacent exons have to be spliceable; instead, one adds a penalty factor in the objective function for cases where the spliceability condition is violated. By doing so, the algorithm will not remove high-scoring exons from a gene model simply because they are not spliceable to other exons, probably caused by other missing exons.

where link(X, Y) is a reward factor when $X = Y$ and $X \neq R_0$, and is zero otherwise, $\mathcal{P}_f(\cdot)$ and $\mathcal{P}_t(\cdot)$ are two penalty factors for a gene missing the initial or terminal exon, respectively, and $\mathcal{L}$ is the minimum distance between two genes ($\mathcal{L} = 1000$ in our current implementation). We now give a dynamic programming algorithm for solving the reference-based gene modeling problem. Let $\{E_1, \ldots, E_N\}$ be the set of all exon candidates (within a pair of gene boundaries) sorted in the increasing order of their right boundaries. The core of the algorithm is a set of recurrences that relate the optimum (partial) solutions ending at an exon E_i to the optimum partial solutions ending at some exons to the left of E_i. We define model(E_i, R_j) to be the optimum value of the objective function (16) under the constraint that the right-most exon in the optimum gene models is E_i and E_i's reference model is R_j. By definition,

$$\max_{i \in [1, N], \, j \in [0, M]} \text{model}(E_i, R_j)$$

corresponds to the optimum solution to the optimization problem (16). For simplicity of discussion, we introduce an auxiliary notation model$_0(E_i, R_j)$, which is defined the same as model(E_i, R_j) except that the $\mathcal{P}_t(\cdot)$ term (in the objective function (16)) is removed for the right-most sublist $\{E_1^D, \ldots, E_{n_D}^D\}$.

Now we give the recursive equations of model(E_i, R_j) and model$_0(E_i, R_j)$. There are two cases we need to consider in calculating these two quantities. We define model$(E_0, R) = 0$ and model$_0(E_0, R) = 0$ for any $R \in \{R_0, \ldots, R_M\}$, for simplicity of discussion.

Case 1. E_i is the first exon of a gene:

$$\text{model}(E_i, R_j) = \max_{p \in [0, i-1], \, q \in [0, M]} \{\text{model}(E_p, R_q) + \text{score}(E_i, R_j) + \mathcal{P}_f(E_i) + \mathcal{P}_t(E_i)$$
$$\text{when } l(E_i) - r(E_p) \geqslant \mathcal{L}, \text{ and } E_i, E_p \text{ not related.}\}$$

$$\text{model}_0(E_i, R_j) = \max_{p \in [0, i-1], \, q \in [0, M]} \{\text{model}(E_p, R_q) + \text{score}(E_i, R_j) + \mathcal{P}_f(E_i)$$
$$\text{when } l(E_i) - r(E_p) \geqslant \mathcal{L}, \text{ and } E_i, E_p \text{ not related.}\}$$

Case 2. E_i is not the first exon of a gene:

$$\text{model}(E_i, R_j) = \max_{p \in [0, i-1], \, q \in [0, M]} \{\text{model}_0(E_p, R_q) + \text{score}(E_i, R_j) + \text{link}(R_q, R_j) + \mathcal{P}_t(E_i)$$
$$\text{when } E_p \text{ is spliceable to } E_i.\}$$

$$\text{model}_0(E_i, R_j) = \max_{p \in [0, i-1], \, q \in [0, M]} \{\text{model}_0(E_p, R_q) + \text{score}(E_i, R_j) + \text{link}(R_q, R_j)$$
$$\text{when } E_p \text{ is spliceable to } E_i.\}$$

These recurrences can be easily proved using a simple induction on i, which we omit. In the general case, model(E_i, R_j) equals the highest value of the two cases, similarly for model$_0(E_i, R_j)$. Using the initial conditions model$(E_0, R) = 0$ and model$_0(E_0, R) = 0$, any model(E_i, R_j) and model$_0(E_i, R_j)$ can be calculated in the increasing order of i using the above recurrences. In the following, we give an efficient implementation of these recurrences.

To implement the recurrences in Case 1 efficiently, we keep a table defined as follows. The table keeps the following information for the optimum (partial) gene model ending at each exon: the right boundary of the exon (*position*), the name of the exon, the score

Table 1
Optimum partial gene scores (case 1)

position	exon_name	gene_score	index_to_maximum
98	exon #1	85	1
105	exon #2	80	1
251	exon #3	173	3
256	exon #4	180	4
1001	exon #5	312	5
1099	exon #6	456	6
1562	exon #7	580	7
⋮	⋮	⋮	⋮

of the (partial) gene model, and the index to the entry that has the maximum *gene_score* among all entries from the top to the current one. The table is listed in the non-decreasing order of *positions*.

It can be seen that to calculate model(E_i, R_j) (similarly model$_0(E_i, R_j)$), we only need to find the entry in this table that is the closest to the bottom under the condition that its distance to E_i is at least $\mathcal{L}$ and its corresponding exon is not related to E_i. To do this, we first get the left boundary L of E's gene segment (it is defined to be $l(E)$ if E does not belong to any gene segment), and search the table for the entry that is closet to the bottom and its *position* is $\leqslant \min\{L, l(E_i) - \mathcal{L}\}$. This can be done in O(log(table_size)), i.e. O(log N) time. After model(E_i, R_j) is calculated, we need to update E_i's entry for each R_j. Each update takes only O(1) time. So the total time used on Case 1 throughout the algorithm is O($N \log(N) + NM$).

To implement the recurrences in Case 2 efficiently using a similar technique is a little more involved due to the requirement of spliceability. Recall that two exons are spliceable if they are at least $\mathcal{I}$ bases apart, their translation frames are consistent, and no in-frame stop can be formed at the joint point when they are appended (the extra conditions for the multiple-gene case can be easily checked and are omitted in our discussion). All these three conditions have to be checked and satisfied when calculating the recurrences in Case 2.

Note that the translation frame consistency condition between two exons E and E',

$$f(E') = (l(E') - r(E) - 1 + f(E)) \bmod 3, \tag{17}$$

can be rewritten as

$$f(E') = (l(E') - 1 - (r(E) - f(E)) \bmod 3) \bmod 3. \tag{18}$$

Hence we can classify all exons into three classes based on the value of

$$\mathcal{F}(E) = (r(E) - f(E)) \bmod 3. \tag{19}$$

If forming in-frame stops was not a concern, we could have three tables for the three classes of exons like the one for Case 1, and calculate model(E_i, R_j) and model$_0(E_i, R_j)$ through searching the table that satisfies the adjacent-exon frame consistency.

Table 2
Optimum partial gene scores (case 2)

position	gene_score	index_to_maximum
10 051	110	1
10 057	100	1
10 302	72	1
10 652	412	4
⋮	⋮	⋮

To deal with the in-frame stop problem, we need to further divide these three classes. Note that when $\mathcal{F}(E) = 0$, E ends with the first base of a codon, and an in-frame stop can be possibly formed when appending E to some exon to its right only if E ends with the letter T (recall the three stops TAA, TAG, TGA); similarly when $\mathcal{F}(E) = 1$, stops can be possibly formed when E ends with either TA or TG; and when $\mathcal{F}(E) = 2$, in-frame stops can be formed only if E ends with a stop codon (E is a terminal exon). Knowing this, we classify all exons E into 7 classes: two classes (T and non-T) for $\mathcal{F}(E) = 0$; three classes (TA, TG and others) for $\mathcal{F}(E) = 1$; and two classes (stop codon and non-stop) for $\mathcal{F}(E) = 2$.

We maintain a table similar to the one in Case 1 for each possible triple $(R, \mathcal{F}, \text{tail})$, where $R \in \{R_0, \ldots, R_M\}$, $\mathcal{F} \in \{0, 1, 2\}$, and tail is one of the 7 cases addressed above. The table keeps an entry for each exon E of this class, and each entry contains three values: E's right boundary (position), the score of optimum partial gene ending at E (i.e. $\text{model}_0(E, R)$), and the index to the entry that has the maximum gene_score among all entries in the table from the top to the current one. The table is listed in the increasing order of its exon's right boundaries (for exons having the same right boundary, we only keep the one with the highest score). We define that each entry of the table for $(R_0, \mathcal{F}, \text{tail})$ keeps the maximum score of the corresponding entries of tables for $(R_i, \mathcal{F}, \text{tail})$ for all $i \in [0, M]$ (recall $R_0 = \emptyset$).

It can be seen that to calculate $\text{model}(E_i, R_j)$ (similarly $\text{model}_0(E_i, R_j)$), we only need to look at R_j's tables and R_0's tables that satisfy the frame consistency condition and the condition that no in-frame stop is formed. For each such table, we find the right entry in the same way as in Case 1 except that this time a search can be done in $O(\mathcal{I})$ time, which is a small constant, i.e. $O(1)$. Note that for each $\text{model}(E_i, R_j)$, there are at most 2×3 tables to look up. After $\text{model}(E_i, R_j)$ is calculated, we need to update the corresponding entries in both R_j and R_0's tables, and each of these updates takes $O(1)$ time. So the total time used on Case 2 is $O(NM)$.

To recover the gene models that achieve

$$\max_{i \in [1, N], \, j \in [0, M]} \text{model}(E_i, R_j),$$

some simple bookkeeping needs to be done, which can be accomplished within the same asymptotic time bound of calculating $\text{model}(E_i, R_j)$. Hence we conclude that

126

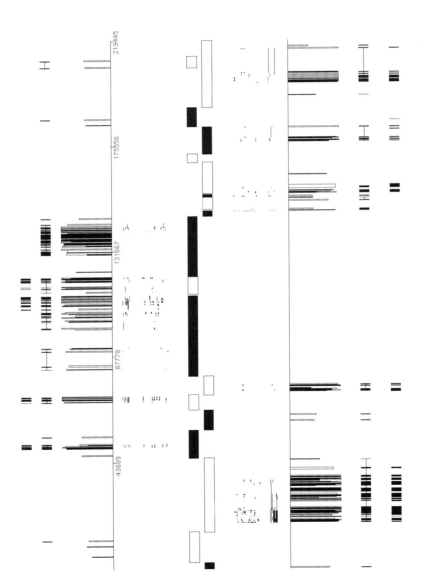

Fig. 7. The solid bars in both the top and the bottom represent the positions of the annotated GenBank exons in the forward and reverse strand, respectively. The solid bars in the next-to-top and next-to-bottom rows represent the exons in the predicted gene models. Each set of bars connected through a line represent one gene. The hollow rectangles represent the predicted GRAIL exons. The short lines (or dots) represent the matched ESTs. The rectangles (alternately hollow and solid) in the middle represent boundaries of one or a number of genes.

the optimization problem (16) can be solved in O($NM + N\log(N)$) time. Figure 7 shows an example of the predicted multiple gene model on the genomic sequence HUMFLNG6PD.

5. *Conclusion*

The performance of multi-agent systems such as GRAIL depends critically on how the information from different agents (algorithms) is combined. Over a dozen exon indicators and correction factors are used in the GRAIL gene recognition process. The relationship between these quantities and the presence of exons is complicated, incomplete and clearly non-linear. To develop an effective mechanism to map these quantities, some of which may not be independent, to exon and non-exonic regions is one of the key goals of our research. By training neural networks with hidden layers on empirical data, GRAIL seems to have captured some of the most essential part of this relationship based on its successful applications to gene recognition by molecular biologists worldwide over the past five years.

By combining the complementary nature of pattern recognition-based exon prediction and the EST gene structural information, we have developed a computer system for automatic identification of gene structures in anonymous DNA sequences. Minimal "inconsistency" between the predicted gene structures and ESTs is the basic rule used in this gene modeling framework. Tests have shown that the predicted gene models are very consistent with the available EST information. With its reliable gene structure prediction supported by the EST information, this computer system should provide molecular biologists with a powerful and convenient tool in analyzing complex genomic sequences.

References

[1] Fickett, J.W. (1982) Recognition of protein coding regions in DNA sequences. Nucleic Acids Res. 10, 5303–5318.
[2] Borodovsky, M., Sprizhitskii, Yu., Golovanov, E. and Aleksandov, A. (1986) Statistical patterns in the primary structures of functional regions in *E. coli*. Mol. Biol. 20, 1390–1398.
[3] Gelfand, M.S. (1990), Computer prediction of exon–intron structure of mammalian pre-mRNAs. Nucleic Acids Res. 18, 5865–5869.
[4] Uberbacher, E.C. and Mural, R.J. (1991) Locating protein-coding regions in human DNA sequences by a multiple sensors-neural network approach. Proc. Natl. Acad. Sci. USA 88, 11261–11265.
[5] Guigo, R., Knudsen, S., Drake, N. and Smith, T. (1992) Prediction of gene structure. J. Mol. Biol. 226, 141–157.
[6] Hutchinson, G.B. and Hayden, M.R. (1992), The prediction of exons through an analysis of spliceable open reading frames. Nucleic Acids Res. 20, 3453–3462.
[7] Snyder, E.E. and Stormo, G.D. (1993) Identification of coding regions in genomic DNA sequences: An application of dynamic programming and neural networks. Nucleic Acids Res. 21, 607–613.
[8] Xu, Y., Mural, R.J. and Uberbacher, E.C. (1994) Constructing gene models from a set of accurately-predicted exons: An application of dynamic programming. Comput. Appl. Biosci. 10, 613–623.
[9] Dong, S. and Searls, D.B. (1994) Gene structure prediction by linguistic methods. Genomics 23, 540–551.

[10] Solovyev, V.V., Salamov, A.A. and Lawrence, C.B. (1994) Predicting internal exons by oligonucleotide composition and discriminant analysis of spliceable open reading frames. Nucleic Acids Res. 22, 5156–5163.

[11] Kulp, D., Haussler, D., Reese, M.G. and Eckerman, F.H. (1996) A generalized hidden Markov model for the representation of human genes in DNA. In: D.J. States, P. Agarwal, T. Gaasterland, L. Hunter and R. Smith (Eds.), Proc. 4th Int. Conf. on Intelligent Systems for Molecular Biology, AAAI Press, Menlo Park, CA, pp. 134–142.

[12] Gelfand, M., Mironov, A. and Pevzner, P. (1996) Gene recognition via spliced alignment. Proc. Natl. Acad. Sci. USA 93, 9061–9066.

[13] Salzberg, S., Chen, X., Henderson, J. and Fasman, K. (1996) Finding genes in DNA using decision trees and dynamic programming. In: Proc. 4th Int. Conf. on Intelligent Systems for Molecular Biology, AAAI Press, Menlo Park, CA, pp. 201–210.

[14] Uberbacher, E.C., Xu, Y. and Mural, R.J. (1996) Discovering and understanding genes in human DNA sequence using GRAIL. Methods Enzymol. 266, 259–281.

[15] Claverie, J.M., Sauvaget, I. and Bougueleret, L. (1990) k-tuple frequency analysis: from intron/exon discrimination to T-cell epitope mapping. Methods Enzymol. 183, 237–252.

[16] Xu, Y. and Uberbacher, E.C. (1997) Automated gene structure identification in large-scale genomic sequences. J. Comp. Biol. 4(3) 325–338.

[17] Costanzo, F., Castagnoli, L., Dente, L., Arcari, P., Smith, M., Costanzo, P., Raugel, G., Izzo, P., Pietronaolo, T.C., Bougueleret, L., Cimino, F., Salvatore, F. and Cortese, R. (1983) Cloning of several cDNA segments coding for human liver proteins. EMBO J. 2, 57–61.

[18] Adams, M.D., Kelley, J.M., Gocayne, J.D., Dubnick, M., Polymeropolous, M.H., Xiao, H., Merril, C.R., Wu, A., Olde, B., Moreno, R.F., Kerlavage, A.R., McCombie, W.R. and Venter, J.C. (1991) Complementary DNA sequencing: expressed sequence tags and human genome project. Science 252, 1651–1656.

[19] Wilcox, A.S., Khan, A.S., Hopkins, J.A. and Sikela, J.M. (1991) Use of 3′ untranslated sequences of human cDNAs for rapid chromosome assignment and conversion to STSs: implications for an expression map of the genome. Nucleic Acids Res. 19, 1837–1843.

[20] Houlgatte, R., Mairage-Samson, R., Duprat, S., Tessier, A., Bentolila, S., Lamy, B. and Auffray, C. (1995) The gene-express index: A resource for gene discovery and the genic map of the human genome. Genome Res. 5, 272–304.

[21] Adams, M.D., et al. (1995) Initial assessment of human gene diversity and expression patterns based upon 83 million nucleotides of cDNA sequence. Nature 377, 3–174.

[22] Aaronson, J.S., Eckman, B., Blevins, R.A., Borkowski, J.A., Myerson, J., Imran, S. and Elliston, K.O. (1996) Towards the development of a gene index to the human genome: an assessment of the nature of high-throughput EST sequence data. Genome Res. 6, 829–845.

[23] Hertz, J., Krogh, A. and Palmers, A.R.G. (1991) Introduction to The Theory of Neural Computation. Addison-Wesley, Reading, MA.

[24] Xu, Y., Mural, R.J. and Uberbacher, E.C. (1996) An iterative algorithm for correcting sequencing errors in DNA coding regions. J. Comp. Biol. 3, 333–344.

[25] Altschul, S.F., Gish, W., Miller, W., Myers, E.W. and Lipman, D.J. (1990) Basic local alignment search tools. J. Mol. Biol. 215, 403–410.

[26] Boguski, M.S., Lowe, T.M. and Tolstoshev, C.M. (1993) dbEST – database for expressed sequence tags. Nat. Genet. 4, 332–333.

S.L. Salzberg, D.B. Searls, S. Kasif (Eds.), *Computational Methods in Molecular Biology*

CHAPTER 8

Modeling dependencies in pre-mRNA splicing signals

Christopher B. Burge

Center for Cancer Research, Massachusetts Institute of Technology,
40 Ames Street, Room E17-526a, Cambridge, MA 02139-4307, USA

List of abbreviations

A	Adenosine	R	Either purine nucleotide (A or G)
B	not-A (C, G or T/U)	snRNA	small nuclear RNA
C	Cytosine	T	Thymidine
CIV	Consensus indicator variable	U	Uracil
D	not-C (A, G or T/U)	V	not-T/U (A, C or G)
G	Guanosine	WMM	Weight matrix model
H	not-G (A, C or T/U)	WAM	Weight array model
HMM	Hidden Markov model	WWAM	Windowed weight array model
IMM	Inhomogeneous Markov model	Y	Either pyrimidine nucleotide (C or T/U)
MDD	Maximal dependence decomposition		

1. Introduction

The discovery in 1977 of split genes in eukaryotic genomes must rank as one of the most surprising scientific findings of this century. Some twenty years later, a number of important questions raised by this discovery still remain, including why introns exist, how they are processed, and how the precise specificity of splicing is achieved. In particular, although a number of theories have been advanced, it is not yet firmly established what purpose(s) introns serve or why they are so widespread in eukaryotic genes. And, although a great deal of progress has been made towards understanding the functioning of the spliceosome, the precise catalytic details of intron removal/exon ligation have not yet been worked out. Finally, it is still not fully understood how the splicing machinery achieves the precise specificity required to distinguish correct splice sites from similar sequences nearby or how the appropriate donor/acceptor pairs are brought together. This chapter addresses the issue of the specificity of pre-mRNA splicing using a computational biology approach. From this perspective, we seek to understand the specificity of splicing by studying the statistical/compositional properties of known splice signal (and non-signal) sequences. As a practical application, the significant compositional features and dependencies observed in such signals are incorporated into probabilistic models which can be used to detect splice signals in genomic DNA, an issue of obvious importance for exon/gene identification efforts.

   This chapter is intended both as a summary of some of the known compositional properties of pre-mRNA splicing signals, as well as a kind of tutorial on the construction and application of simple probabilistic models of biological signal sequences. Although all of the examples given here relate to splicing signals, many of the techniques described could equally well be applied to other types of nucleic acid signals such as those involved in transcription, translation, or other biochemical processes. Particular emphasis is placed on the use of simple statistical tests of dependence between sequence positions. In some cases, observed dependencies can give clues to important functional constraints on a signal, and it is demonstrated that incorporation of such dependencies into probabilistic models of sequence signals can lead to significant improvements in the accuracy of signal prediction/classification. Some caveats are also given (with examples) concerning the dangers of: (1) constructing models with too many parameters; and (2) overinterpreting observed correlations in sequence data.
   The next three sections briefly review pre-mRNA splicing, summarize the probabilistic approach to signal classification, and describe several standard types of discrete probabilistic models. Subsequent sections describe datasets of available gene sequences and illustrate particular types of models as applied to human acceptor and donor splice signals. The discussion progresses from relatively simple weight matrix models (WMMs) to certain natural generalizations of WMMs, and concludes with the fairly complex maximal dependence decomposition (MDD) method. Along the way, a number of statistical properties of splice signal sequences are described, and possible explanations for the observed compositional features are discussed. In some cases, references to related experimental results are given, but the treatment of the extensive experimental literature on splicing is by no means comprehensive. This chapter is loosely based on the fourth chapter of my doctoral dissertation [1] and also overlaps with parts of the paper (with Karlin) which introduced the GENSCAN gene identification program [2]. Although many of the tables and figures used are borrowed or adapted from refs. [1,2], for completeness a few new tables and figures have also been constructed which serve to illustrate data which was referred to but not explicitly shown in these sources.

2. Review of pre-mRNA splicing

Introns are removed from eukaryotic pre-mRNAs in the nucleus by a complex process catalyzed by a 60 S particle known as the spliceosome (e.g. ref. [3]). The spliceosome is composed of five small nuclear RNAs (snRNAs) called U1, U2, U4, U5 and U6, and numerous protein factors. Splice site recognition and spliceosomal assembly occur simultaneously according to a complex sequence of steps outlined below (see Fig. 1). Two of the most comprehensive reviews of pre-mRNA splicing are those by Staley and Guthrie [3] and Moore et al. [4]; see also ref. [5] and, for a review of alternative splicing, ref. [6]. The first step appears to be recognition of the donor (5′) splice site at the exon/intron junction: a substantial amount of genetic (e.g. ref. [7,8]) and biochemical evidence [9] has established that this occurs primarily through base pairing with U1 snRNA over a stretch of approximately nine nucleotides encompassing the last two or three exonic nucleotides and the first five or six nucleotides of the intron. The

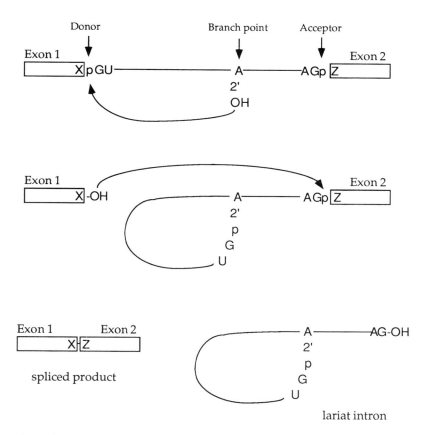

Fig. 1. The essential steps in the splicing of a pre-mRNA containing two exons and one intron are illustrated. The letter "p" is used to represent the two phosphodiester linkages which are broken in the course of splicing; the letters "OH" represent hydroxyl groups, e.g., the 2' hydroxyl of the branch point adenosine. The conserved donor (GU), acceptor (AG) and branch point (A) nucleotides are indicated by the corresponding letters; X and Z represent the last nucleotide of exon 1 and the first nucleotide of exon 2, respectively.

second step in spliceosomal assembly involves recognition of the branch point/acceptor site. This process is more complex, involving binding of U2 auxiliary factor (U2AF) and possibly other proteins to the pyrimidine-rich region immediately upstream of the acceptor site, which directs U2 snRNA binding to the branch point sequence approximately 20 to 40 bp upstream of the intron/exon junction [3]. The U2 snRNA sequence 3' GGUG 5' has been shown to base pair with the branch point signal, consensus 5' YYRAY 3', with the unpaired branch point adenosine (A) bulged out of the RNA duplex [10]. Subsequently, a particle containing U4, U5 and U6 snRNAs is added, U5 snRNA possibly interacting with the terminal exonic nucleotides, leading eventually to formation of the mature spliceosome [11].

Splicing itself occurs by two sequential transesterification reactions. First, an unusual 2'–5' phosphodiester bond (RNA branch) is formed between the 2' hydroxyl of an adenosine (A) near the 3' end of the intron (the branch point) and the guanosine (G) at the 5' end of the intron, resulting in cleavage at the 5' or donor splice site (exon/intron junction). In the second step, the 3' or acceptor splice site (intron/exon junction) is cleaved and the two exons are ligated together, causing the intron to be released as an "RNA lariat" which is rapidly degraded *in vivo*. After cap formation, polyadenylation, and excision of all introns, the resulting processed mRNA is exported to the cytoplasm for translation.

Despite fairly extensive knowledge of the factors involved in splice site recognition (e.g. ref. [12]), the precise mechanisms by which the proper splice sites are distinguished from similar "pseudo-sites" nearby is not well understood, although much of the specificity for constitutive splicing appears to be determined by signals at or near the splice junctions. However, it should be emphasized that a number of signals involved in both constitutive and regulated (alternative) splicing have recently been discovered which are located at some distance from the splice junctions, in both exons and introns. Examples include: the class of purine-rich exonic splicing enhancer signals (e.g. ref. [13]); intronic signals involved in developmentally regulated splicing of fibronectin [14]; and intronic signals involved in the constitutive splicing of vertebrate "micro-exons" [15]. In this chapter, we consider only the signals present at and near the splice junctions themselves, i.e. the donor splice signal, poly-pyrimidine/acceptor signal, and (to a limited extent) the branch point.

3. Signal prediction

The general problem of identifying occurrences of a particular biological signal of fixed length λ in the genomic DNA of a particular organism is now considered. For concreteness, the example of finding donor splice signals of length $\lambda = 9$ bp in human genomic DNA will often be used in this section, but most of the discussion applies to any well localized nucleic acid signal. Given a long genomic sequence, one can always break it into (overlapping) λ bp segments, essentially reducing the problem to that of classifying each λ-length DNA fragment. What is needed, then, is simply a decision rule which says "yes" (donor) or "no" (not donor) for each of the 4^9 possible DNA sequences of length 9. Now suppose that we had perfect knowledge of the statistical composition of actual donor sites (+) and non-donor sites (−) in the human genome, i.e. that we knew the frequency $p^+(X)$ with which each ninemer X is used as a donor splice site, and also the corresponding ninemer frequencies $p^-(X)$ in general (non-donor) locations in the genome. Obviously, if $p^+(X) > 0$ and $p^-(X) = 0$ for a particular ninemer X, then we can say with confidence that X is a donor site, and we can be equally certain that sites for which $p^-(X) > 0$ and $p^+(X) = 0$ are not donor sites. (The case $p^+(X) = 0$ and $p^-(X) = 0$ corresponds to a pattern which never occurs in the genome, so no decision rule is needed.) The only interesting question is therefore what to do when both $p^+(X)$ and $p^-(X)$ are positive, i.e. in the case of a ninemer which might or might not be a donor site.

Intuitively, one would tend to guess "donor" when $p^+(X)$ is much greater than $p^-(X)$ and the opposite when $p^-(X) > p^+(X)$, but the precise rule to use depends in general on the relative importance one places on "false positives" (predicted donor sites which are not) versus "false negatives" (missed donor sites). A classical result known as the

Neyman–Pearson lemma (ref. [16], but see also ref. [17]) says essentially that optimal decision rules depend only on the ratio $r(X) = p^+(X)/p^-(X)$ and are monotone in the value of this ratio. In other words, optimal decision rules predict "donor" when $r > C$ and "not donor" when $r < C$ for a particular constant C whose choice depends on the relative importance (weight) one places on false positives versus false negatives. (The case $r = C$, if it occurs, is handled as described in ref. [17].) From this point of view, the essential idea in probabilistic modeling of biological signals is to construct approximations $\hat{p}^+()$ and $\hat{p}^-()$ of the distributions $p^+()$ and $p^-()$ described above and then to use the ratio $\hat{r}(X) = \hat{p}^+(X)/\hat{p}^-(X)$ to discriminate true from false signals. In practice, one often works with the signal "score" $\hat{s}(X) = \log \hat{r}(X)$ rather than the signal ratio itself. A decision rule based on comparing the ratio $\hat{r}(X)$ to the constant C is equivalent to a rule which compares $\hat{s}(X)$ to $\log C$, since any sequence with signal ratio $\hat{r}(X) > C$ will have score $\hat{s}(X) > \log C$ and conversely (because log is a monotone increasing function). Some authors have used the quantity $\hat{s}'(X) = \log \hat{p}^+(X)$ instead of the score $\hat{s}(X)$ defined above. This is equivalent to use of the score $\hat{s}(X)$ in the special case when the non-signal model $\hat{p}^-$ assigns the same probability to all sequences X, since in this case the scores $\hat{s}(X)$ and $\hat{s}'(X)$ differ only by a constant. In some situations this may be a reasonable assumption (e.g. for an organism with nearly "random" base composition), but in general the use of the score $\hat{s}(X)$ is likely to be somewhat more accurate.

The simplest way of estimating the distribution $p^+()$ described above would be simply to set the probability $\hat{p}^+(X)$ equal to the observed frequency of the pattern X in an appropriate set of known signal sequences. (From this point on, we will write $p^+()$ and $p^-()$ instead of $\hat{p}^+()$, $\hat{p}^-()$ to simplify the notation.) This approach is almost never used in practice, however, since it requires estimation of a huge number of parameters (essentially equal to the number of possible DNA sequences of length λ) from what is generally a relatively small number of known signal sequences. For example, the number of available non-redundant human donor splice sites is on the order of 10^3 (see below), far less than the number of possible ninemers ($4^9 \approx 2.6 \times 10^5$), making such an approach impractical. The remainder of this chapter is devoted primarily to a discussion of several alternative types of signal/non-signal models which are simple enough so that all necessary parameters can be reliably estimated from a reasonable number of sequences.

4. Probabilistic models of signal sequences

Several simple types of models which assign a probability to each possible DNA sequence of some fixed length λ will now be introduced. For signals of variable length (e.g. for signals which might have insertions/deletions), more complex hidden Markov models (HMMs) may be appropriate – see chapter 4 by Krogh, elsewhere in this volume. Further discussion of biological signal models can also be found in the review by Gelfand [18]. The simplest model one can imagine for generating a signal sequence of length λ would be to label a tetrahedral die (singular of dice) with the letters A, C, G, T and then to simply roll the die λ times to generate the sequence. By appropriately weighting (biasing) the die, one could capture the overall base composition of the signal, but not the position-specific biases or patterns which are commonly observed in real biological signals. For this reason, such models are rarely used in practice. Instead, a more general independence model corresponds to using a particular set of λ biased or weighted dice rather than a single die, so that the probability of generating the particular sequence $X = x_1, x_2, \ldots, x_\lambda$ is given by

$$p(X) = p^{(1)}(x_1)p^{(2)}(x_2)\cdots p^{(\lambda)}(x_\lambda) = \prod_{i=1}^{\lambda} p^{(i)}(x_i),$$

where $p^{(i)}(x)$ is the probability of generating nucleotide x at position i of the signal (the frequency with which the ith die comes up x). With such a model one can account for features such as the requirement for particular nucleotide(s) at particular position(s) (e.g. the invariant AG in acceptor sites), or in general for any particular pattern of position-specific base preferences. Typically, the position-specific probabilities $p^{(i)}(x)$ are estimated from observed position-specific frequencies $f_x^{(i)}$ from a set of known signal sequences (see below). Biological signal models of this type, often referred to as "weight matrix models" (WMMs), were first introduced in the early 1980s [19,20]) and have been widely used ever since.

Weight matrix models have the important advantages that they are simple, easy to understand, and easy to use. In addition, the WMM is probably the best type of model to use when relatively few (say, dozens up to a few hundred) signal sequences are available. An important limitation of WMMs, however, is the assumption of independence between positions. Examples are given below in which the independence assumption is clearly violated, and more complicated models which incorporate observed dependencies often outperform WMMs in terms of signal prediction/classification. A natural generalization of weight matrices which allows for dependencies between adjacent positions is a (first-order) inhomogeneous Markov model (IMM or WAM – see below), in which the first nucleotide is generated at random (e.g. by dice roll), but each subsequent nucleotide is generated according to a probability distribution which depends on the previous nucleotide in the signal. Under a first-order Markov model, the probability of generating the sequence $X = x_1, x_2, \ldots, x_\lambda$ is given by

$$p_{\text{WAM}}(X) = p^{(1)}(x_1)p^{(2)}(x_2|x_1)p^{(3)}(x_3|x_2)\cdots p^{(\lambda)}(x_\lambda|x_{\lambda-1}) = p^{(1)}(x_1)\prod_{i=2}^{\lambda} p^{(i)}(x_i|x_{i-1}),$$

where $p^{(i)}(z|y)$ is the conditional probability of generating nucleotide z at position i given nucleotide y at position $i-1$, which is typically estimated from the corresponding conditional frequency (see below). This type of model, referred to by Zhang and Marr [21] as a "weight array model" (WAM), was used by these authors to model the donor splice signal in the yeast *Schizosaccharomyces pombe* – WAM models of the human donor and acceptor splice sites will be described in the next sections. The splice signal models used by the gene identification program VEIL [22], though embedded in a more complex HMM framework, are also quite similar to the WAM model described above. Here, the terms "weight array model" (WAM) introduced by Zhang and Marr and the more precise "inhomogeneous Markov model" (IMM) will be used interchangeably.

4.1. The directionality of Markov models

This section is intended partly to familiarize the reader with typical Markov model calculations and also to examine a common perception about Markov models. Specifically, Markov models are often seen as being "asymmetric" or "directional" since the first ("leftmost" or 5′) nucleotide is distinguished from the others and because each nucleotide depends on the one to its "left" (5′) but not on the one to its "right" (3′). It is argued below that this assessment is somewhat misleading since "left-to-right" and "right-to-left" inhomogeneous Markov models derived from the same data give identical results. Specifically, consider a "left-to-right" Markov model L with transition probabilities estimated in the standard way, i.e. setting $p^{(i)}(z|y) = f_{y,z}^{(i)}/f_y^{(i-1)}$, where $f_{y,z}^{(i)}$ is the observed frequency of the dinucleotide y, z at position $i-1, i$ in the dataset used. For such a model, the probability of generating the sequence X is given by

$$P_L(X) = p^{(1)}(x_1) \prod_{i=2}^{\lambda} p^{(i)}(x_i|x_{i-1}) = f_{x_1}^{(1)} \prod_{i=2}^{\lambda} \frac{f_{x_{i-1},x_i}^{(i)}}{f_{x_{i-1}}^{(i-1)}} = \frac{\prod_{i=2}^{\lambda} f_{x_{i-1},x_i}^{(i)}}{\prod_{i=2}^{\lambda-1} f_{x_i}^{(i)}}.$$

Now consider deriving (from the same data) a "right-to-left" Markov model R in which the last (rightmost) nucleotide is generated first (with probability $q^{(\lambda)}(x_\lambda) = f_{x_\lambda}^{(\lambda)}$) and then each nucleotide x_{i-1} is generated conditionally on the nucleotide to the right, x_i, with probability $q^{(i-1)}(x_{i-1}|x_i)$. If the transition probabilities are again estimated from the corresponding conditional frequencies, i.e. setting $q^{(i-1)}(z|y) = f_{z,y}^{(i)}/f_y^{(i)}$, then the probability of generating the sequence X under the model R is given by:

$$P_R(X) = q_{x_\lambda}^{(\lambda)} \prod_{i=\lambda}^{2} q^{(i-1)}(x_{i-1}|x_i) = f_{x_\lambda}^{(\lambda)} \prod_{i=\lambda}^{2} \frac{f_{x_{i-1},x_i}^{(i)}}{f_{x_i}^{(i)}} = \frac{\prod_{i=2}^{\lambda} f_{x_{i-1},x_i}^{(i)}}{\prod_{i=2}^{\lambda-1} f_{x_i}^{(i)}},$$

which is identical to the probability previously calculated under the L model. If two models give the same probability for all sequences, then for all practical purposes they are equivalent, so the apparent directionality of Markov signal models is, in a sense, illusory. Therefore, it is probably better to think of a Markov model as a model which "incorporates dependencies between adjacent positions" rather than as a model in which the "next" symbol is generated from the "previous" symbol.

A general k-order Markov model ($k = 2, 3, \ldots$) is similar to the first-order Markov model described above except that the probability of generating the nucleotide x_i depends not just on the previous nucleotide, x_{i-1}, but on the previous k nucleotides, $x_{i-k}, \ldots, x_{i-1}$. In principle, a k-order Markov model will more accurately reflect the higher-order dependencies present in a signal than simple weight matrix or first-order Markov models, but sufficient data must be available to reliably estimate the larger number of parameters of such models. (This issue is dealt with in more detail below.) It is not hard to show that general k-order Markov models are "the same in both directions" in the sense described above for first-order models. Specific examples of these types of models (and certain extensions and generalizations) are given in later sections of this chapter, as applied to human donor and acceptor splice signals. A general introduction to Markov models (homogeneous and inhomogeneous) can be found in chapter 3 of ref. [23].

5. Datasets

All of the examples given in this chapter are taken from two sets of human gene sequences which were constructed for the training and testing of the splice signal models used by GENSCAN [2]. The learning set $\mathcal{L}$ is a set of 380 non-redundant genes which was derived from a previous set of GenBank (Release 89, 1995) sequences constructed by David Kulp (University of California, Santa Cruz) and Martin Reese (Lawrence Berkeley National Laboratories) for use as a common dataset for the training and testing of gene prediction methods (see [http://www-hgc.lbl.gov/inf/genesets.html]). Specific details of dataset construction and a list of GenBank accession numbers are given in appendix A of ref. [1]. Each sequence in this set contains a single human gene sequenced at the genomic (as opposed to cDNA) level and containing at least the initial ATG through the stop codon. Certain additional consistency constraints were also enforced, e.g. that there should be no in-frame stop codons in the annotated coding region and that splice signals should match the minimal consensus (GT for donor sites, AG for acceptor sites). The rationale for the latter condition is that, although a handful of splice sites which do not obey the "universal" GT–AG rule are known, sequences whose annotated splice junctions lack these consensus signals probably represent annotation errors at least as often as bona fide non-canonical splice signals. This restriction excludes introns with AT–AC terminal dinucleotides, many of which are spliced by the "second spliceosome" [24].

This set was culled of redundant or homologous entries by comparison at the protein level with the program BLASTP [25] by Kulp and Reese. The final set comprises a total of 238 multi-exon genes, containing a total of 1254 introns, hence 1254 donor and acceptor splice sites. It should be emphasized that use of a dataset which has been culled of redundant and homologous genes is absolutely essential for many of the statistical tests described below, which are based on an assumption of independence between positions which would clearly be violated by the presence of two or more identical or near-identical sequences. The test set $\mathcal{T}$ is a set of 65 non-redundant human multi-exon gene sequences (described in appendix B of ref. [1]) which was based on a later version of the Kulp/Reese dataset derived from GenBank Release 95 (1996). This set, totaling slightly more than 600 kilobases of genomic DNA, contains a total of 338 introns, hence 338 donor/acceptor splice signals. Importantly, this set is completely independent of the learning set in that it contains no proteins which are more than 25% similar to any protein of the learning set.

6. Weight matrix models and generalizations

This section describes specific models of the human acceptor/branch point signal which were constructed from the dataset $\mathcal{L}$ described above. Table 1 displays the base composition at specific positions relative to the intron/exon junction for the 1254 acceptor splice sites of the learning set. Intron positions are labeled $-40, -39, \ldots$, up to -1 for the last intron nucleotide: the first three exon positions are labeled $+1, +2, +3$. These positions divide fairly naturally into three regions. First, the region $[-40, -21]$ in which the branch point adenosine typically resides is characterized by nearly random (equal)

Table 1a

Base composition around intron/exon junctions[a]; branch point region I, [−40, −30] (50–57% Y)

Position	−40	−39	−38	−37	−36	−35	−34	−33	−32	−31	−30
A%	21	22	22	20	22	24	21	21	20	22	23
G%	29	27	25	26	25	22	23	22	22	21	23
C%	24	25	28	28	26	28	28	29	29	29	29
T%	26	26	26	27	26	26	28	28	29	28	25
Y%	50	52	54	54	52	55	56	57	57	57	55

[a] The observed frequency, expressed as a percentage, of each DNA base at each position in the aligned set of all 1254 annotated acceptor splice sites in genes of the learning set, $\mathcal{L}$ (see text). Positions −40,...,−1 are the last forty bases of the intron; positions +1, +2, +3 are the first three bases of the succeeding exon. The letter Y indicates the sum of the frequencies of the two pyrimidine nucleotides (C and T) at the given position. (Modified from Burge 1997 [1].)

Table 1b

Base composition around intron/exon junctions[a]; branch point region II, [−29, −21] (58–64% Y)

Position	−29	−28	−27	−26	−25	−24	−23	−22	−21
A%	22	21	21	22	23	21	23	20	20
G%	20	20	18	20	16	17	18	17	16
C%	30	30	31	30	31	30	29	31	34
T%	28	28	30	28	31	33	30	32	30
Y%	58	59	61	58	61	63	59	63	64

[a] See footnote in Table 1a.

Table 1c

Base composition around intron/exon junctions[a]; pyrimidine-rich region, [−20, −7] (65–82% Y)

Position	−20	−19	−18	−17	−16	−15	−14	−13	−12	−11	−10	−9	−8	−7
A%	20	16	15	14	14	12	9	9	8	8	8	8	8	9
G%	16	18	18	18	15	12	13	13	12	12	13	13	12	10
C%	31	32	32	31	35	37	35	34	34	33	33	38	41	41
T%	34	33	35	37	35	39	42	45	46	47	46	42	39	41
Y%	65	66	66	68	71	76	78	79	80	80	80	80	80	82

[a] See footnote in Table 1a.

Table 1d
Base composition around intron/exon junctions[a]; acceptor site region, [−6, +3]

Position	−6	−5	−4	−3	−2	−1	+1	+2	+3
A%	6	7	22	4	100	0	25	25	27
G%	6	6	22	0	0	100	52	22	24
C%	44	38	33	74	0	0	13	21	27
T%	44	48	22	21	0	0	9	32	23
Y%	88	87	55	96	0	0	23	53	50

[a] See footnote in Table 1a.

usage of the four nucleotide types, with a weak bias toward pyrimidine nucleotides (Y = C or T/U) which increases in strength as one proceeds in a 5′ to 3′ direction. Second, the region [−20, −7] where U2AF and/or other pyrimidine-tract-associated proteins typically bind, is characterized by a pronounced bias toward pyrimidine nucleotides, which increases monotonically from 65% to over 80% immediately upstream of the acceptor site. Third, the region [−6, +3] around the acceptor splice site itself, is thought to interact with U5 snRNA, U1 snRNA and possibly other factors. Positions −6 and −5, which are overwhelmingly biased toward pyrimidines, probably represent the typical 3′ terminus of the poly-pyrimidine tract, while the remaining positions exhibit the consensus pattern CAGG, flanked on either side by poorly conserved (unbiased) positions. Such compositional biases at the acceptor site were first observed by Breathnach and Chambon [26] and first systematically tabulated by Mount [27].

For comparison purposes, both WMM and WAM models (see above) of the pyrimidine-tract/acceptor region [−20, +3] were constructed from the acceptor sites annotated in the learning set, together with corresponding non-signal models derived from the non-signal/non-coding portions of the genes of $\mathcal{L}$. These models were then tested on the test set $\mathcal{T}$ of 65 human gene sequences as follows. Each (overlapping) segment X of length 23 bp from the sequences of $\mathcal{T}$ was considered as a possible acceptor site and assigned a score, $s(X) = \log_2(p^+(X)/p^-(X))$, by both models as described previously. Since any segment lacking the requisite AG dinucleotide at positions −2, −1 has probability zero, such segments have score −∞ under both models. The score distribution for the remaining approximately 46 000 segments which did have the required AG is shown for both models (and a third model to be described later) in Fig. 2, divided into true (annotated) acceptor sites and "pseudo-acceptor sites" (all other positions with AG at −2, −1). Comparison of Figs. 2a and 2b shows that, while both models are clearly able to separate true from false acceptor sites to a degree, the WAM model gives somewhat better discrimination (see also Table 3 below).

What accounts for the improved discrimination of the WAM model relative to the simpler WMM model of the acceptor site? The differences between these two models are most easily seen by comparing their signal ratios, i.e. by looking at the ratio of $r_{\mathrm{WAM}}(X) = p^+_{\mathrm{WAM}}(X)/p^-_{\mathrm{WAM}}(X)$ to $r_{\mathrm{WMM}}(X) = p^+_{\mathrm{WMM}}(X)/p^-_{\mathrm{WMM}}(X)$. Plugging in

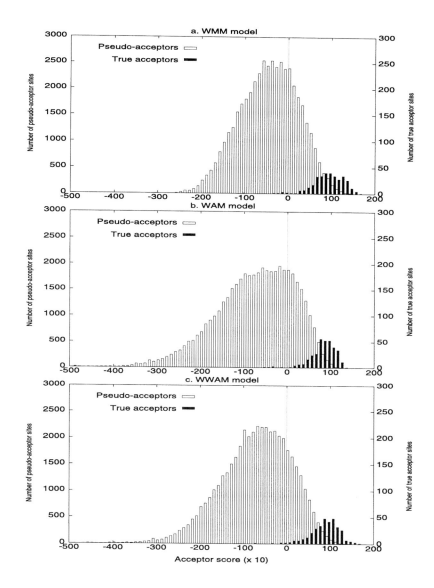

Fig. 2. The three acceptor splice signal models described in the text are compared in terms of their score histograms for true and false acceptor splice signals in the test set $\mathcal{T}$ of 65 human gene sequences (see text). Each position of each sequence in the set $\mathcal{T}$ was assigned a "score" by each model, defined as the logarithm base two of the signal ratio, $p^+(X)/p^-(X)$ (see text). Separate score distributions were then plotted for true acceptor sites (black bars) and "pseudo-acceptor" sites (open bars). Here, pseudo-acceptors are positions which have the requisite AG at positions $-2, -1$ but which are not annotated as acceptor splice sites (non-AG positions have score $-\infty$ under all three models and so are not plotted.) The height of each bar corresponds to the number of true or false sites observed at the given score level: note that separate scales were used for true (right scale) and false (left scale) acceptor sites. Note also that the x-axis is in units of score times ten rather than score. (Modified from Burge 1997 [1].)

the formulas given previously for $p^+()$ and $p^-()$ for each model gives (after a bit of rearrangement) the following simple form:

$$r_{\text{WAM}}(X)/r_{\text{WMM}}(X) = \frac{\prod_{i=2}^{\lambda} f_{x_{i-1},x_i}^{(i)}}{(\prod_{i=2}^{\lambda-1} f_{x_i}^{(i)})(\prod_{i=1}^{\lambda} f_{x_i}^{(i)})} \bigg/ \frac{\prod_{i=2}^{\lambda} g_{x_{i-1},x_i}^{(i)}}{(\prod_{i=2}^{\lambda-1} g_{x_i}^{(i)})(\prod_{i=1}^{\lambda} g_{x_i}^{(i)})} = \prod_{i=2}^{\lambda} \frac{r_{x_{i-1},x_i}^{(i)}}{\rho_{x_{i-1},x_i}^{(i)}}.$$

Here, $f_{x_i}^{(i)}$ etc. are frequencies observed in true acceptor signals, while $g_{x_i}^{(i)}$ etc. are the corresponding values observed in non-signal sequences. The term

$$r_{j,k}^{(i)} = \frac{f_{j,k}^{(i)}}{f_j^{(i-1)} f_k^{(i)}}$$

is the "positional odds ratio" of the dinucleotide j, k ending at position i of the signal, which is a measure of how much the frequency of j, k at position $i-1, i$ differs from what would be expected if the nucleotides at positions $i-1$ and i were independent. The terms

$$\rho_{j,k}^{(i)} = \frac{g_{j,k}^{(i)}}{g_j^{(i-1)} g_k^{(i)}}$$

are the corresponding odds ratios derived from non-signal positions, which will be essentially equal to the "global odds ratio" for the dinucleotide j, k in bulk genomic DNA (see ref. [28]).

6.1. Positional odds ratios

From the factorization above it is clear that only if the deviations from independence between adjacent positions (as measured by the dinucleotide positional odds ratios) are substantially different at particular positions in the acceptor splice signal from those observed generally in genomic DNA will the WAM and WMM models give significantly different results. To see which dinucleotides/positions account for the observed differences in discrimination, the positional odds ratios of all dinucleotides were determined at each position in the pyrimidine-tract region, $[-20, -5]$ (and, for future reference, in the branch point region, $[-40, -21]$) and compared to the corresponding global odds ratios measured in the genomic sequences of the set $\mathcal{L}$. These data are displayed for YR and YY dinucleotides ($Y = C$ or T, $R = A$ or G) in Fig. 3a.

Several features of these data are apparent. First, although many dinucleotides exhibit deviations from independence (positional odds ratios substantially above or below one), most of these ratios are quite similar to those observed globally in human genomic DNA. The dinucleotide CG, for example, is very strongly avoided, with odds ratios in the range of approximately 0.25 to 0.45, but this range is centered on the typical genomic value of 0.36. Thus, it appears that the dinucleotide odds ratios are in some way an intrinsic property of the genome (see also refs. [28,29]), which tend to gravitate toward typical genomic values even in the presence of strong selective pressures, e.g. the

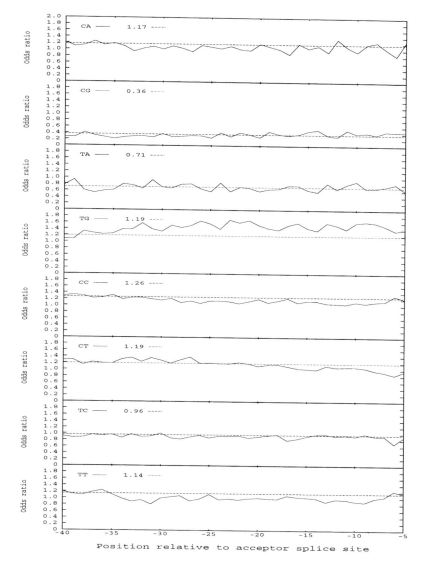

Fig. 3a. Distribution for YR and YY dinucleotides (Y=C or T, R=A or G). Solid lines show the positional odds ratios (see text) of each dinucleotide at each position in the range [−40, −5] relative to the intron/exon junction in the annotated acceptor splice signals of the genes of the learning set $\mathcal{L}$ (see text). Dashed lines show the corresponding "global odds ratio" values for non-coding, non-signal positions in the genomic DNA of learning set genes. (Modified from Burge 1997 [1].)

presumably quite powerful selection for pyrimidines in the region [−20, −5]. The only doublet in this figure which exhibits positional odds ratios consistently different from the global value is TG, with ratios typically in the range of 1.4 to 1.6, higher than the global

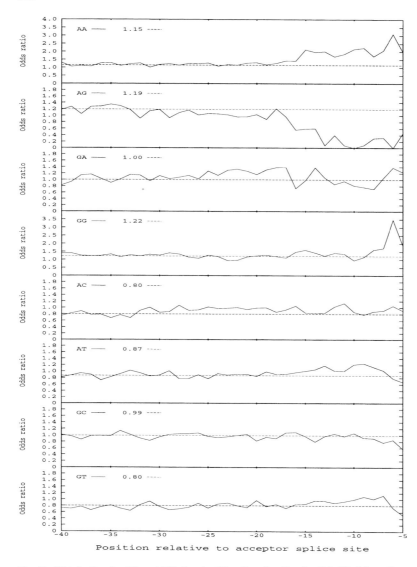

Fig. 3b. Distribution for RR and RY dinucleotides. See also Fig. 3a. (Modified from Burge 1997 [1].)

average value of 1.19. The reason for this excess of TG dinucleotides is not immediately apparent (but see below for a possible explanation).

The positional odds ratios and global odds ratios for RR and RY dinucleotides are displayed in Fig. 3b. Interestingly, three of the four RR doublets exhibit very dramatic over- or under-representation relative to their typical genomic values in the pyrimidine-tract region [−20, −5], but not in the branch point region. (These biases are so strong, in fact, that for AA and GG the y-axis had to be rescaled to display the full range of values observed – see figure.) In particular, the AG doublet is dramatically avoided in this region and the AA and GG doublets are strongly favored at certain positions. To appreciate just how low the frequency of AG is in this region, note (Table 1) that the product $f_A^{(i-1)}f_G^{(i)}$ is typically around 0.01 (1%) in this region, so that the positional odds ratios of less than 0.4 observed for AG over much of this region imply that its frequency is not more than 0.004 (0.4%), i.e. almost completely absent (see also ref. [30]). The strength of this avoidance suggests that occurrence of an AG in the region [−20, −5] might be extremely deleterious for splicing. Consistent with previous experimental results (e.g. refs. [31,32,7]), the most likely explanation is that, in the course of spliceosome assembly, the branch point is chosen first, followed by some sort of scanning mechanism which chooses the first available AG dinucleotide downstream (3′) of the branch point, so that presence of an AG in the region [−20, −5] would lead to incorrect acceptor site choice (often with direct consequences for the translation product of the incorrectly spliced mRNA). Selection against AG may also explain the unusually high frequencies of the two dinucleotides, AA and GG, which are a single transition mutation away from AG. In principle, this type of effect could also account for the unusually high frequencies of TG (one transversion away from AG) mentioned above, but this does not explain why the other doublets (AC, AT, CG) which are one transversion away from AG are not also over-represented. Nor, of course, does the AG avoidance hypothesis explain why AA should be over-represented over most of the pyrimidine-tract region while GG is over-represented only over the last four positions, [−8, −5] (Fig. 3b). In summary, while both the WMM and WAM models capture the bias toward pyrimidine nucleotides in [−20, −5] and the preferred pattern CAGG at [−3, +1], the improved discrimination observed for the WAM model apparently relates to its ability to capture the biases away from AG and towards AA, GG and TG dinucleotides in the pyrimidine-rich region.

6.2. The branch point region

Interestingly, the branch point region [−40, −21] exhibits very few biases either in terms of nucleotide (Table 1) or dinucleotide composition (Fig. 3) and, for this reason, extension of the WMM and WAM models described above to encompass this region gave little improvement in discrimination of true from false acceptor splice sites (data not shown). This lack of compositional bias at particular positions is presumably due, at least in part, to the variable location of the branch point signal in the region [−40, −21] (and sometimes further upstream), which would tend to average out any biases which might otherwise be observed. One approach to modeling this signal would be to locate and align the branch point signals from each sequence, as was done for the acceptor (and donor) splice sites on the basis of the CDS feature annotation. However, this approach is not feasible in practice because sequencing laboratories very rarely perform the extra experiments required to localize the branch point and so branch points are almost never annotated in GenBank (or EMBL) sequences. Nor is searching for the branch point consensus (YYRAY or variants) a reliable way of locating branch points, e.g. Harris and Senapathy found only a very weak tendency toward the branch point consensus in this region [33]. Consistent with these findings, only 30% of the introns in the set $\mathcal{L}$ contained a YYRAY pentamer in the appropriate region [−40, −21]. Thus, any model of the acceptor signal which required the presence of even such a weak consensus pattern would tend to miss most true sites. Nevertheless, mutations which affect the branch point signal can have a significant effect on pre-mRNA splicing both *in vitro* [34] and *in vivo* [35], and have even been associated with certain human diseases (e.g. refs. [36,37]), so it would clearly be desireable to incorporate aspects of this signal into models of the acceptor region. One approach would be to develop a higher-order inhomogeneous Markov (IMM/WAM) model to capture biases in triplet, tetramer or even pentamer composition at particular positions in the branch point region, since the branch point signal is at least five to seven nucleotides in length. While such a model might work in principle, serious problems arise in estimating the increased number of parameters in such models. For example, construction of a second-order Markov model requires the estimation of four transition probabilities conditional on each doublet at each position. For the set of 1254 acceptor sequences from $\mathcal{L}$, most doublets occur about $n = 1254/16 = 78$ times per position, and some doublets occur significantly less often, e.g. TA occurs about $0.71 \times 78 = 55$ times per position (Fig. 3a) and CG has even lower counts. Unfortunately, these are not sufficient data to reliably estimate the transition probabilities, as is discussed below.

6.3. Parameter estimation error

If a sample of n nucleotides is randomly chosen from a large population (of nucleotides), then the count n_i of nucleotide type i ($i =$A, C, G or T) will be distributed binomially, with mean $\mu_i = np_i$ and variance $\sigma_i^2 = np_i(1 - p_i)$, where p_i is the (true) frequency of nucleotide i in the original population. A convenient measure of the typical estimation error made in using the observed fraction $f_i = \frac{n_i}{n}$ to estimate p_i is the "coefficient of variation", CV_i, defined as

$$CV_i = \frac{\sigma_i}{\mu_i} = \frac{\sqrt{np_i(1 - p_i)}}{np_i} = \sqrt{\frac{1 - p_i}{np_i}}.$$

Note that this error increases with decreasing values of p_i (i.e. for lower frequency nucleotides), but decreases (of course) with increasing sample size, n. Values of this error measure for typical ranges of p_i and n are given in Table 2. It can be seen that

Table 2
Parameter estimation error (coefficient of variation) as a function of symbol probability, p_i, and sample size, n[a]

p_i	Sample size, n						
	30	50	100	175	300	500	1000
0.50	18.3%	14.1%	10.0%	7.6%	5.8%	4.5%	3.2%
0.25	31.6%	24.5%	17.3%	13.1%	10.0%	7.7%	5.5%
0.15	43.5%	33.7%	23.8%	18.0%	13.7%	10.6%	7.5%
0.10	54.8%	42.4%	30.0%	22.7%	17.3%	13.4%	9.5%

[a] A measure of the error in the estimation of a Markov transition probability, $P\{x_i|W\}$, by the corresponding conditional frequency, f_{Wx_i}/f_W, is given as a function of n, the count of the pattern W which is being conditioned on, and p_i, the true (conditional) probability of the nucleotide x_i. The measure used is the "coefficient of variation" defined in the text. (From Burge 1997 [1].)

estimation errors are quite large for sample sizes of about one hundred or less, but become tolerable (say, in the range of 10 to 20% or so) at around $n = 175$ to 300, depending on the value of p_i. From above, the sample size n available to estimate the (second-order) Markov transition probabilities conditional on a dinucleotide at a given position i is about 78, while the corresponding sample size for a first-order WAM is about $4 \times 78 = 312$. The resulting difference in parameter accuracy is the most likely explanation for the result (not shown) that a second-order WAM acceptor model (derived from the set $\mathcal{L}$) actually performed less well than the simpler first-order WAM model described previously. This example illustrates the dangers of constructing models which are more complex (have more parameters) than can be reliably approximated by the available data.

6.4. A windowed weight array model

In an attempt to bypass this type of sample-size limitation, an alternative approach was used, in which data from a "window" of adjacent signal positions are pooled to form what was termed a "windowed weight array model" (WWAM) or "windowed inhomogeneous Markov model" (WIMM) (see chapter 4 of ref. [1]). In this approach, the second-order transition probabilities at a given position i are estimated by averaging the conditional frequencies observed at i with those observed in a window of nearby positions centered at the position i. The basic assumption is that nearby positions will have sufficiently similar triplet biases so that the averaged conditional frequencies will give a more accurate estimate of the true conditional probability than could be achieved using simply the observed conditional frequency (based on quite a small sample) at the given position. For the dataset $\mathcal{L}$ of 1254 acceptor sites, a window size of five was chosen, so that transition probabilities for positions i in the branch point region $[-40, -21]$ were estimated from the data at positions $i - 2, i - 1, i, i + 1, i + 2$. This results in typical sample sizes $n = 5 \times 1254/16 = 392$ per position, which should be sufficient to give fairly reliable parameter estimates according to the data in Table 2. This branch region model (and a corresponding negative model derived from non-signal sites) was then "fused" to the previously derived first-order WAM model of the pyrimidine-tract/acceptor site, to give a hybrid WWAM model. This model was then tested on the set $\mathcal{T}$ as for the previous models; results are shown in Fig. 2c. Interestingly, the WWAM model appears to give better separation of true acceptors from pseudo-sites than either of the other models; the three models are compared in a somewhat more systematic fashion below.

A standard way to assess the discriminatory power of a set of models is to compare the number of "false positives" (FP) observed at several different levels of "sensitivity". For a particular sensitivity level p, the score threshold s_p is chosen as the minimum score for which a fraction p of true sites have score $\geqslant s_p$. Thus, the number of true positives (TP) at this level is approximately pN, where N is the total number of true sites. The number of false sites with score $\geqslant s_p$ is then determined: improved discrimination corresponds to fewer false positives at a given level of sensitivity. Performance can then be measured either by the total number (or percentage) of false positives (FP) or by the specificity, $Sp = TP/(TP + FP)$, at a given sensitivity threshold. These values are tabulated for the three models at selected sensitivity levels in Table 3. The results show quite clearly that

Table 3
Specificity versus sensitivity for acceptor splice signal models[a]

Model	Sensitivity level							
	20%		50%		90%		95%	
	FP	Sp	FP	Sp	FP	Sp	FP	Sp
WMM	68	50.7%	629	21.6%	3,983	7.1%	6,564	4.7%
WAM	65	51.5%	392	30.9%	3,322	8.4%	5,493	5.5%
WWAM	48	58.6%	343	33.8%	3,170	8.8%	5,397	5.6%

[a] Accuracy statistics are given for the three acceptor splice signal models described in the text, as tested on the test set T of 65 human gene sequences. Each position of each sequence in the set T was assigned a "score" by each model, as discussed in the text. For each sensitivity level p, the score cutoff s_p is defined as the minimum score for which at least p% of true acceptor sites scored above the level s_p. True positives (TP) is the number of real acceptor sites with score $\geq s_p$; false positives (FP) is the number of non-acceptor site positions with score $\geq s_p$; and Sp is the ratio of true positives over predicted positives, i.e. $Sp = TP/(TP + FP)$, at the given sensitivity level. (From Burge 1997 [1].)

WWAM > WAM > WMM in terms of discriminatory power, and this same ordering holds at all levels of sensitivity.

What accounts for the improved discriminatory ability of the WWAM branch point/acceptor model? Previously, the improved performance of the first-order Markov (WAM) model relative to the (zero-order) independence or WMM model was analyzed by comparing the observed positional dinucleotide frequencies to their "expected" values under independence, as measured by the dinucleotide positional odds ratios. The analogous test is to compare the observed positional triplet frequencies, $f_{x,y,z}^{(i)}$, at positions i in the branch point region (from which the second-order WWAM parameters were estimated) to their expected values, $E_{x,y,z}^{(i)}$, under a first-order Markov (WAM) model, i.e. to consider the "triplet positional odds ratios"

$$r_{x,y,z}^{(i)} = f_{x,y,z}^{(i)}/E_{x,y,z}^{(i)} = f_{x,y,z}^{(i)} / (f_x^{(i-2)} \cdot \frac{f_{x,y}^{(i-1)}}{f_x^{(i-2)}} \cdot \frac{f_{y,z}^{(i)}}{f_y^{(i-1)}}) = \frac{f_{x,y,z}^{(i)} f_y^{(i-1)}}{f_{x,y}^{(i-1)} f_{y,z}^{(i)}}.$$

To plot these values for all 64 DNA triplets at all positions, as was done for dinucleotides in Fig. 3, would take up a great deal of space; instead, a more concise representation of the major trends in triplet biases was sought.

Consistent patterns of bias toward or away from particular triplets may be seen by considering the signs of the differences $\Delta_i = r_{x,y,z}^{(i)} - 1$ for particular triplets x, y, z at positions i in the branch point region, $[-40, -21]$. Since the probability of observing all signs positive (or all signs negative) is approximately $(\frac{1}{2})^{20}$, which is less than 10^{-6}, to observe such a pattern is significant even considering that 64 such tests were performed. Therefore, a triplet which has all $\Delta_i > 0$ is considered "significantly favored" and one with all $\Delta_i < 0$ is "significantly avoided". Of the 64 possible DNA triplets, four (CAG, CTG, TAA and TTT) were significantly favored according to the above definition and five (CAA, CTA, CTT, TAG and TTG) were significantly avoided. The favored triplets CTG and TAA are probably related to the branch point signal itself, forming portions of the YYRAY consensus. The bias toward CAG was somewhat unexpected, but might (speculatively) represent an alternative branch point pattern, YYARY, since under certain circumstances either of the purine nucleotide positions in the branch signal can function as the branch nucleophile [10]. If so, then perhaps the avoidance of the triplet TAG might reflect a particular preference of YYARY branch points for C over T at the second position. The finding that TTT is favored while CTT is avoided is particularly intriguing, since both are YYY triplets which might form portions of the polypyrimidine tract in cases where this signal is located in, or extends into, the region [−40, −21]. The most likely explanation is probably that such distal polypyrimidine tracts are particularly sensitive to the relative usage of T(U) versus C, with a preference for uninterrupted runs of T, an idea for which there is a certain amount of experimental support [38,39]. The explanation for the other avoided triplets is not immediately apparent, although it is interesting to observe that each of the avoided triplets is a single transition mutation away from at least one favored triplet. In summary, the improved discriminatory power of the WWAM model versus the other two models can be attributed to its ability to capture certain triplet biases which appear to be related to the branch point and polypyrimidine-tract signals, as well as other biases of uncertain origin.

7. Measuring dependencies between positions

In the previous section, several models of the acceptor splice signal were introduced, their performance was compared, and the compositional features contributing to the differences in performance were analyzed by studying the patterns of over- and under-representation of di- and trinucleotides. In an important respect, however, this analysis was incomplete, since none of the models considered allows for the possibility of "long-range" interactions between positions further apart than two or three nucleotides. To address this issue, we must first have a way of measuring the amount of dependence between arbitrary positions in a signal. A standard test of dependence between two discrete-valued random variables Y and Z, which may take on values $y_1, \ldots, y_m$ and $z_1, \ldots, z_n$, respectively, is the chisquare test, in which the joint distribution of the variables Y and Z is compared to that expected under the null hypothesis of independence using the measure

$$X^2 = \sum_{i,j} \frac{(O_{i,j} - E_{i,j})^2}{E_{i,j}} \, .$$

Here, $O_{i,j}$ is the observed count of the event that $Y = y_i$ and $Z = z_j$; $E_{i,j}$ is the value of this count expected under the null hypothesis that Y and Z are independent (calculated from the product of the observed frequencies of the events $Y = y_i$ and $Z = z_j$); and the sum is taken over all possible pairs of values, i, j. Under the null hypothesis of independence, X^2 will have (for sufficiently large datasets) a χ^2 distribution with $(m-1)(n-1)$ degrees of freedom.

To use this test in the context of a biological signal, one must choose appropriate pairs of variables to study. The most obvious choice is to examine dependencies between the variables N_i and N_j (which may take on the four possible values A, C, G, T), indicating the nucleotides at positions i and j of the signal, i.e. to ask whether there is an association

Table 4a

Measuring dependencies between positions in a signal: contingency table for nucleotide variables; N_{-2} versus N_{+6} in donor splice signal[a,b]

N_{-2}	N_6								
	A		C		G		T		All
	O	E	O	E	O	E	O	E	O
A	136	(114)	144	(139)	182	(154)	292	(345)	754
C	17	(24)	26	(29)	18	(32)	99	(73)	160
G	19	(26)	33	(32)	29	(35)	93	(79)	174
T	19	(25)	29	(30)	28	(34)	90	(75)	166
All	191		232		257		574		1254

[a] Two ways of measuring the amount of dependence between two positions in a nucleic acid signal sequence are illustrated, using the comparison of positions −2 and +6 in the human donor splice signal as an example. Position −2 is the penultimate nucleotide of the exon (consensus: A); +6 is the sixth nucleotide of the succeeding intron (consensus: T). Here, N_i is a variable indicating the identity of the nucleotide at position i of the signal; K_i is a consensus indicator variable (CIV), which is defined to be one if N_i matches the consensus at i and zero otherwise. Part **a** shows a comparison of the variables N_{-2} and N_{+6} in the set of all 1254 donor splice signals in genes of the learning set $\mathcal{L}$ (see text), using the standard contingency table representation. For each pair of nucleotides X, Y, the observed count ("O") of the event that $N_{-2} = X$ and $N_{+6} = Y$ is given first, followed by the expected value ("E"), in parentheses. The expected value is calculated in the standard way as the product of the row and column sums divided by the total number of observations, N ($= 1254$). Part **b** shows a similar contingency table for the variables K_{-2} versus N_{+6}. Chisquare statistics X^2 (see text) are given in footnote b.
[b] $X^2 = 46.4$ ($P < 0.001$, 9 df).

between occurrence of particular nucleotide(s) at position i and occurrence of other nucleotide(s) at position j in the same signal sequence. An example of such a comparison, for positions −2 versus +6 in the set of 1254 donor splice signals in genes of the set $\mathcal{L}$ is illustrated, using the standard 4×4 contingency table representation, in Table 4a. The value of $X^2 = 46.4$ observed indicates a significant degree of dependence between the positions at the level $P < 0.001$. Examination of the contingency table data shows that most of the dependence is a result of a negative association between A at position −2 and T at position +6, with a corresponding increased occurrence of T at +6 when N_{-2} is any base other than A. There is also a somewhat weaker positive association between occurrence of A at position −2 and A or G at +6, and again the opposite is observed when N_{-2} is C, G or T. (Possible interpretations of dependencies in donor splice signals are discussed in the next section.) Overall, the most notable feature of this table is that the distribution of nucleotides at position +6 appears to depend only on whether N_{-2} is or is not A (the consensus at position −2) and not on the exact identity of N_{-2} in the event that it is not A.

This phenomenon can be quantified by measuring the dependence of N_{+6} on the "consensus indicator variable" (CIV) K_{-2}, which has the value one (say) if N_{-2} matches the consensus at position −2 (A) and zero otherwise. (The symbol K was chosen for the CIV rather than C to avoid confusion with the symbol for the nucleotide cytosine.) This comparison is illustrated using a similar contingency table representation in Table 4b.

Table 4b

Measuring dependencies between positions in a signal: contingency table for consensus indicator versus nucleotide variables (K_{-2} versus N_{+6}) in donor splice signal[a,b]

K_{-2}	N_6								
	A		C		G		T		All
	O	E	O	E	O	E	O	E	O
A	136	(114)	144	(139)	182	(154)	292	(345)	754
[CGT]	55	(76)	88	(92)	75	(102)	282	(228)	500
All	191		232		257		574		1254

[a] See footnote a in Table 4a.
[b] $X^2 = 42.9$ ($P < 0.001$, 3 df).

Strikingly, the chisquare statistic for this 2×4 table is $X^2 = 42.9$ ($P < 0.001$), almost as high as that observed in the 4×4 table, even though the number of degrees of freedom is only three instead of nine. This result underscores the observation made previously that N_{+6} depends only on whether N_{-2} matches the consensus. (Another indication that the donor splice site "does not care" about the identity of N_{-2} when it is not A is, of course, the fact that the frequencies of the three non-A nucleotides C, G and T are nearly equal at position -2.) Importantly, this phenomenon of "consensus indicator dependence" (CID) is not unique to the particular pair of positions chosen, but appears to hold generally for almost all pairs of positions in the donor and acceptor splice signals (data not shown), and may hold also for other types of nucleic acid signals. (Possible explanations for the CID phenomenon are discussed in the next section.) As a consequence, K_i versus N_j tests may represent a more powerful way to detect significant dependencies between positions in a signal than the more commonly used N_i versus N_j comparisons. Another significant disadvantage of N_i versus N_j comparisons is that, for positions i and j with strongly biased composition, the contingency table expected values may become so small (say < 10) that the χ^2 test is no longer reliable. Clearly, when lower frequency nucleotides are pooled in a K_i versus N_j comparison, this problem becomes less acute. For these reasons, only K_i versus N_j tests are considered in the remainder of this chapter.

7.1. Dependencies in acceptor splice signals

The dependence structure of the acceptor splice signal was investigated by constructing K_i versus N_j contingency tables for all pairs $i \neq j$ in the range $[-20, +1]$ – positions in the pyrimidine-tract region were considered to have consensus $Y = C$ or T (see Table 1). Those pairs for which significant dependence was detected at the level $P < 0.001$ are shown as dark gray squares in Fig. 4. (The stringent P-value cutoff was used to compensate for the relatively large number of tests performed.) The distribution of dark squares in the figure is highly non-random, with two primary trends observed: (1) a near-diagonal bias; and (2) a horizontal bar of squares near the bottom of the figure. The dark squares immediately above and below the diagonal indicate interactions between adjacent positions in the pyrimidine-tract region; the fact that so many are observed provides another perspective

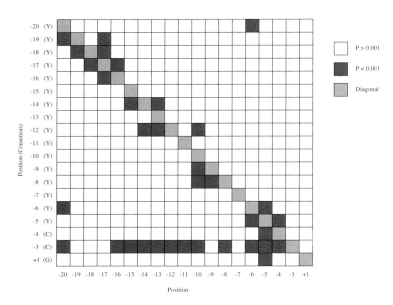

Fig. 4. Statistically significant interactions (dependencies) between positions in the 1254 acceptor splice signals of the learning set $\mathcal{L}$ (see text) are illustrated. For each pair of positions i (y-axis) and j (x-axis) with $i \neq j$, a K_i versus N_j contingency table was constructed (see Table 4b) and the value of the chisquare statistic X^2 was calculated. Dark gray squares indicate pairs with significant chisquare values at the level $P < 0.001$ (3 df); all other squares are white, except those on the diagonal ($i = j$), which are colored light gray for visual reference.

on why the WAM model, which incorporates interactions between adjacent positions, gave better discrimination than the simpler WMM model, which does not. Closer examination of the data showed that in most cases this dependence takes the form of a positive association between pyrimidine nucleotides at adjacent positions. Such a dependence could be explained on the basis that the factors which interact with the pyrimidine tract may bind preferentially to contiguous blocks of pyrimidines without intervening purines (e.g. refs. [38,39]), so that the strength of the selection for pyrimidines at a given position i is stronger when the nucleotides at positions $i-1$ and/or $i+1$ are pyrimidines. This idea could perhaps be further investigated through more detailed computational analysis of available acceptor sequences.

The dark horizontal bar at position -3 on the y-axis is more mysterious since it corresponds to interactions between positions quite far apart (up to 16 nucleotides) and also because it is highly asymmetric in that there is no corresponding vertical bar at position -3 on the x-axis. This asymmetry reflects the surprising result that, while the nucleotides N_j at many positions *are* significantly dependent on the value of the consensus indicator, K_{-3}, the nucleotide variable N_{-3} is *not* significantly dependent on consensus indicator variables K_j in the pyrimidine-tract region. How can this be? The data in Table 5 shed some light on this apparent paradox. First, it is clear that occurrence of C at position -3 is positively associated with occurrence of C at positions in the pyrimidine-tract region, but negatively associated with occurrence of T at these positions, explaining

Table 5

Long-range interactions in acceptor splice signals: usage of C and T in the pyrimidine tract as a function of consensus indicator variable K_{-3} [a]

Position	C at -3			[TAG] at -3			X^2
	C%	T%	Y%	ΔC	ΔT	ΔY	
-20	32	31	64	-9	$+10$	$+1$	23.7
-19	34	31	66	-9	$+8$	-1	11.6
-18	33	32	65	-5	$+10$	$+5$	14.7
-17	33	34	67	-6	$+10$	$+4$	15.4
-16	39	31	70	-13	$+16$	$+3$	45.5
-15	39	36	75	-9	$+12$	$+3$	21.8
-14	39	38	77	-15	$+18$	$+3$	43.9
-13	36	42	78	-8	$+11$	$+3$	18.1
-12	36	43	79	-6	$+10$	$+4$	17.7
-11	36	43	79	-12	$+17$	$+5$	31.7
-10	36	43	79	-10	$+14$	$+4$	25.9
-9	40	40	80	-9	$+18$	$+9$	12.1
-8	44	34	78	-14	$+20$	$+6$	39.6
-7	43	39	82	-8	$+17$	$+9$	7.1
-6	48	40	88	-15	$+14$	-1	25.4
-5	42	44	86	-13	$+13$	0	26.4

[a] The influence of the consensus indicator variable K_{-3} on the pyrimidine composition of the region $[-20, -5]$ in human acceptor splice signals is illustrated. Data are from the set of 1254 acceptor signals from genes of the learning set, $\mathcal{L}$ (see text). Column one gives the position in the intron relative to the intron/exon junction (e.g., the conserved AG is at positions $-2, -1$). Columns 2–4 give the observed percentage frequencies of C, T and Y ($=$C or T) at each position in the subset of acceptor sites which have the consensus nucleotide (C) at position -3. The values listed in columns 5, 6 and 7 represent the differences between the observed frequencies of C, T and Y in acceptor sites which do not match the consensus at -3 and the values listed in columns 2, 3 and 4, respectively. For each position j, column 8 gives the value of the chisquare statistic X^2 calculated as described in the text for a K_{-3} versus N_j comparison.

the high X^2 values observed for many K_{-3} versus N_j comparisons. However, these two effects largely cancel each other out, leaving only a very weak (apparently not statistically significant) negative association between occurrence of pyrimidines at positions j in the pyrimidine tract and occurrence of C at position -3, explaining the absence of significant interactions between the variables K_j ($-20 \leqslant j \leqslant -5$) and N_{-3}.

Of course, the data described above only explains *how* the asymmetry can exist, but not *why* it exists, i.e. why the relative usage of C versus T in the pyrimidine-tract region appears to be so strongly affected by the presence or absence of the consensus (C) at position -3. Given the strengths of the statistical interactions observed in Fig. 4 and Table 5, it is tempting to speculate about possible molecular mechanisms involving physical interactions (perhaps mediated by spliceosomal proteins) between the nucleotides at -3 and those in the pyrimidine tract, for example. However, it must always be kept in mind that correlation by itself does not imply causation. Specifically, while an

Table 6
Isochore effects on acceptor splice signal composition[a]

Position	Group I[b]		Group II[c]		Group III[d]		Group IV[e]	
	C%	T%	C%	T%	C%	T%	C%	T%
−20	21	42	29	41	36	30	36	26
−19	20	46	34	36	34	32	42	21
−18	19	50	32	37	35	29	40	24
−17	18	51	30	41	36	32	41	25
−16	18	49	34	40	42	31	46	22
−15	23	51	35	45	44	36	46	27
−14	17	60	32	48	42	37	49	27
−13	17	63	33	45	40	40	44	32
−12	23	56	30	52	41	39	43	36
−11	19	63	30	51	42	38	39	37
−10	21	59	27	55	42	40	43	34
−9	27	50	36	44	44	38	44	36
−8	26	54	36	49	49	31	50	26
−7	25	55	39	42	42	38	56	30
−6	22	66	40	49	54	33	60	27
−5	23	62	31	55	43	45	55	34
−3	54	38	67	27	83	13	90	9

[a] The influence of local (C+G)% composition on the pyrimidine composition of positions [−20, −5] and −3 in human acceptor splice signals is illustrated. The genes of the learning set $\mathcal{L}$ (see text) were partitioned into four roughly equal-sized groups, designated I–IV, on the basis of the (C+G)% composition of the GenBank (genomic) sequence. For each group the range of (C+G)% composition, the number of genes in the group, and the total number of acceptor splice sites (introns) in the group are listed in footnotes b–e, respectively. The table gives the observed percentage frequency of C and T nucleotides at each position in the genes of each group.
[b] C+G%, 32–45; number of genes, 59; total number of acceptor splice sites (introns), 317.
[c] C+G%, 45–51; number of genes, 60; total number of acceptor splice sites (introns), 286.
[d] C+G%, 51–58; number of genes, 60; total number of acceptor splice sites (introns), 304.
[e] C+G%, 58–72; number of genes, 59; total number of acceptor splice sites (introns), 347.

observed correlation between two variables A and B may indicate a direct causal/physical relationship, it is equally possible to result simply from dependence of both A and B on some third variable, C. In this instance, an argument will be made that the observed association between K_{-3} and N_j ($-20 \leqslant j \leqslant 5$) is most simply explained on the basis of the dependence of both variables on local compositional/mutational biases rather than on some previously unknown physical interaction related to splicing.

From Table 1, the alternative to C at position −3 is almost always T (never G and rarely A). Therefore, the association described above can be described either as a positive association between C at −3 and C at positions −20,...,−5 and/or as a positive association between T at −3 and T at positions −20,...,−5. This description highlights the possibility that the observed dependencies might result simply from compositional heterogeneity of the dataset of genes used, reflecting the known compositional heterogeneity of the human genome [40]. Specifically, if some genes were subjected to mutational pressure biased toward C + G as a result of their location in a C + G-rich "isochore" (see ref. [40]), while others were subjected to mutational pressure towards A + T, such heterogeneity alone might lead to measurable positive associations (auto-correlations) between the occurrence of T at one position and T at other positions in the same sequence, for example. (See also ref. [28] for a more thorough discussion of this general phenomenon.) To investigate this potential explanation, the genes of the learning set $\mathcal{L}$ were partitioned according to the (C + G)% content of the GenBank sequence (assuming that this value will be somewhat reflective of the nature of local mutational pressures) into four subsets of roughly equal size, and the acceptor site composition of these sets was compared (Table 6). The data show quite clearly that both the C versus T composition of position -3 and the C versus T composition of all positions in the pyrimidine-tract region are strongly dependent on the (C + G)% composition of the local genomic region, supporting the hypothesis proposed above.

To ask whether this compositional effect can "explain" the observed dependencies, the dependence between K_{-3} and N_j ($-20 \leqslant j \leqslant 5$) was measured for the sequences of each compositional subset separately. The results were that *no* significant chisquare values were observed in any of the sets for any position j in the range $[-20, -5]$, i.e. partitioning based on (C + G)% composition completely eliminates the observed dependence. Importantly, this effect does not appear to be simply a consequence of the reduction in sample size since significant dependencies between some *adjacent* positions still remained after partitioning into compositional subsets (not shown). Thus, the available evidence supports the idea that the K_{-3} versus N_j dependence is primarily due to mutational forces acting differentially on different (C + G)% compositional regions of the genome rather than on factors directly related to the specificity of splicing. For this reason, further models accounting for these dependencies were not developed. To summarize, there are two primary types of dependencies in acceptor splice signals: an "adjacent pyrimidine bias" in the region $[-20, -5]$, which is putatively related to the affinities of pyrimidine-tract binding proteins; and a long-range dependence between K_{-3} and pyrimidine-tract nucleotides which appears not to be directly related to splicing.

7.2. Dependence structure of the donor splice signal

A similar analysis of the dependence structure of the donor splice signal is now described, leading to a new model of this signal. The donor splice signal comprises the last 3 exonic nucleotides (positions -3 to -1) and the first 6 nucleotides of the succeeding intron (positions $+1$ through $+6$), with consensus sequence [c/a]AG!GT[a/g]AGt (the splice junction is indicated by !; uppercase letters indicate nucleotides with frequency $> 50\%$ – see Fig. 5). As discussed previously, the GT dinucleotide at positions $+1, +2$ is essentially invariant, with only a small number of exceptions known (but see ref. [24]). The dependence structure of this signal was investigated by testing for dependencies between K_i and N_j variables at all pairs of variable positions $i \neq j$ as above. The results (Table 7) show a much greater density and more complex pattern of dependencies than was observed for the acceptor signal (Fig. 4). In particular, while dependencies between

154

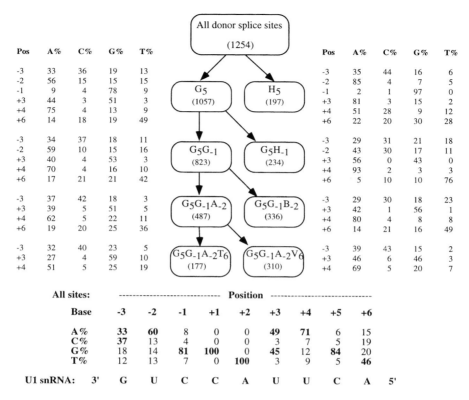

Fig. 5. Subclassification of the donor splice signals of the learning set $\mathcal{L}$ by the MDD procedure is illustrated. Each rounded box represents a subset of donor sites corresponding to a pattern of matches/mismatches to the consensus nucleotide(s) at a set of positions. For example, G_5G_{-1} is the set of donor sites with consensus nucleotide (G) at both positions +5 and −1. Positions −3, −2, −1 are the last three bases of the exon; positions +1, ..., +6 are the first six bases of the succeeding intron. Here, H indicates A, C or U; B indicates C, G or U; and V indicates A, C or G. The number of sites in each subset is given in parentheses beneath the pattern description; the percentage frequencies of the four nucleotides at each variable position are indicated for each subset immediately adjacent to the corresponding box. Data for the entire set of 1254 donor sites are given at the bottom of the figure, with the frequencies of consensus nucleotides shown in bold face. The sequence near the 5′ end of U1 snRNA which has been shown to base pair with the donor signal is shown below in 3′ to 5′ orientation. (From Burge and Karlin 1997 [2]).

acceptor site positions were relatively sparse, almost three fourths (31/42) of the i,j pairs in the donor site exhibit significant dependence! And, while two primary trends appeared to explain most observed dependencies in the acceptor site, no clear trend is apparent in Table 7 other than a tendency for (almost) every position to depend on (almost) every other position. More specifically, every donor site pair i,j which is physically adjacent (i.e. such that $|i-j|=1$) exhibits significant dependence in the donor site, while less than half of all nearest-neighbor tests were significant in the acceptor site. Most interesting, however, is the observation that a large number of *non*-adjacent pairs also exhibit significant dependence, even pairs quite far apart. For example, extremely high χ^2 values

Table 7

Dependencies between positions in human donor splice signals: chisquare statistic for consensus indicator K_i versus nucleotide N_j [a]

Position i	Consensus	Position j							Row sum
		-3	-2	-1	$+3$	$+4$	$+5$	$+6$	
-3	c/a	–	**61.8**	14.9	5.8	**20.2**	11.2	**18.0**	**131.8**
-2	A	**115.6**	–	**40.5**	**20.3**	**57.5**	**59.7**	**42.9**	**336.5**
-1	G	15.4	**82.8**	–	13.0	**61.5**	**41.4**	**96.6**	**310.8**
$+3$	a/g	8.6	**17.5**	13.1	–	**19.3**	1.8	0.1	**60.5**
$+4$	A	**21.8**	**56.0**	**62.1**	**64.1**	–	**56.8**	0.2	**260.9**
$+5$	G	11.6	**60.1**	**41.9**	**93.6**	**146.6**	–	**33.6**	**387.3***
$+6$	t	**22.2**	**40.7**	**103.8**	**26.5**	17.8	**32.6**	–	**243.6**

[a] For each pair of positions $\{i,j\}$ with $i \neq j$, a 2×4 contingency table was constructed for the consensus indicator variable K_i (see text) versus the variable N_j identifying the nucleotide at position j. The consensus nucleotide(s) at each position i are shown in the second column: the invariant positions $+1, +2$ (always GT) are omitted. For each contingency table, the value of the chisquare statistic X^2 (see text) was calculated and is listed in the table above. Those values exceeding 16.3 ($P < 0.001$, 3 df) are displayed in boldface. The last column in the table lists the sum of the values in each row, which is a measure of the dependence between K_i and the vector of nucleotides at the six remaining positions $j \neq i$. All values exceeded 42.3 ($P < 0.001$, 18 df) and so are displayed in boldface: the largest value, for K_{+5}, is indicated by *. (From Burge and Karlin 1997 [2].)

of 96.6 and 103.8 are exhibited in K_{-1} versus N_{+6} and K_{-6} versus N_{-1} comparisons, respectively, even though these positions are separated by five intervening bases.

8. Modeling complex dependencies in signals

The observations made at the end of the previous section raise two important questions: (1) Are the complicated dependencies in donor splice sites related to splicing or to something else? and (2) How can a probabilistic model be constructed which accounts for such a complicated pattern of dependencies? This section is devoted primarily to a discussion of these two related topics. First, we ask whether the observed pattern of dependencies might be explained on the basis of local variations in $(C+G)\%$ composition, as appeared to be the case for the K_{-3} versus N_j dependence in acceptor signals. The position-specific nucleotide frequencies in the donor splice signals of learning set genes, partitioned on the basis of $(C+G)\%$ composition as before, are shown in Table 8. Comparison of the data in Table 8 with that in Table 6 shows quite clearly that local $(C+G)\%$ composition has a much less significant impact on donor sites than acceptor sites. In particular, while the frequencies of C and T at most positions in the pyrimidine-tract region differ by 20–30% or more between genes of groups I and IV (Table 6), the variability in donor splice sites (Table 8) is not more than 10% at most positions. (The most notable exception is position $+3$, consensus R = A or G, for which the relative usage of G versus A is strongly associated with the $C+G$ content of the gene.) Further evidence that $C+G$ composition

Table 8
Isochore effects on donor splice signal composition[a]

Base	Group	Position								
		−3	−2	−1	+1	+2	+3	+4	+5	+6
A	I	34	66	14	0	0	74	75	11	20
	II	39	61	4	0	0	52	74	6	18
	III	30	63	7	0	0	38	71	6	14
	IV	30	52	7	0	0	35	65	3	10
G	I	19	11	74	100	0	21	9	79	12
	II	16	13	87	100	0	41	9	83	22
	III	16	11	83	100	0	57	14	87	26
	IV	22	19	81	100	0	60	17	88	22
C	I	33	10	2	0	0	2	4	3	12
	II	35	10	4	0	0	4	7	5	14
	III	42	12	5	0	0	2	9	4	19
	IV	38	18	4	0	0	3	10	6	27
T	I	15	13	11	0	100	3	12	7	56
	II	10	16	5	0	100	3	9	6	45
	III	12	14	6	0	100	2	6	4	41
	IV	10	10	7	0	100	2	8	2	41
Consensus		c/a	A	G	G	T	a/g	A	G	t

[a] The influence of local $(C+G)\%$ composition on human donor splice signal composition is illustrated. The percentage frequency of each nucleotide at each position in the donor splice signal is given for each of the four $(C+G)\%$ compositional subsets of the learning set $\mathcal{L}$ described in the footnote to Table 6. The consensus nucleotide(s) at each position is given at the bottom of the table. Positions $-3, -2, -1$ are the last three nucleotides of the exon; positions $+1, \ldots, +6$ are the first six nucleotides of the succeeding intron.

does not explain most of the dependencies in donor splice signals came from the observation that X^2 values for many of the same K_i versus N_j comparisons remained statistically significant when measured within particular compositional subsets of genes (not shown). If, instead, the dependencies are related to the specificity of pre-mRNA splicing (as will be argued below), then incorporation of such dependencies into a probabilistic model of the donor signal promises to provide improvements in predictive accuracy.

But what type of model should be used? To this point, three principal types of signal models have been described, but none seems particularly well suited to the present situation. Specifically, the independence assumption made by a WMM model is clearly inappropriate as is the assumption, made in derivation of the WWAM model, that nearby positions have similar composition. A first-order Markov (WAM) donor model is a partial solution, since it can account for the strong nearest-neighbor dependencies observed in Table 7, but such a model still fails to incorporate potentially important longer-range interactions. Therefore it appears that a new type of model is required. One possible solution to this problem is an approach called maximal dependence decomposition

(MDD), which seeks to account for the most significant non-adjacent as well as adjacent dependencies in a signal (see ref. [1], chapter 4) using an iterative subdivision of the sequence data. Other types of models which consider complex dependencies will be discussed in the last section.

8.1. Maximal dependence decomposition

The goal of the MDD procedure is to generate, from an aligned set of signal sequences of moderate to large size (i.e. at least several hundred or more sequences), a model which captures the most significant dependencies between positions (allowing for non-adjacent as well as adjacent dependencies), essentially by replacing unconditional WMM probabilities by appropriate conditional probabilities provided that sufficient data are available to do so reliably. Given a dataset $\mathcal{D}$ of aligned sequences of length λ, the first step is to assign a consensus nucleotide or nucleotides at each position. Then, for each pair of positions i, j with $i \neq j$, the chisquare statistic $X_{i,j}^2$ is calculated for the K_i versus N_j contingency table as described previously (e.g. Table 4b). If no significant dependencies are detected (for an appropriately stringent P-value), then a simple WMM model should be sufficient. If significant dependencies are detected, but they are exclusively or predominantly between adjacent positions, then a WAM model may be appropriate. If, however, there are strong dependencies between non-adjacent as well as adjacent positions (as was observed in Table 7), then the MDD procedure is carried out.
(1) Calculate, for each position i, the sum $S_i = \sum_{j \neq i} X_{i,j}^2$ (the row sums in Table 7), which is a measure of the amount of dependence between the variable K_i and the nucleotides at all other positions of the signal. (Under the null hypothesis of pairwise independence between positions, S_i will have approximately a χ^2 distribution with $3(\lambda - 1)$ degrees of freedom.)
(2) Choose the value i_1 such that S_{i_1} is maximal and partition $\mathcal{D}$ into two subsets: $\mathcal{D}_{i_1}$, all sequences which have the consensus nucleotide(s) at position i_1; and $\mathcal{D}_{\bar{i}_1} (= \mathcal{D} \setminus \mathcal{D}_{i_1})$, all sequences which do not. (The name of the method derives from this step, of course.)
Now repeat the first two steps on the subsets $\mathcal{D}_{i_1}$ and $\mathcal{D}_{\bar{i}_1}$ and on subsets thereof, and so on, yielding a binary "tree" (often called a decision tree) with at most $\lambda - 1$ levels (see Fig. 5). Interested readers are referred to Salzberg's chapter 10 in this volume for a tutorial on decision tree classifiers and some interesting applications to exon/intron discrimination. This process of subdivision is carried out successively on each branch of the tree until one of the following three conditions occurs:
(i) the $(\lambda - 1)$th level of the tree is reached (so that no further subdivision is possible);
(ii) no significant dependencies between positions in a subset are detected (so that further subdivision is not indicated); or
(iii) the number of sequences in a resulting subset falls below a preset minimum value M so that reliable conditional probabilities could not be determined after further subdivision.
Finally, separate WMM models are derived for each subset of the tree (WAM models could be used instead if sufficient data were available), and these are combined to form a composite model as described below.

The results of applying the MDD procedure to the 1254 donor sites of the learning set are illustrated in Fig. 5. The initial subdivision was made based on the consensus (G) at position +5 of the donor signal (see Table 7), resulting in subsets G_5 and H_5 (H meaning A, C or U) containing 1057 and 197 sequences, respectively. (In this section, U is used instead of T in order to emphasize that the interactions discussed below occur at the RNA level.) Based on the data of Table 2, the value $M = 175$ was chosen as a reasonable minimum sample size (giving typical parameter estimation errors in the range of 7–23%), so the set H_5 is not divided further. The subset G_5 is sufficiently large, however, and exhibits significant dependence between positions (data not shown), so it was further subdivided according to the consensus (G) at position -1, yielding subsets G_5G_{-1} (i.e. all donor sites which have G at position +5 and G at position -1) and G_5H_{-1} (all donors with G at position +5 and non-G nucleotides at position -1), and so on. Thus, the essential idea of the method is to iteratively analyze the data, accounting first for the most significant dependence present, and then for dependencies which remain after previously chosen dependencies have been accounted for by subdivision of the data.

The composite model for generation of signal sequences is then essentially a recapitulation of the subdivision procedure, as described below.
(0)   The (invariant) nucleotides N_1 ($=$G) and N_2 ($=$T) are generated.
(1) N_5 is generated from the original WMM for all donor sites combined.
(2a) If $N_5 \neq$ G, then the (conditional) WMM model for subset H_5 is used to generate the nucleotides at the remaining positions in the donor site.
(2b) If $N_5 =$ G, then N_{-1} is generated from the (conditional) WMM model for the subset G_5.
(3a) If ($N_5 =$ G and) $N_{-1} \neq$ G, then the WMM model for subset G_5H_{-1} is used.
(3b) If ($N_5 =$ G and) $N_{-1} =$ G, N_{-2} is generated from the model for G_5G_{-1}.
(4)   ... and so on, until the entire 9 bp sequence has been generated.

The result is a model quite different from any previously described in this chapter in which the dependence structure of the model differs for different sequences. For instance, in the case that N_5 is a nucleotide other than G, the MDD probabilities of nucleotides at other positions $i \neq 5$ depend only on the fact that N_5 was not G. In the case that $N_5 =$ G, on the other hand, the remaining nucleotides depend also on whether N_{-1} is or is not G and possibly on the identities of nucleotides at other positions. Thus, the model actually represents a combination of first-, second-, third- and even some fourth-order dependencies.

The performance of the MDD donor site model is compared to WMM and WAM models of this signal in Fig. 6, as measured on the test set $\mathcal{T}$ containing 338 true donor splice sites. Given the strong dependencies observed between adjacent positions (Table 7), it is not surprising to see that the WAM model gives better separation than the simpler WMM model. What is most striking about the figure, however, is the much greater separation between typical scores for true and false donor sites observed for the MDD model, indicating that this model is apparently much better at distinguishing (and assigning very low scores to) sites with poor potential as donor signals. To further compare the three models, the number of false positives (FP) was also determined at various levels of sensitivity (Table 9), as described previously for the acceptor site models. This comparison shows the clear superiority of the MDD model over the other two at all

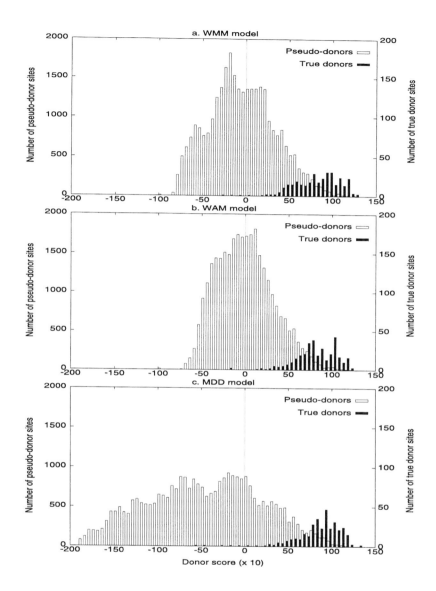

Fig. 6. The three donor splice signal models described in the text are compared in terms of their score histograms for true and false donor splice signals in the test set T of 65 human gene sequences (see text). Each position of each sequence in the set T was assigned a "score" by each model, defined as the logarithm base two of the signal ratio, $p^+(X)/p^-(X)$ (see text). Separate score distributions were then plotted for true donor sites (black bars) and "pseudo-donor" sites (open bars). Here, pseudo-donors are positions which have the requisite GT at positions +1, +2 but which are not annotated as donor splice sites. The height of each bar corresponds to the number of true or false sites observed at the given score level: note that separate scales were used for true (right scale) and false (left scale) donor sites. For convenience, the x-axis is in units of score times ten rather than score. (Modified from Burge 1997 [1]).

Table 9
Specificity vs sensitivity for donor splice signal models[a]

Model	Sensitivity level							
	20%		50%		90%		95%	
	FP	Sp	FP	Sp	FP	Sp	FP	Sp
WMM	68	50.0%	368	32.0%	2954	9.4%	4185	7.1%
WAM	79	49.6%	350	33.0%	2160	12.4%	4153	7.2%
MDD	59	54.3%	307	36.0%	1985	13.4%	3382	8.7%

[a] Accuracy statistics are given for the three donor splice signal models described in the text, as tested on the test set T of 65 human gene sequences. Each position of each sequence in the set T was assigned a "score" by each model, as discussed in the text. Accuracy statistics were then calculated as described for acceptor signal models in the footnote to Table 3. (From Burge 1997 [1].)

levels of sensitivity. However, as was the case for acceptor sites (Table 3), even the best model is still not sufficiently accurate to allow reliable prediction of isolated splice sites in genomic sequences. This finding suggests that a significant portion of the specificity of splicing is determined by factors other than the sequences at and immediately adjacent to the splice sites. Such factors might include, for example, the relative locations of potential splice signals in the pre-mRNA, or additional signals such as exonic or intronic splicing enhancers (e.g. refs. [13–15]). It should also be emphasized that much higher predictive accuracy can be achieved when other types of information such as compositional differences between coding and non-coding regions, frame/phase compatibility of adjacent exons and so on are used in an integrated gene model, e.g. refs. [41,2].

Aside from the improvement in predictive ability demonstrated above, the MDD procedure may also lend insight into the mechanism of donor splice signal recognition. Specifically, the data of Fig. 5 suggest some fairly subtle properties of the U1:donor signal interaction, namely:

(i) a *5'/3' compensation effect*, in which matches to consensus nucleotides at nearby positions on the same side of the intron/exon junction are positively associated, while poor matching on one side of the splice site is almost always compensated by stronger matching on the other side;

(ii) an *adjacent base-pair effect*, in which base pairs at the edge of the donor splice site appear to form only in the presence of adjacent base pairs; and

(iii) a G_3 *preference effect*, in which G is preferred at position +3 only for a subclass of strongly U1-binding donor sites.

The evidence for each of these effects is summarized below.

5'/3' compensation effect. First, G_{-1} is almost completely conserved (97%) in H_5 donor sites (those with a non-G nucleotide at position +5) versus 78% in G_5 sites, suggesting that absence of the G:C base pair with U1 snRNA at position +5 can be compensated for by a G:C base pair at position -1, with a virtually absolute requirement for one of these two G:C base pairs (only 5 of 1254 donor sites lacked both G_5 and G_{-1}). Second, the H_5 subset exhibits substantially higher consensus matching at position -2 ($A_{-2} = 85\%$ in H_5 versus 56% in G_5), while the G_5 subset exhibits stronger matching at positions +4 and +6. Similar compensation is also observed in the G_5G_{-1} versus G_5H_{-1} comparison:

the G_5H_{-1} subset exhibits substantially higher consensus matching at positions +6 (76% versus 42%), +4 (93% versus 70%) and +3 (100% R_3 versus 93%). Another example of compensation is observed in the $G_5G_{-1}A_{-2}$ versus $G_5G_{-1}B_{-2}$ comparison, with the $G_5G_{-1}B_{-2}$ subset exhibiting increased consensus matching at positions +4 and +6, but somewhat lower matching at position −3. Finally, the dependence between positions −2 and +6 described previously (Table 4) also falls into the category of a 5′/3′ compensation phenomenon. The most probable explanation for these effects relates to the energetics of RNA helix formation, in which adjacent base pairs with U1 snRNA (resulting from adjacent consensus matches) contribute greater stability due to favorable base stacking interactions. Conversely, when one side of the splice site matches poorly to the consensus, selection for matches on the other side might be increased in order to provide a sufficiently stable interaction with U1. Another possibility is that this compensation effect results from constraints acting at steps subsequent to the U1:donor interaction. In particular, the two sides (exon and intron) of the donor signal may be recognized separately by U5 and U6 snRNAs, respectively (e.g. ref. [4]), and it could be that at least one (but not necessarily both) of these interactions must be fairly strong for splicing to take place.

Adjacent base-pair effect. H_5 splice sites have nearly random (equal) usage of the four nucleotides at position +6, implying that base pairing with U1 at position +6 does not occur (or does not aid in donor recognition) in the absence of a base pair at position +5. The almost random distribution of nucleotides at position −3 of $G_5G_{-1}B_{-2}$ donor sites also suggests that base pairing with U1 snRNA at position −3 does not occur or is of little import in the absence of a base pair at position −2. These results again support the idea that the U1:donor interaction consists of a block of adjacent base pairs rather than isolated base pairs interrupted by small loops or bulges.

G_3 preference effect. Comparison of the relative usage of A versus G at position +3 in the various subsets reveals several interesting features. Perhaps surprisingly, G is almost as frequent as A at position +3 (45% versus 49%) in the entire set of donor sites, despite the expected increased stability of an A:U versus G:U base pair at position +3. Only in subset H_5 is a dramatic preference for A over G at position +3 observed (81% versus 15%), suggesting that only in the absence of the strong G:C base pair at position +5 does the added binding energy of an A:U versus G:U base pair at position +3 become critical to donor site recognition by U1 snRNA. On the other hand, in the most strongly consensus-matching donor site subset, $G_5G_{-1}A_{-2}U_6$, there is actually a strong preference for G_3 over A_3 (59% versus 27%)! Two possible explanations for this observation seem reasonable: either (1) there is selection to actually weaken the U1:donor interaction in these strongly matching sites so that U1 snRNA can more easily dissociate from the donor site to permit subsequent steps in splicing; or (2) G_3 is preferred over A_3 at some step in splicing subsequent to donor site selection (but this effect is only apparent when the strong constraints of U1 binding are satisfied by consensus matches at many other positions). It is also possible that this phenomenon may be related in some way to the previous observation (Table 8) that N_3 composition is strongly affected by local (C+G)% composition.

In summary, the MDD model not only provides improved discrimination between true and false donor sites by accounting for several potentially important non-adjacent as well as adjacent interactions, but may also give some insight into how the donor site is recognized. It may be of interest in the future to apply this method to other biological

162

signals, e.g. transcriptional or translational signals in DNA/RNA or perhaps even protein motifs. In many cases, however, use of this approach will have to be postponed until sufficiently large sets of sequences have accumulated so that complex dependencies can be reliably measured.

9. Conclusions and further reading

Previous sections have described a number of compositional features of pre-mRNA splicing signals and several types of probabilistic models of such signals based on presumed or observed patterns of dependence. In some cases, statistical properties of signal sequences can give important clues to the nature of the biochemical mechanisms which recognize and process such signals. Particular examples described above include the avoidance of AG in the 20 bases immediately upstream of acceptor sites; significant favoring of certain triplets and avoidance of others in the region near the branch point signal; and a number of properties of the donor splice signal, most notably the $5'/3'$ compensation effect, in which matches to the consensus on one side of the exon/intron junction are positively associated while mismatches on one side tend to be compensated for by stronger matching on the other side. In other cases, such as the K_{-3} versus N_j dependence described in acceptor sites, observed statistical interactions appear to result not from factors directly related to splicing, but instead from local compositional/mutational effects, emphasizing the need to consider a range of possible interpretations of observed correlations in sequence data. Another important conclusion of this work is that incorporation of observed dependencies in signals into more complex probabilistic models, such as the WWAM acceptor model and MDD donor model described above, can lead to significant improvements in the accuracy of signal prediction/classification, but that overly complex models with too many parameters perform less well.

Aside from the MDD approach described in the previous section, several other approaches to signal prediction which allow for potentially complex dependencies between positions in a signal sequence are available, each with its own particular strengths and weaknesses. Notable examples include HMMs, loglinear or logitlinear models, discriminant analysis and (artificial) neural networks, each of which is discussed briefly below. Of these approaches, the HMM model described by Henderson et al. [22] is most similar to the IMM/WAM models described here, and these authors also observed dependencies between adjacent positions in splice signals similar to those described in this chapter. Logitlinear (and loglinear) models are a fairly classical technique from statistics (e.g. chapter 2 of ref. [42]) which have recently been applied with promising results to the problem of identifying donor and acceptor splice signals in plant pre-mRNAs by Kleffe et al. [43]. Discriminant analysis is another approach with a long history in the statistical literature which has been applied in different forms to the problem of splice signal identification by Solovyev et al. [44] and Sirajuddin et al. [45], among others. Neural networks are an approach from machine learning which are also capable of accounting for complicated dependencies in a signal: notable applications of neural networks to pre-mRNA splicing signals include the work of Brunak et al. [46] and the recent improvements of this method described by Reese et al. [47].

All of these methods assign some sort of "score" to sequence segments which is used for prediction/classification into true versus false signals. However, in contrast to the methods described here, for some of these other approaches (e.g. neural networks) the scores do not have an explicit interpretation in terms of probabilities, $p^+(X)$ and $p^-(X)$, under signal and non-signal models. Thus, although many of these models make use of similar types of compositional properties of signal sequences, the scores obtained by different methods are in many senses non-equivalent. A consequence is that, although one can compare the accuracy of two approaches in terms of particular training and test sets of sequences, scores derived by different methods cannot in general be used interchangeably in gene finding algorithms, for example. A final point to be made is that, although each author has his or her favorite types of methods and particular methods give demonstrably better results than others in particular applications, no single type of model is likely to be appropriate for all types of biological signals.

The continued progress of genomic sequencing efforts presents significant opportunities and challenges for the computational analysis of sequence signals such as those involved in pre-mRNA splicing. On the one hand, the accumulation of larger numbers of experimentally verified splice signal sequences will present the opportunity to study more complex statistical properties of such signals, and the results may lead to new perspectives on splicing mechanisms. On the other hand, generation of huge quantities of unannotated genomic DNA sequences will also present significant challenges to computational biologists to develop accurate and efficient means of predicting important biological properties such as promoter, splice site and gene locations. In this chapter, a probabilistic approach to these sorts of problems has been advocated: a particular advantage of this approach is that the relative performance of different models can generally be compared in terms of relatively simple statistical properties such as the "positional odds ratios" described above. An important challenge for the future is to develop more flexible and sensitive approaches to the analysis of available sequence data which may allow detection of more subtle biological features. In the longer term, it may even be possible to construct realistic models of such complex biological processes as transcription and pre-mRNA splicing *in silico.*

Acknowledgments

I gratefully acknowledge Dr. Luciano Brocchieri and Dr. Steven Salzberg for helpful comments on the manuscript, and Dr. Samuel Karlin for his encouragement and support. This work was supported in part by NIH Grants 5R01HG00335-09 and 2R01GM10452-32 and NSF Grant DMS-9403553-002.

References

[1] Burge, C. (1997) Identification of genes in human genomic DNA, PhD thesis. Stanford University, Stanford, CA.
[2] Burge, C. and Karlin, S. (1997) J. Mol. Biol. 268, 78–94.

[3] Staley, J.P. and Guthrie, C. (1998) Cell 92, 315–326.

[4] Moore, M.J., Query, C.C. and Sharp, P.A. (1993) In: R.F. Gesteland and J.F. Atkins (Eds.), The RNA World. Cold Spring Harbor Lab. Press, Plainview, NY, pp. 305–358.

[5] Nilsen, T.W. (1998) In: R. Simons and M. Grunberg-Manago (Eds.), RNA Structure and Function. Cold Spring Harbor Lab. Press, Cold Spring Harbor, NY, pp. 219–307.

[6] Wang, J. and Manley, J.L. (1997) Curr. Opin. Genet. Dev. 7, 205–211.

[7] Zhuang, Y. and Weiner, A.M. (1990) Gene 90, 263–269.

[8] Siliciano, P.G. and Guthrie, C. (1988) Genes Dev. 2, 1258–1267.

[9] Heinrichs, V., Bach, M., Winkelmann, G. and Luhrmann, R. (1990) Science 247, 69–72.

[10] Query, C.C., Moore, M.J. and Sharp, P.A. (1994) Genes Dev. 8, 587–597.

[11] Konarska, M.M. and Sharp, P.A. (1987) Cell 49, 763–774.

[12] Will, C.L. and Lührmann, R. (1997) Curr. Opin. Cell Biol. 9, 320–328.

[13] Watakabe, A., Tanaka, K. and Shimura, Y. (1993) Genes Dev. 7, 407–418.

[14] Huh, G.S. and Hynes, R.O. (1994) Genes Dev. 8, 1561–1574.

[15] Carlo, T., Sterner, D.A. and Berget, S.M. (1996) RNA 2, 342–353.

[16] Neyman, J. and Pearson, E.S. (1933) Phil. Trans. Roy. Soc. Ser. A 231, 289–337.

[17] Berger, J.O. (1980) Statistical Decision Theory and Bayesian Analysis. Springer, New York.

[18] Gelfand, M.S. (1995) J. Comput. Biol. 2, 87–115.

[19] Staden, R. (1984) Nucleic Acids Res. 12, 505–519.

[20] Stormo, G.D., Schneider, T.D., Gold, L. and Ehrenfeucht, A. (1982) Nucleic Acids Res. 10, 2997–3011.

[21] Zhang, M.Q. and Marr, T.G. (1993) Comput. Appl. Biosci. 9, 499–509.

[22] Henderson, J., Salzberg, S. and Fasman, K.H. (1997) J. Comput. Biol. 4, 127–141.

[23] Taylor, H.M. and Karlin, S. (1984) An Introduction to Stochastic Modeling. Academic Press, San Diego, CA.

[24] Sharp, P.A. and Burge, C.B. (1997) Cell 91, 875–879.

[25] Altschul, S.F., Gish, W., Myers, E.W. and Lipman, D.J. (1990) J. Mol. Biol. 215, 403–410.

[26] Breathnach, R. and Chambon, P. (1981) Annu. Rev. Biochem. 50, 349–383.

[27] Mount, S.M. (1982) Nucleic Acids Res. 10, 459–472.

[28] Karlin, S. and Burge, C. (1995) Trends Genet. 11, 283–290.

[29] Nussinov, R. (1981) J. Biol. Chem. 256, 8458–8462.

[30] Senapathy, P., Shapiro, M.B. and Harris, N.L. (1990) Meth. Enzymol. 183, 252–278.

[31] Reed, R. (1989) Genes Dev. 3, 2113–2123.

[32] Smith, C.W., Porro, E.G., Patton, J.G. and Nadal-Ginard, B. (1989) Nature 342, 243–247.

[33] Harris, N.L. and Senapathy, P. (1990) Nucleic Acids Res. 18, 3015–3019.

[34] Hornig, H., Aebi, M. and Weissmann, C. (1986) Nature 324, 589–591.

[35] Noble, J.C., Prives, C. and Manley, J.L. (1988) Genes Dev. 2, 1460–1475.

[36] Webb, J.C., Patel, D.D., Shoulders, C.C., Knight, B.L. and Soutar, A.K. (1996) Hum. Mol. Genet. 5, 1325–1331.

[37] Maslen, C., Babcock, D., Raghunath, M. and Steinmann, B. (1997) Am. J. Hum. Genet. 60, 1389–1398.

[38] Coolidge, C.J., Seely, R.J. and Patton, J.G. (1997) Nucleic Acids Res. 25, 888–896.

[39] Roscigno, R.F., Weiner, M. and Garcia-Blanco, M.A. (1993) Nucleic Acids Res. 25, 888–896.

[40] Bernardi, G. (1995) Ann. Rev. Genet. 29, 445–476.

[41] Burset, M. and Guigó, R. (1996) Genomics 34, 353–367.

[42] Bishop, Y.M.M., Fienberg, S.E. and Holland, P.W. (1975) Discrete Multivariate Analysis: Theory and Practice. MIT Press, Cambridge, MA.

[43] Kleffe, J., Hermann, K., Vahrson, W., Wittig, B. and Brendel, V. (1996) Nucleic Acids Res. 24, 4709–4718.

[44] Solovyev, V.V., Salamov, A.A. and Lawrence, C.B. (1994) Nucleic Acids Res. 22, 5156–5163.

[45] Sirajuddin, K., Nagashima, T. and Ono, K. (1993) Comput. Appl. Biosci. 11, 349–359.

[46] Brunak, S., Engelbrecht, J. and Knudsen, S. (1991) J. Mol. Biol. 220, 49–65.

[47] Reese, M.G., Eeckman, F.H., Kulp, D. and Haussler, D. (1997) J. Comput. Biol. 4, 311–324.

S.L. Salzberg, D.B. Searls, S. Kasif (Eds.), *Computational Methods in Molecular Biology*

CHAPTER 9

Evolutionary approaches to computational biology

Rebecca J. Parsons

Department of Computer Science, University of Central Florida, P.O. Box 162362,
Orlando, FL 32816-2362, USA. Phone: 1+407-823-5299; fax: 1+407-823-5419; rebecca@cs.ucf.edu

1. Introduction

Many of the problems posed by the current explosion of biological sequence data require tremendous computational resources to solve exactly. These resources are not always readily available, though, and are certainly not to be found in many molecular biology laboratories. However, these problems are still important for molecular biologists to be able to solve. Hence, alternative computational approaches are constantly being explored. Evolutionary algorithms are one possible tool for addressing such problems. These algorithms approach computational problems using the techniques of survival of the fittest and natural selection to evolve solutions to particular problem instances. These approaches are attractive for attacking computationally challenging problems and computational problems that are not yet well understood. Caution must be used, however, since these approaches require an understanding of how the different components of the algorithms interact with the problem and how these interactions affect the performance of the evolutionary approach.

Computer scientists traditionally classify problems based on their computational complexity. Many of the interesting problems arising in the analysis of biological sequence data and elsewhere are, at best, in the class of problems termed NP-hard (see chapter 2 for a description of NP-completeness and NP-hardness). In general, problems are NP-hard because the number of possible alternatives that must be examined grows exponentially with the size of the input. For example, pairwise sequence alignment is not NP-hard. The complexity of this problem grows in proportion to the product of the length of the two strings. The problem addressed in this paper, the sequence assembly problem for shotgun sequencing of DNA, is NP-complete. The growth of the complexity is exponential in the number of fragments because solutions to the assembly problem are orderings of the fragment set. There are an exponential number of possible orderings, requiring an exponential amount of time to determine which ordering is the best. Thus, for a problem with 50 fragments, an optimal algorithm would require in excess of 3570 years to find the solution. Since a typical shotgun assembly project can have over 10 000 fragments [1], it is clear that an exact solution is infeasible. Instead, we must use an approximation algorithm that will give us a solution that is almost always good enough.

Many different approximation algorithms have been used on NP-hard problems, including greedy algorithms, heuristic algorithms, Monte Carlo methods, simulated annealing (inspired by the physical annealing process) and evolutionary approaches [2–4]. This chapter examines the evolutionary approaches, each inspired in some way by the processes of natural evolution.

1.1. Optimization problems

Many sequence analysis problems are optimization problems; we want to find the best answer to a particular question. The formulation of an optimization problem has two primary components: the search space and the fitness function. The search space for an optimization problem is simply the set of all possible solutions, or answers, to the question. The fitness function is the measure used to determine how good a particular answer is; therefore, the fitness function determines which answer is the best. An optimization problem, then, becomes a problem of searching the space of all possible answers and finding the best one.

Analogous to the genotype–phenotype distinction known in the genetics community, a particular solution to a problem is different from a representation of that solution. These differences are sometimes meaningless. In particular, if there is a one-to-one correspondence between solutions and their encodings as solutions, the differences become less relevant. In many practical problems, there are competing representation schemes for the problem. Additionally, there might be multiple representations for the same solution under a particular representation scheme. The particular representation choice determines the search space for the problem.

The fitness function component drives the optimization process. The function specifies, indirectly, the right answer or answers. The representation choice affects the difficulty of finding the answer for a particular fitness function, as the same representation choice can be good or bad depending on the fitness function. When looking at a problem, then, the fitness function and the type of optimization method to be employed affect the choice of a representation scheme.

Once a problem is characterized within this optimization framework, there are several possible approaches that can be employed. A complete characterization of all optimization algorithms is far beyond the scope of this chapter. Much research in the field of operations research is devoted to these sorts of problems. Some general approaches to optimization problems include greedy algorithms [4], simulated annealing [5,6], tabu search [7,8], Lagrangian relaxation [9,10], artificial neural networks [11,12], and the evolutionary approaches discussed here. Reeves [2] contains an overview of several of these techniques.

This chapter reviews the different evolutionary approaches to optimization problems. Then, we present our results on applying one of these approaches, genetic algorithms, to the problem of DNA sequence assembly. Section 2 introduces the various evolutionary approaches that have been applied to optimization problems. This section focuses on genetic algorithms, since that is the technique applied here. Section 3 describes the DNA fragment assembly problem as we have characterized it. Section 4 describes the process involved in developing the genetic algorithm we used for the fragment assembly problem. Section 5 introduces some modifications to the genetic algorithm that can address certain problems with the particular genetic algorithm selected. In section 6 we describe the direction our research is taking with respect to genetic algorithms and DNA sequence analysis.

2. Evolutionary computation

The way species have evolved is a form of optimization of an admittedly unknown, varying and complex fitness function. All individuals of a species are potential solutions to this fitness function. The theory of evolution states that nature uses the processes of natural selection, survival of the fittest, random mutation, and the passing of genetic material from parents to children as a mechanism to search for individuals that are better able to survive in the natural environment. Evolutionary algorithms use these same notions to evolve solutions to problems in artificial environments.

Evolutionary algorithms have been applied in many different contexts, including planning, scheduling, traveling salesman, fragment assembly, classification of genetic sequences, immune system modeling, computer security, and modeling of predator/prey relationships, to name just a few. Evolutionary algorithms do not need a lot of problem-specific information to perform adequately. However, practice has shown that exploiting problem-specific information in the selection of the problem representation and the operators can be quite useful. This point will be explored further in section 4.

2.1. Landscape of evolutionary approaches

Evolutionary algorithms explore the search space of a problem by working from individuals, with each individual representing an encoding of a potential solution. Evolutionary algorithms modify individuals using artificial operators inspired by natural mechanisms. Then, these individuals compete on the basis of their value under the fitness function. Selected individuals reproduce and live into the next generation. The major classes of evolutionary algorithms are genetic algorithms [13,14], evolutionary strategies [15] and evolutionary and genetic programming [16,15].

These different evolutionary approaches have many features in common. They are mostly population based, meaning that they operate over several individuals at the same time. In addition, they all use the fitness of an individual in some way as a means of selection. Finally, all use operators that are inspired by genetics. There are many variants within each of the major classes of algorithms, some of which appear to belong more appropriately to a different class. However, the basic differences among the approaches can be broadly organized as follows:
- representation choice;
- how the operators are used, and which operators are used;
- fixed- or variable-sized individuals; and
- deterministic versus probabilistic selection.

Table 1 summarizes the characteristics of the major evolutionary approaches. We will look at each of the approaches briefly, before focusing for the remainder of the chapter on genetic algorithms.

Table 1
Landscape of evolutionary approaches

	Rep. choice	Operators used	Size	Selection mechanism
Genetic algorithms	Bit string	Crossover*, mutation	Fixed	Probabilistic
Genetic programming	Syntax trees	Crossover*, mutation	Variable	Probabilistic
Evolutionary strategies	Vector of floats	Mutation*, crossover	Fixed	Deterministic

* denotes the emphasized operator.

2.1.1. Genetic algorithms

More than 20 years ago, Holland [13] introduced genetic algorithms as a way of understanding adaptive systems. The genetic algorithm (GA) formulation he introduced is often referred to as the canonical genetic algorithm. His focus was on understanding adaptive systems and examining the role and importance of the crossover operator in such systems. Indeed, his original vision for genetic algorithms did not include utilizing them as an approach to optimization. Genetic algorithms were popularized by Goldberg, whose approachable book on genetic algorithms has become a standard reference [14].

The primary features of the canonical genetic algorithm, which are explained in more detail below, are the following:
- bit-string representation;
- fitness-proportionate, probabilistic selection;
- heavy reliance on crossover;
- random, low probability mutation;
- the Schema Theorem and the Building Block Hypothesis.

In general terms, the genetic algorithm works by taking two relatively good individuals and combining them, using crossover, to create two new individuals. Under the assumption that the parents are good based on some good pieces in their genomes, these good features should combine, in at least some offspring, resulting in individuals which include the good features of both parents. Of course, other offspring will be inferior, but these inferior individuals will have a low probability of surviving. The general goal of the crossover operator is to exploit the good genetic material in the population. Random mutation allows the genetic algorithm to explore new regions of the search space by introducing new genetic material into the population.

The Schema Theorem and the Building Block Hypothesis form the core of the theoretical underpinnings of the genetic algorithm. The Building Block Hypothesis formalizes the informal notions just described about how a genetic algorithm searches for good solutions. A building block is a short, contiguous piece of an individual that has good fitness. The Building Block Hypothesis states that the genetic algorithm works by combining smaller building blocks in different parents together to form larger and better building blocks. Crossover is seen as the general mechanism for convergence in the genetic algorithm.

The Schema Theorem addresses the issue of how trials are allocated to different parts of the search space. Holland defined the notion of a schema to represent a region of the search space that shares some characteristic. Schemes are strings over a three character alphabet, consisting of 0, 1, and ∗, with the ∗ as a wild card character. For example, the schema 0∗∗0 represents all length 4 strings over the alphabet 0,1 that both begin and end in a 0. The Schema Theorem shows that the genetic algorithm allocates an increasing number of individuals (trials) to a schema with above average fitness. Thus, the dynamics of the genetic algorithm drive the population towards a schema with higher fitness values. Goldberg provides a more complete, yet still approachable, discussion of the Schema Theorem and its implications for genetic algorithms [14].

The processing of the genetic algorithm proceeds as follows.
(1) Randomly initialize the population of n individuals with binary strings of length l.
(2) Repeat until the population converges.
   (a) Apply the crossover operator and the mutation operator to the individuals in the population. Children replace their parents in the population.
   (b) Evaluate the fitness of all individuals in the population.
   (c) Compute the average fitness f^* over all individuals in the population.
   (d) For individual i, the expected number of copies of i in the next generation is f_i/f^*. Using these expectations, create a new population of n individuals.

Each individual in the population is a string of bits (0 or 1). An interpretation function is applied to this individual to map it to the abstract solution space. The interpretation function may be as simple as translating the bit string to a series of integers or floating point numbers. Thus, the interpretation function recovers the particular solution from the individual. The fitness function often implicitly includes the interpretation function, which decodes the solution from an individual. In the genetic algorithm, individuals are selected at random to reproduce. The reproduction/selection phase selects individuals for the next generation using a probability distribution based on the relative fitness of the individual. This selection is probabilistic; a less-fit individual still has a non-zero probability of surviving into the next generation. The genetic algorithm primarily uses crossover to drive the population changes; standard crossover rates are 60–80%, meaning that most of the individuals in a population participate in crossover and are replaced by their children. Mutation typically is applied at very low rates, for example 0.1–0.5%.

The canonical genetic algorithm applies operators blindly. Selection relies only on relative fitness, not on the details of the individual. Crossover and mutation affect randomly selected bits in the individual and randomly selected individuals.

Mutation in the canonical genetic algorithm is quite simple. A bit is chosen at random to be altered, and this bit position is flipped (0 becomes 1, or 1 becomes 0). Crossover combines the genetic material from two parents to create two children. In the simplest form of crossover, one-point crossover, a bit position, i, is selected. For parent individuals $p_1, p_2, \ldots, p_i, p_{i+1}, \ldots p_L$ and $q_1, q_2, \ldots, q_i, q_{i+1}, \ldots q_L$, the crossover operators creates, using crossover point i, the children $p_1, p_2, \ldots, p_i, q_{i+1}, \ldots, q_L$ and $q_1, q_2, \ldots, q_i, p_{i+1}, \ldots, p_L$. Figure 1 demonstrates this operation graphically.

As these descriptions show, the canonical genetic algorithm exploits no problem-specific heuristics in the application of the operators. Consequently, it is easy to analyze the canonical genetic algorithm from a theoretical perspective, and it is applicable to any problem in this form. This approach allows exploration of an unknown problem; the genetic algorithm can uncover properties of the search space and the problem that can be exploited in a more problem-specific way. However, as discussed in later sections,

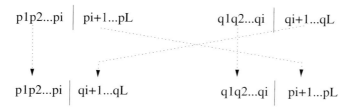

Fig. 1. Single point crossover at location *i*. This operation creates two children from two parents.

this generality can render the canonical genetic algorithm too inefficient for practical applications.

Many of the applications of genetic algorithms to biological problems utilize variants of the standard genetic algorithm to increase the efficiency of the genetic algorithm, the quality of the solution, or both. Many use a more complex representation, in conjunction with specialized operators. Several variations of the selection mechanisms, notably the steady-state genetic algorithm [17,18], have also been used. In computational biology, genetic algorithms have been applied to physical mapping [19,20], and protein structure prediction [21–23].

2.1.2. Evolutionary strategies

Evolutionary strategies were first introduced in Germany by Schwefel and Rechenberg, at approximately the same time Holland introduced genetic algorithms in the United States (Bäck discusses these works in English; the original citations are in German [15]). Evolutionary strategies developed independently of genetic algorithms, although there are some striking similarities and the approaches are converging.

The initial formulation of evolutionary strategies was not population based and focused solely on mutation. In an evolutionary strategy, there is a parent individual, consisting of a vector of real numbers. Mutation is applied to each component of the vector using a normal distribution, with mean zero and the same standard deviation. If the child's fitness exceeds the parent's, the child replaces the parent; otherwise, the process continues. This formulation of evolutionary strategies is termed (1+1)-ES, denoting a population size of one and one child being produced from which to select the new population. Rechenberg has proposed a multi-membered evolutionary strategy, termed $(\mu+1)$-ES, where μ is the number of parents which are combined to form one offspring. The resulting offspring is mutated and then replaces the worst parent, assuming the offspring's fitness is better than that of the worst parent. This strategy has been generalized by Schwefel to the (μ, λ)-ES and $(\mu + \lambda)$-ES. In $(\mu + \lambda)$-ES, μ parents generate λ offspring, with the next generation being chosen from the set of parents and offspring. In (μ, λ)-ES, parents are not eligible to move into the next generation. Most research currently focuses on the (μ, λ)-ES approach.

The primary differences, then, between genetic algorithms and evolutionary strategies are the representation, the emphasized operator, and the selection mechanism. The canonical genetic algorithm uses only bit representations, while evolutionary strategies uses only real-valued vectors. Genetic algorithms emphasize crossover and use mutation sparingly, while evolutionary strategies emphasize mutation, but do use some form of recombination, in all but the (1+1)-ES formulation. Selection in genetic algorithms is probabilistic while it is deterministic in evolutionary strategies. One very interesting characteristic of the evolutionary strategies approach is the introduction of the standard deviations for mutation. Thus, the way mutation occurs is also subject to evolutionary mechanisms.

Evolutionary strategies have been applied to many different kinds of problems, including component design [24], network optimization [25] and visual systems [26].

2.1.3. Genetic programming

The automatic generation of computer programs that solve a particular problem is a goal for many researchers. Genetic programming uses the same basic evolutionary mechanisms as genetic algorithms to address this need. The particular formulation of evolving programs addressed here is that developed by Koza [16]. L.J. Fogel [27,28], D.B. Fogel [29,30] and Cramer [31] proposed similar approaches prior to Koza's work, but the genetic programming approach of Koza currently receives the most attention.

The basic approach in genetic programming is the same as that for genetic algorithms. A population of individuals is altered using the genetic operators of crossover and mutation. Selection is on the basis of the fitness of the individuals. The goal is to find a solution that performs well, based on the fitness function. The differences, however, have a significant impact on how one approaches a genetic programming application, the tasks for which one uses genetic programming, and how much is understood about genetic programming.

The first obvious difference between genetic programming and genetic algorithms is that the individuals are program trees. When one sets up a genetic programming application, the set of primitive functions that are available to an individual, the data domains for these functions, and the different mechanisms for combining these functions must all be chosen. The fitness function for an individual includes a set of input/output pairs that characterize a piece of the desired program behavior. The individual program is evaluated on the test set and its input/output behavior on the test set determines its fitness value.

Crossover and mutation operators for genetic programming must be chosen to maintain legal trees and to account for the biases in random selection arising from the changing size of individuals. Different kinds of selection mechanisms such as rank-based selection are often employed in genetic programming applications [17,18]. Genetic programming has been applied to several different protein classification problems [32,33] as well as in other settings.

2.2. A primer on genetic algorithms

Above we described the canonical genetic algorithm. Next, we generalize this notion of a genetic algorithm to match more closely with the practice of genetic algorithms. The major components of a genetic algorithm are the fitness function, the representation mapping, the genetic operators, the selection mechanisms, and the control parameters. Each of these components must be chosen with the others in mind, as each has the ability to affect adversely the performance of the genetic algorithm for a particular problem.

172

2.2.1. Fitness function

The fitness function embodies the essential aspects of the problem to be solved. The genetic algorithm strives to find the best input value for the given fitness function. Different characteristics of the fitness function affect how easy or difficult the problem is for a genetic algorithm. Research into deceptive functions [34] discusses different aspects of the interaction between fitness functions and search strategies. This component is least likely to be subject to a choice on the part of the genetic algorithm designer, although our own research has examined the use of different fitness functions for the same problem [35,36].

Within the bounds of the problem, however, there are some properties of fitness functions that are desirable from the perspective of the genetic algorithm and over which the designer may have some choice. A crucial factor is the degree to which the fitness functions discriminate between individuals. An extreme example of a bad fitness function is one that assigns zero to all individuals except the best one, which gets a significant fitness value. The genetic algorithm receives no information from the fitness function about the relative merits of all the other individuals and thus must effectively do a random search through the space. Additionally, it is desirable for individuals with significant shared characteristics to have similar fitness values. Finally, as the discussion surrounding deceptive functions has shown, the fitness function should point the genetic algorithm towards the correct value, rather than away from it.

2.2.2. Representation

Choosing a representation for a problem is a critical design decision. Our early work began with a poor representation choice and the results show the degree to which this can affect the performance of the genetic algorithm [35]. The chosen representation for the problem defines the search space. This space should obviously include the desired (right) answer. In addition, the representation should be appropriate for the problem, and specifically should help preserve the building blocks of the problem. Representation choices include the ordering of different parameters in an individual, and the use of bits, or floating point or integers numbers as the basic unit of the representation. Although the canonical genetic algorithm uses only bit strings as individuals, the practice of genetic algorithms utilizes more complex representations, such as vectors of floating point or integer numbers, permutations, or alternatively coded integers [17,14].

2.2.3. Genetic operators and selection mechanism

The canonical genetic algorithm includes a crossover and a mutation operator. Each of these operate blindly on the bit representation. Many genetic algorithm applications use specialized operators that are tailored to the representation or to the problem For example, significant work has been done on finding operators that work on permutation representations [37]. These operators are designed to create permutations from permutations; standard crossover and mutation operators are not guaranteed to generate a legal permutation from two parent permutations. It is also possible, as we did in our work, to design operators that exploit some feature of the problem. The genetic algorithm for protein structure prediction is one such application [21].

2.2.4. Control parameters

The genetic algorithm has numerous parameter settings, including population size, crossover rate, and mutation rate. Proper setting of these parameters can significantly improve performance. Unfortunately, there is little concrete advice on how to set these parameters, although folk wisdom provides some guidance [14]. In section 5.1 we describe the application of statistical techniques to this question [38].

2.3. Putting the pieces of the genetic algorithm together

The interaction of each of the genetic algorithm components affects the ability of the genetic algorithm to search the space of available solutions. For example, too small a population leads to premature convergence as it is not possible to maintain enough diversity in the population. A representation that interacts poorly with the operators in the retention of building blocks renders useless the mechanisms the genetic algorithm uses to explore the search space. A fitness function that does not adequately discriminate between different solutions, providing an essentially flat landscape, gives the genetic algorithm insufficient information on which to base its decisions. Therefore, the design of an efficient genetic algorithm to solve a particular problem necessitates some understanding of how the individual components will work together.

Recalling the Schema Theorem and the Building Block Hypothesis, a genetic algorithm performs an effective search by locating the good building blocks in a population, with these good building blocks being identified through the fitness function of the whole individual. Individuals with these good building blocks increase their presence in the population. Meanwhile, the crossover operator combines the (presumably) good building blocks from two different individuals in a population, creating individuals that combine the genetic characteristics of both parents. In this way, larger and better building blocks should be formed from the parents' smaller building blocks. Mutation helps to maintain diversity, and the probabilistic nature of the algorithm broadens the space of exploration, thus increasing the likelihood of avoiding local extrema. Therefore, it is obvious why all the components of the genetic algorithm must be considered together. Their interaction is the primary driver of effective performance of the genetic algorithm.

However, a genetic algorithm is also capable of explicating just how these components should work together. Because of the general natural of the representation and the operators, the genetic algorithm can be used as a tool for exploring the nature of a problem and for uncovering the critical interactions required for effective solution by a genetic algorithm or by another approach. Since genetic algorithms can function with no domain-specific knowledge, they are quite useful in learning enough about a particular problem to allow for the formulation of a better solution. Indeed, this chapter documents our use of the genetic algorithm in this manner. We began with a particular representation choice that was ineffective for our problem. A change in representation and operators gave us a genetic algorithm that performed well on smaller problems but that did not scale well. However, that genetic algorithm also gave us clues as to the additional operators that were required to increase the performance of the genetic algorithm.

Thus, the complex interactions of the genetic algorithm components and the generality of the approach are both a strength and a weakness. Proper understanding of the approach

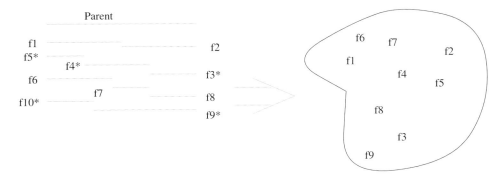

Fig. 2. DNA sequence assembly with $2\times$ coverage. The asterisk denotes a fragment from the anti-parallel strand. After the fragmentation process, strandedness, orientation and placement information is lost.

allows one to avoid the weakness and exploit the strength of the genetic algorithm approach.

3. Assembling the DNA sequence jigsaw puzzle

The news is filled with successful application of DNA sequence analysis. The analysis of human DNA sequences has resulted in the discovery of genes which indicate increased susceptibility to diseases and others which indicate increased resistance to diseases. These discoveries each require the sequencing of numerous strands of DNA. Thus, it is easy to see the importance of making this process as efficient as possible. Improving sequencing involves both laboratory processes and computational processes. Large scale sequencing [39,40] attempts to take numerous short strands of DNA from a longer sequence and from them recover the longer sequence.

Unfortunately, current DNA sequencing techniques are not able to sequence DNA segments of the lengths required. Gel electrophoresis methods can reliably sequence fragments up to approximately 600 bases in length. This size is at least two orders of magnitude too small. Chromosomal DNA can be millions (in bacteria) or hundreds of millions (in higher eukaryotes) of bases long. So, another approach is needed. One such approach is called shotgun sequencing [41]. Essentially, this approach first makes multiple copies of the DNA in question, breaks up these copies into smaller pieces, sequences those smaller pieces and finally puts the pieces back together to yield the complete sequence. Figure 2 illustrates this process for $2\times$ coverage of the original sequence (i.e., each base is included in two different fragments). The $*$ by some of the fragments indicate they derive from the anti-parallel strand of the DNA. The strand orientation is lost when the fragments are produced.

The sequencing process uses as its basic assumption that highly similar fragments necessarily come from the same region of the parent DNA sequence. Similar fragments are assumed to overlap each other and are thus placed near each other in the assembly. In more detail, the basic sequencing process has the following steps.
(1) Replicate the DNA to create some number k of identical copies.

(2) Apply different enzymes to the different copies of the parent to split the long pieces into smaller fragments. This process separates the two strands of the parent sequence and loses the orientation of those separate strands. It also loses the relative placement of the individual fragments.

(3) Sequence each fragment. This step is not error-free.

(4) For each pair of fragments, compare the sequences of the two fragments to determine their similarity.

(5) Using this similarity information, order the fragments in a way that optimizes the placement of the fragments and respects the similarity.

(6) Apply an alignment algorithm to combine the individual fragment sequences and the order information from the previous step, resulting in a consensus sequence for the original parent strand.

The algorithm described in this chapter focuses on the ordering step. The input we consider is a set of fragment identifiers with a given base sequence and a computed similarity score, sometimes referred to as the overlap strength, for each pair of fragments. The higher the similarity score for a pair, the more these two fragments resemble each other. The desired result is a "sensible" ordering of the fragments that accounts for the similarities demonstrated by the fragment overlap strengths.

There are several other problems, some very familiar, that are related to the fragment ordering problem. From a computational perspective, the physical mapping problem is the same as the fragment ordering problem with simply a different similarity metric [42,43]. Outside the realm of computational biology, the familiar traveling salesperson problem (TSP) [44] is quite similar to the fragment assembly problem. An instance of the traveling salesperson problem includes a set of cities to be visited by the salesperson and the distance (cost) associated with traveling between any two pairs of cities. The objective is to find a route that visits each city exactly once and that is of the lowest cost. A route is simply an ordering of the cities. If one replaces fragments with cities and the similarities with the distances, the only difference between these problems is that the fragment ordering problem is a maximization problem, since we want to maximize the similarity of adjacent fragments.

TSP has received a great deal of attention from the optimization community and many excellent heuristic algorithms exist to solve instances of this problem [45]. However, these heuristics rely on properties of the distance function in TSP that do not hold for the fragment assembly problem, making these heuristics inapplicable. Several different techniques have been applied to the fragment ordering problem, including greedy algorithms, exact and heuristic algorithms, and simulated annealing [46–52]. A survey of different methods is given in Burks et al. [53]. The next section describes a series of genetic algorithms developed to address this ordering problem.

4. Design for DNA sequence assembly

As described earlier, designing a genetic algorithm for a particular problem requires selection of the representation, the fitness function, the genetic operators and the settings for the parameters. The fitness function we used is a simple one; the fitness of an individual is the sum of the pairwise overlap strengths for adjacent fragments. For a fragment set of size n, with the fragment in position i denoted f_i and the pairwise overlap strength of two fragments given by the function $w(f_i, f_j)$, the fitness function value is computed as follows:

$$\sum_{i=0}^{n-1} w(f_i, f_{i+1}).$$

This function has the advantage of being quick to compute (it takes time proportional to the size of the fragment set). However, it does nothing directly to penalize failing to utilize a particular adjacency, and thus does not necessarily properly discriminate between individuals. There is still debate as to the validity of the underlying assumptions of sequencing (see Myers [54]). The problem with this formulation arises as a result of repeated sequences of DNA. Under this model of assembly, all fragments from any of the repeated sequence will be placed overlapping one another, even though some are far removed from the others in the actual parent sequence. The simple fitness function suffices for these investigations, since this function captures the current practice of sequence assembly.

4.1. Choosing the right representation

From the perspective of the DNA sequence assembly problem, a solution is an ordering of the fragments. This ordering should account for the overlaps present between the fragments and should allow for the creation of the assembly through the alignment process. As should be clear from the description of the process, there are in general several orderings for the same fragment set that are usable as results. Thus, we do not have to find the best answer; instead we search for an acceptable ordering.

An ordering of the fragment set requires that all fragments be represented in the ordering exactly once. So, our solution is some permutation of the fragment identifier set. An obvious representation choice, then, is to use an array of fragment identifiers with legal solutions being permutations of the set of possible identifiers. However, the standard genetic algorithm operators are not closed for permutations. It is easy to see that both mutation and crossover, when applied to legal permutations, can easily generate individuals that are not permutations.

Thus, we have three choices.

(1) Use the array representation with the standard operators and have the fitness function penalize solutions that are not permutations.

(2) Use the array representation and specialized operators that are closed for permutations.

(3) Use a different representation such that the standard operators are guaranteed to generate permutations.

A different representation, referred to as both the random keys and the sorted order representation, has been applied to other permutation problems [55–57]. However, it has been shown that different classes of permutation problems require different representation choices [37]. This work has shown that some permutation problems

Individual	1 1 1 0	0 0 1 0	1 0 0 1	1 0 1 1	0 0 1 1	0 0 1 1
Decimal Number	14	2	9	6	11	3
Sort Order	5	1	3	2	4	
Intermediate Layout	2	4	3	5	1	
Final Layout	3	5	1	2	4	

Fig. 3. Example of the sorted order representation.

emphasize adjacencies, some absolute ordering and some relative ordering. As there were indications that the DNA sequence assembly problem had components of all of these classes, we wanted to explore the application of the sorted order representation to this problem.

The advantage of this representation is that every bit string of length $k \times (n + 1)$, for $2^k \geqslant n$, represents a legal permutation of the fragments numbered from 1 to n. The appropriate permutation is recovered by interpreting the bit string as $n + 1$ different keys, each of length k bits. The first n keys are sorted; the last key determines the starting position for the ordering. The position of a key in the individual determines the fragment to which that key applies for this bit string. The location of the key for a fragment in the sorted order determines where the fragment is placed in a preliminary ordering. As shown in Fig. 3, if the key value in position j of the individual appears in position i of the sorted list, then fragment j is in position i of the intermediate layout. The last key value in the individual determines the first individual in the final ordering; the ordering is rotated to make this fragment first. In the example, the key value in the fourth position, six, is in the second position in the sorted order. As a result, fragment four appears in the second position in the intermediate layout. The final key value, three, specifies that the third position becomes the starting position in the ordering.

The results of this experiment are described in more detail in [35]. To summarize, this representation choice was a poor one. For the sequence assembly problem, the most important information about a good solution, the building blocks, are adjacencies. A good ordering contains valuable adjacencies. Thus, our choice of representation and operators should preserve and exploit adjacency information. As we were attempting to explain the poor performance of the genetic algorithm, we discovered that the standard crossover operator, when applied to these permutations using the sorted order representation, was very disruptive of the adjacency information. This representation preserves relative ordering information, not adjacency information. Thus, this representation was increasing the probability that good building blocks would be broken down as opposed to preserved and combined to form larger building blocks.

A B C D E F and B C F E A D

Adjacencies	
A	B, D, E
B	A, C*
C	B*, D, F
D	C, E, A
E	D, F*, A
F	E*, C

yield

A E F C B D

Fig. 4. Permutation crossover operator. Shared adjacencies are denoted by asterisks.

4.2. Permutation-specific operators

Thus, we made a different choice. We went back to the standard permutation representation and used operators that are specifically designed to preserve and exploit adjacency information. These efforts, described in more detail in Parsons et al. [36], were much more successful. The initial attempts with this representation used a simple swap operation as the mutation operator. Specifically, each time mutation was to be applied, two fragments positions were chosen at random. These two fragments swapped positions unconditionally. For example, if the individual represented the ordering A–B–C–D–E, a swap of positions 2 and 4 would yield the individual A–B–E–D–C. It is easy to see that this operator is incapable of creating an individual that is not a permutation.

The permutation crossover operator is more complex. The strategy of the operator, however, is relatively simple. The goal is to maintain adjacencies while combining the information of two parents. (More complete descriptions can be found in Starkweather et al. [37] and Parsons et al. [36].) To accomplish this goal, the crossover function records the adjacency information from both parents. Then, the initial position of one of the parents is chosen as the initial position of the offspring. The crossover function selects the fragment to place next to fragment f in the offspring using the following rules.

(1) If both parents have the same fragment f' adjacent to fragment f, and fragment f' has not yet been placed in the individual, pick f'.
(2) Otherwise, if there is some fragment f' that is adjacent in one of the parents and has not yet been placed in the offspring, pick it.
(3) If there is no unplaced fragment adjacent to f in either parent, pick an unplaced fragment f'. The heuristic used is to pick the fragment with the most unplaced adjacent fragments in an attempt to reduce the number of further random picks that would be necessary. This same heuristic is used to break any other ties.

This process continues until all fragments have been placed into the offspring individual somewhere. Figure 4 demonstrates this process for two sample parents. The new individual starts with the first fragment from the first parent, fragment A. Fragment A has no shared adjacencies, but two of its adjacent fragments, B and E, both do. We choose E as the next fragment; E has a shared adjacency to F, so that is placed next. Fragment F has as its remaining unplaced adjacency fragment C, so that is placed into the individual. C's only unplaced adjacency is B, so that is placed. Finally, D is chosen since it is unplaced.

Because the rules specify unplaced fragments only, this operator is guaranteed to generate a legal permutation. In addition, it is easy to see how adjacency information is maintained. The priority of the operator is to exploit as much as possible the adjacency information present in the parents. The operator first uses those adjacencies present in both parents, then those present in at least one parent. Only if no adjacency information can be used is a random choice made.

The advantages of this operator from the perspective of the building blocks are obvious. However, this operator is clearly more time consuming than the standard crossover operator. Since crossover is applied a significant number of times in a genetic algorithm run, the efficiency of the operator is important. However, the improvement that this operator made in the convergence of the genetic algorithm to good solutions easily made up for the additional time spent in the crossover operation. The genetic algorithm solved the smaller data set readily and repeatedly. However, the genetic algorithm's performance did not scale well to larger data sets. The next stage of work involved analyzing the kinds of improper solutions found by the genetic algorithm. The next section describes these efforts.

5. Problem-specific operators

The use of an operator appropriate to an adjacency-biased permutation problem made a significant difference in the performance of the genetic algorithm. However, the genetic algorithm was still not able to scale to data sets of realistic size. In examining the behavior of the genetic algorithm and the solutions found, two recurring problems with the solutions became clear: large blocks in the wrong place, and large blocks in the reverse order. An example of the reverse block problem is shown in Table 2. The fragment ordering should be A–B–C–D–E–F–G. However, the genetic algorithm initially constructs small building blocks and then combines these. With the blocks A–B, E–D–C, and F–G, the correct solution A–B–C–D–E–F–G could have been constructed. However, in this ordering, the middle contig E–D–C is in the reverse orientation from the rest of the fragments.

This situation evolves because the genetic algorithm continually connects fragments based on their relative overlap and grows this contiguous stretch of fragments, referred

Table 2
Illustration of reversed contigs

	Overlap strengths							Possible solutions
	A	B	C	D	E	F	G	
A	250	175	75	0	0	0	0	
B	175	300	200	0	0	0	0	A-B E-D-C F-G
C	75	200	700	75	50	0	0	
D	0	0	75	100	75	0	0	should be
E	0	0	50	75	500	400	125	
F	0	0	0	0	400	620	300	A-B-C-D-E-F-G
G	0	0	0	0	125	300	420	

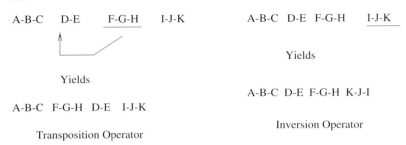

Fig. 5. Illustration of transposition and inversion operators.

to as a *contig*. However, because the overlap score is symmetrical and the strandedness of the fragments is lost, there is no bias for the block to be in one direction or the other until the rest of the fragments are organized into blocks. At this point, the fragments that span these blocks will determine the proper orientation. Visual inspection finds these locations, but there is no single operator that will make the required change in one step. Rather, correcting the error would require a number of different low probability swaps done in just the right way. As this is an unlikely scenario, the genetic algorithm becomes trapped on this local peak. The block placement problem evolves in a similar fashion, since the genetic algorithm concentrates first on getting the individual adjacencies right, and therefore does not consider the overall block ordering.

To address these two problems, we have developed two problem-specific operators, called transposition and inversion [36]. Each of these operators works on contig boundaries, which are defined as follows.

Definition: *Contig.* A contig in an individual is a stretch of adjacent fragments $f_i, f_{i+1}, \ldots, f_j$ such that the overlap strengths $w(f_k, f_k + 1)$ for $i \leqslant k \leqslant j - 1$ are non-zero and the overlap strengths $w(f_{i-1}, f_i)$ and $w(f_j, f_{j+1})$ are zero.

Referring again to Fig. 2, the contigs are A–B, E–D–C, and F–G. The correct solution has a single contig A–B–C–D–E–F–G.

The transposition operator randomly selects a contig to move, and moves it to a random contig boundary in the individual. The inversion operator randomly selects a contig and inverts that contig within the individual. Figure 5 shows both of these operators.

The performance with these new operators is shown in Table 3. This table shows the data for two different data sets. The first data set, POBF, contains 177 fragments and is a human apolipoprotein [58] sequence, with accession number M15421 in GenBank,

Table 3
Performance results for GA with crossover rate 0.3, swap rate 0.2, transposition rate 0.28, inversion rate 0.28, sigma scaling factor 2.0, elitist strategy

Data set name	Population size	Number of generations	Number of trials	Number of contigs
POBF	1500	13 K	5.9 Mil	1
AMCG	2500	5.6 K	2.3 Mil	13

consisting of 10 089 bases. The data set AMCG is the initial 40% (20 100) of the bases from LAMCG, the complete genome of bacteriophage lambda, accession numbers J02459 and M17233 [59]; the set contains 352 fragments. The test run for the second data set had ceased to improve, so the run was terminated. The genetic algorithm found the right ordering for the POBF data set, but it took a significant amount of time, and the indications were that the scaling behavior was unacceptable.

5.1. How to set the parameters

Introducing the two new operators added two more parameters to the genetic algorithm, the application rate for the new operators. The next logical step was an exploration of the space of parameter settings for the rates of the different operators. Initial investigation led us to a lower rate of crossover than is the norm [36]. To refine these values, the statistical techniques of design of experiments and response surface methodology were employed [60,61]. The details of this investigation appear in Parsons and Johnson [38].

The process began with a full-factorial design using every combination of a low setting and a high setting for each of the four operator rate parameters: crossover, swap, transposition and inversion. Multiple runs at each of these combinations were made, and several interesting trends arose. The first conclusion was that low rates of crossover performed significantly better than higher rates. Perhaps the most startling conclusion, though, was the lack of correlation between the different rate settings. Finally, although the presence of the transposition and inversion operators had already been shown to be crucial, the performance was insensitive to changes in the rate. As expected, low rates of swap gave better performance.

The next stage of this approach is to use a steepest descent to alter the rates using the decrements yielded by the full-factorial analysis. After a few runs, we found a large "mesa" of good parameter settings. The size of the mesa indicated that these parameter settings should be relatively robust. This indication was demonstrated as we applied these parameter settings to the larger data sets that our genetic algorithm had not yet been able to solve completely or efficiently. The results on the larger data sets and the parameter setting are shown in Table 4. The genetic algorithm found the correct solution to both problems. The scaling in the number of trials is far less worrying, although it is still not ideal. Most importantly, though, with relatively small population sizes and lower rate settings than expected, the genetic algorithm solved realistic DNA sequence assembly problems with no manual intervention.

Table 4
Performance results for GA with rates set using DOE/RSM. Crossover rate 0.1, swap rate 0.04, transposition rate 0.38, inversion rate 0.28, sigma scaling factor 2.0, elitist strategy

Data set name	Population size	Number of generations	Number of trials	Number of contigs
POBF	200	3393	500 117	1
AMCG	400	6786	2 Mil	1

6. *Good versus best answers*

The final issue that we have examined in our extensive study of this problem is that of convergence. As is typically the case with genetic algorithms, this genetic algorithm rapidly finds good solutions to the problem at hand. This solution is far better than average and is usually quite good. However, it is often not the best solution possible, and in some cases it is not even an acceptable one. The time from a good solution to the best solution is often much longer than desirable. This behavior was clearly manifested in the performance studies for this genetic algorithm.

The endgame problem, as it is sometimes called, is not unique to genetic algorithms [62]. The problem arises at the end of this problem for two reasons. First, by the later stages, the optimization process has found the right area of the search space. Being in the right place in the search space implies that greedy application of operators can be effective. "Greedy" approaches always take the step that has the greatest immediate benefit. The usual failure of these approaches is that they converge too quickly to the best local solution without seeing how this local area differs from the global space. However, once the right local area has been found, greedy operators are generally very efficient at selecting the best point in that area. Second, the behavior of the contig operators changes during the run. This result occurs because the individual contigs get larger as the individuals improve and the number of contigs declines. These two changes require different approaches to improve the performance of the genetic algorithm on large problems.

We have so far experimented with one greedy operator, a greedy swap operator. This operator randomly selects a fragment, f, in the ordering and then finds the fragment with which it has the strongest overlap, f'. An analysis of the local areas of each of these fragments determines whether f is moved to be next to f', or if f' is moved. We have so far only applied this operator late in the run. Use of this operator, however, appears to significantly decrease the time required to go from a good solution to the right solution. Early application of this operator would significantly alter the dynamics of the genetic algorithm, causing it to behave simply as a greedy algorithm. More experimentation is clearly required to determine when and how often to apply such a strategy.

The approach to the contig operators is more complex. The issue here is that fragments may be in an acceptable location without being in the right location within the interior of the contigs. This situation arises because of the redundancy in the data set. As was discussed earlier, the first step in the laboratory processing is to replicate the parent DNA sequence. It is not uncommon for there to be seven or more copies of the same stretch of DNA represented by different fragments in the data set. All these fragments share some similarity, and thus all these adjacent fragments will have non-zero overlaps and will thus be considered interior to the contig. However, while an overlap of 50, for example, is clearly different than random similarity, placing this fragment next to one with which it has a similarity score of 250 is likely a more sensible placement. However, the contig operators will not disrupt the interior of a contig. Indeed, we initially made the selection of the part of the individual these operators would effect completely random and found that this was not useful [36].

As a result, we are beginning to experiment with adaptively relaxing the definition of a contig boundary. Early in the run, the contig boundaries are found by looking for a zero overlap at both ends of the contig. Later in the run, these boundaries are probabilistically allowed to appear at non-zero, yet smaller, overlap strengths. These results are still preliminary, but are encouraging. Rearrangement of the interior of a contig produces a better arrangement of the fragments at the ends of the contig. There are several issues to be explored here, including when to begin the blurring of the boundaries, how quickly to blur them, and by how much.

7. Future directions

Many other directions remain to be explored, both for this genetic algorithm and for genetic algorithms in general. With regards to the DNA sequence assembly problem, the choice of fitness function is still problematic. Laboratory test data indicates strongly that reliance solely on the similarity metric employed here to sequence laboratory test sets will be inadequate. There is still significant human reasoning that goes into placing fragments when their computed similarity score is not different than random. Until this information can be included in the similarity metric, these algorithms will be unable to handle real-world data. The issue is not the size of the data set but the accuracy and completeness of the information used by the algorithm.

Another issue with the fitness function is the compression of the solutions resulting from the significant regions of repeated DNA that appear in real sequence data. Myers [54] has proposed a different model for assembly that addresses this issue. Burks et al. [63] have proposed using additional laboratory data, such as physical mapping data, to disambiguate these fragments. Clearly additional work is needed to capture more clearly the appropriateness of a particular ordering.

From the perspective of genetic algorithms, the use of adaptive rates is clearly important. For large problems, the dynamic behavior of the genetic algorithm changes, sometimes significantly, during the course of a run. As a result, it seems obvious that there could be significant opportunities to improve genetic algorithm performance by altering the rates and perhaps even the suite of operators applied during the different phases of the run. Several authors are currently looking at this issue, notably Booker [64].

Finally, our study of this particular problem has pointed out significant deviations from the standard theory of genetic algorithm behavior. The existing genetic algorithm theory centers on simple operators and bit string representations. Many of these theories are demonstrably invalid in the case of the permutation operators and representations. We and others are exploring different models of behavior for the permutation genetic algorithms. Whitley and Yoo have looked into this problem using an infinite population assumption [65]. We are examining it in the case of finite populations.

8. Conclusions

The latter sections of this chapter described the path we took in tuning a genetic algorithm for a particular problem. This path has given us insight into both the behavior of permutation genetic algorithms and the specifics of the DNA sequence assembly problem. The work has produced different approaches to resolving some of the deficiencies of genetic algorithms in addressing practical problems. Our hope is to incorporate these insights into a more general understanding of the behavior of genetic algorithms in the more complex situations in which they are now being applied.

Evolutionary algorithms offer a significant opportunity for addressing problems which have no current acceptable deterministic or heuristic solution. These algorithms utilize the basic principles of survival of the fittest and natural selection to locate better solutions to a problem using the information from an existing population of solutions. These techniques are quite general, and thus are useful for exploring properties of problems. As this chapter demonstrates, though, improved understanding of the problem can lead to much better representation and operator choices and thus to significant performance improvements.

References

[1] Fleischmann, R., Adams, M., White, O., Clayton, R., Kirkness, E., Kerlavage, A., Bult, C., Tomb, J.-F., Dougherty, B., Merrick, J., McKenney, K., Sutton, G., FitzHugh, W., Fields, C., Gocayne, J., Scott, J., Shirley, R., Liu, L.-I., Glodek, A., Kelley, J., Weidman, J., Phillips, C., Spriggs, T., Hedblom, E., Cotton, M., Utterback, T., Hanna, M., Nguyen, D., Saudek, D., Brandon, R., Fine, L., Fritchman, J., Fuhrmann, J., Geoghagen, N., Gnehm, C., McDonald, L., Small, K., Fraser, C., Smith, H. and Venter, J. (1995) Science 269, 496–512.
[2] Reeves, C.R., Ed. (1993) Modern Heuristic Techniques for Combinatorial Problems. Halsted Press, New York.
[3] Davis, L., Ed. (1987) Genetic Algorithms and Simulated Annealing. Morgan Kaufmann, San Francisco, CA.
[4] Cormen, T., Leiserson, C. and Rivest, R. (1990) Introduction to Algorithms. MIT Press, Cambridge, MA.
[5] Kirkpatrick, S., Gellat, C.D. and Vecchi, M. (1983) Science 220, 671–680.
[6] Metropolis, N., Rosenbluth, A., Teller, A. and Teller, E. (1953) J. Chem. Phys. 21, 1087–1091.
[7] Glover, F. (1986) Comp. Oper. Res. 5, 533–549.
[8] Hansen, P. (1986) Presented at the Congress on Numerical Methods in Combinatorial Optimization, Capri, Italy.
[9] Held, M. and Karp, R. (1970) Oper. Res. 18, 1138–1162.
[10] Held, M. and Karp, R. (1971) Math. Programming 1, 6–25.
[11] Rumelhart, D.E., Hinton, G. and McClellend, J. (1986) In: D.E. Rumelhart and J. McClellend (Eds.), PDP – Explorations in the Microstructure of Cognition, Vol. 1, Foundations. MIT Press, Cambridge, MA, pp. 318–362.
[12] Hertz, J., Krogh, A. and Palmer, R. (1991) Introduction to the Theory of Neural Computation. Addison Wesley, Reading, MA.
[13] Holland, J. (1975) Adaptation in Natural and Artificial Systems. University of Michigan Press, Ann Arbor, MI.
[14] Goldberg, D.E. (1989) Genetic Algorithms in Search, Optimization, and Machine Learning. Addison Wesley, Reading, MA.
[15] Bäck, T. (1996) Evolutionary Algorithms in Theory and Practice: Evolution Strategies, Evolutionary Programming and Genetic Algorithms. Oxford University Press, Oxford.
[16] Koza, J.R. (1992), Genetic Programming: On the Programming of Computers by Natural Selection. MIT Press, Cambridge, MA.
[17] Davis, L., Ed. (1991) The Genetic Algorithms Handbook. Van Nostrand Reinhold, Princeton, NJ
[18] Whitley, D. (1989) In: J. Shaffer (Ed.), Proc. 3rd Int. Conf. on Genetic Algorithms, San Mateo, CA. Morgan Kaufmann, San Francisco, CA, pp. 116–121.

185

[19] Fickett, J. and Cinkosky, M. (1993) In: C.R. Cantor and R. Robbins (Eds.), Proc. 2nd Int. Conf. on Bioinformatics, Supercomputing, and Complex Genome Analysis. World Scientific, Singapore, pp. 273–285.
[20] Platt, D. and Dix, T. (1993) In: T. Mudge, V. Milutinovic and L. Hunter (Eds.), Proc. 26th Hawaii Int. Conf. on System Sciences, Vol. I: Systems Architecture and Biotechnology. IEEE Computer Society Press, Los Alamitos, CA, pp. 756–762.
[21] Schulze-Kremer, S. and Tiedemann, U. (1993) In: J.-G. Ganascia (Ed.), Int. Joint Conf. on Artificial Intelligence. Artificial Intelligence and Genome Workshop 26, AAAI Press, Menlo Park, CA, pp. 119–141.
[22] Dandekar, T. and Argos, P. (1992) Prot. Eng. 5(7), 637–645.
[23] Unger, R. and Moult, J. (1993) J. Mol. Biol. 231, 75–81.
[24] Parmee. I.C. and Vekeria, H.D. (1997) In: T. Bäck (Ed.), Proc. 7th Int. Conf. on Genetic Algorithms (ICGA97). Morgan Kaufmann, San Francisco, CA, 529–536.
[25] Schweitzer, F., Ebeling, W., Rose, H. and Weiss, O. (1996) Parallel Problem Solving from Nature – PPSN IV, Vol. 1141 of Lecture Notes in Computer Science. Springer, Berlin, pp. 940–949.
[26] Lohmann, R. (1991) Parallelism, Learning and Evolution, 1989, Vol. 565 of Lecture Notes in Computer Science. Springer, Berlin, pp. 500–508.
[27] Fogel, L. (1962) Indus. Res. 4, 14–19.
[28] Fogel, L., Owens, A. and Walsh, M.J. (1966) Artificial Intelligence through Simulated Evolution. Wiley, New York.
[29] Fogel, D. (1991) System Identification through Simulated Evolution: A Machine Learning Approach to Modeling. Ginn Press, Needham Heights, MA.
[30] Fogel, D. (1992) Evolving Artificial Intelligence, PhD Thesis. University of California, San Diego, CA.
[31] Cramer, M. (1985) In: J. Grefenstette (Ed.), Proc. 1st Int. Conf. on Genetic Algorithms and Their Applications. Lawrence Erlbaum Associates, Mahwah, NJ, pp. 183–187.
[32] Handley, S. (1994) In: J. Zurada II, R.J. Marts and C.J. Robinson (Eds.), Proc. 1994 IEEE World Congress on Computational Intelligence, Vol. 1. IEEE Press, New York, pp. 474–479.
[33] Handley, S.G. (1994) In: R. Altman, D. Brutlag, P. Karp, R. Lathrop and D.B. Searls (Eds.), 2nd Int. Conf. on Intelligent Systems for Molecular Biology, Stanford University, Stanford, CA, USA. AAAI Press, Menlo Park, CA, pp. 156–159.
[34] Goldberg, D.E. (1989) Complex Syst. 3, 153–171.
[35] Parsons, R., Forrest, S. and Burks, C. (1993) In: L. Hunter, D.B. Searls and J. Shavlik (Eds.), Proc. 1st Int. Conf. on Intelligent Systems in Molecular Biology. AAAI Press, Menlo Park, CA, pp. 310–318.
[36] Parsons, R., Forrest, S. and Burks, C. (1995) Mach. Learning 21(1/2), 11–33.
[37] Starkweather, T., McDaniel, S., Mathias, K., Whitley, D. and Whitley, C. (1991) In: R.K. Belew and L.B. Booker (Eds.), 4th Int. Conf. on Genetic Algorithms. Morgan Kaufmann, San Francisco, CA, pp. 69–76.
[38] Parsons, R.J. and Johnson, M.E. (1997) Am. J. Math. Manag. Sci. 17, 369–396.
[39] Hunkapiller, T., Kaiser, R., Koop, B. and Hood, L. (1991) Science 254, 59–67.
[40] Hunkapiller, T., Kaiser, R. and Hood, L. (1991) Curr. Opin. Biotech. 2, 92–101.
[41] Venter, J.C., Ed. (1994) Automated DNA Sequencing and Analysis Techniques. Academic Press, San Diego, CA.
[42] Waterman, M. (1995) Introduction to Computational Biology: Maps, Sequences and Genomes, Interdisciplinary Statistics Series. Chapman and Hall, London.
[43] Goldstein, L. and Waterman, M. (1987) Adv. Appl. Math. 8, 194–207.
[44] Lawler, E., Rinnooy Kan, A. and Shmoys, D., Eds. (1985) The Traveling Salesman Problem. Wiley, New York.
[45] Lin, S. and Kernighan, H.W. (1973) Oper. Res. 21, 498–516.
[46] Sutton, G., White, O., Adams, M. and Kerlavage, A. (1995) Genome Sci. Technol. 1, 9–19.
[47] Huang, X. (1992) Genomics 14, 18–25.
[48] Staden, R. (1980) Nucleic Acids Res. 8, 3673–3694.
[49] Tarhio, J. and Ukkonen, E. (1988) Theor. Comput. Sci. 57, 131–145.

[50] Kececioglu, J. (1991) Exact and approximation algorithms for DNA sequence reconstruction, PhD Thesis. University of Arizona, Tucson, AZ, TR 91-26, Department of Computer Science.

[51] Peltola, H., Soderlund, H., Tarhio, J. and Ukkonnen, E. (1983) In: R. Mason (Ed.) Inf. Process. 83. Elsevier, Amsterdam, pp. 59–64.

[52] Turner, J. (1989) Inf. Comput. 83, 1–20.

[53] Burks, C., Engle, M., Forrest, S., Parsons, R., Soderlund, C. and Stolorz, P. (1994) In: J.C. Venter (Ed.), Automated DNA Sequencing and Analysis Techniques. Academic Press, San Diego, CA.

[54] Myers, G. (1994) An alternative formulation of sequence assembly. Presented at DIMACS Workshop on Combinatorial Methods for DNA Mapping and Sequencing.

[55] Bean, J.C. (1992), Genetics and random keys for sequencing and optimization, Technical Report 92-43. University of Michigan, Ann Arbor, MI.

[56] Syswerda, G. (1989) In: J. Shaffer (Ed.), Proc. 3rd Int. Conf. on Genetic Algorithms, San Mateo, CA. Morgan Kaufmann, San Francisco, CA, pp. 2–9.

[57] Schaffer, J.D., Caruana, R., Eshelman, L.J. and Das, R. (1989) In: J. Shaffer (Ed.), Proc. 3rd Int. Conf. on Genetic Algorithms, San Mateo, CA. Morgan Kaufmann, San Francisco, CA, pp. 51–60.

[58] Carlsson, P., Darnfors, C., Olofsson, S.-O. and Bjursell, G. (1986) Gene 49, 29–51.

[59] Sanger, F., Coulson, A., Hill, D. and Petersen, G. (1982) J Mol. Biol. 162, 729–773.

[60] Box, G.E., Hunter, W.G. and Hunter, J.S. (1978) Statistics for Experimenters, An Introduction to Design, Data Analysis and Model Building. Wiley, New York.

[61] Box, G.E. and Draper, N.E. (1987) Empirical Model-Building and Response Surfaces. Wiley, New York.

[62] Bohachevsky, I., Johnson, M. and Stein, M. (1992) J. Comp. Graph. Stat. 1(4), 367–384.

[63] Burks, C., Parsons, R. and Engle, M. (1994) In: R. Altman, D. Brutlag, P. Karp, R. Lathrop and D.B. Searls (Eds.), Proc. 2nd Int. Conf. on Intelligent Systems in Molecular Biology. AAAI Press, Menlo Park, CA, pp. 62–69.

[64] Booker, L.B. (1987) In: L. Davis (Ed.), Genetic Algorithms and Simulated Annealing. Morgan Kaufmann, San Francisco, CA, pp. 61–73.

[65] Whitley, D. and Yoo, N. (1995) Foundations of Genetic Algorithms, Vol. 3. Morgan Kaufman, San Francisco, CA.

S.L. Salzberg, D.B. Searls, S. Kasif (Eds.), *Computational Methods in Molecular Biology*
© 1998 Elsevier Science B.V.   All rights reserved

CHAPTER 10

Decision trees and Markov chains for gene finding

Steven L. Salzberg

The Institute for Genomic Research, 9712 Medical Center Drive, Rockville, MD 20850, USA.
Phone: +1 301-315-2537; Fax: +1 301-838-0208; Email: salzberg@cs.jhu.edu or salzberg@tigr.org
and Department of Computer Science, Johns Hopkins University, Baltimore, MD 21218, USA

Introduction

Decision trees are an important and widely used technique for the classification of data. Recently, they have been adapted for use in computational biology, especially for the analysis of DNA sequence data. This chapter describes how decision trees are used in the gene-finding system MORGAN [1], and how they might be used for other problems of sequence analysis. First I present a tutorial introduction to decision trees, describing what they are and how they can be created automatically from data. Readers already familiar with decision tree algorithms can skip this tutorial section and go directly to the section describing the applications of decision trees in DNA sequence analysis. The tutorial also contains pointers to several places on the Internet where the interested reader can find and download decision tree software.

1. A tutorial introduction to decision trees

The basic structure of a decision tree was explained in chapter 2. In short, a decision tree is a collection of nodes and edges, where each node tests some feature or features of the data. These trees are used as classifiers by passing an example down from the root node, which is a specially designated node in the tree (we usually draw the "root" of the tree at the top). An example is tested at each node, and it proceeds down the left or right branch depending on the outcome of the test. We assume here that all tests are binary or "yes–no" questions, and each node therefore has two child nodes below it. We can also build trees where a test has many possible answers, and a node will have one child for each answer. An simple example of a decision tree is shown in Fig. 1.

The nice thing about decision trees is that they contain the knowledge for making complex decisions in a very small data structure. Frequently we find that trees with just a handful of nodes can perform very well at classifying databases with thousands of examples. In order to get such good performance, we have to be pretty clever at choosing a good set of features to measure on the data, and this often involves detailed knowledge of the problem. Once the features have been specified, the decision tree *induction* algorithm does the rest of the work for us, building the tree from scratch. The purpose of this tutorial section is to explain in simple terms how this tree-building algorithm works.

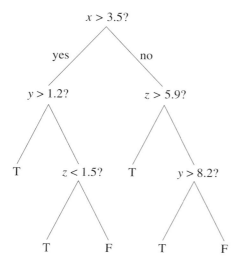

Fig. 1. Sample decision tree. Examples are passed down the tree beginning at the top, where a "yes" result on any test means that an example should be passed down to the left. The tests are applied to any aspect of the examples that is deemed to be useful for the classification task. The leaf nodes contain class labels, here listed simply as true (T) and false (F).

1.1. Induction of decision trees

Decision tree algorithms have been described in many places, for example Quinlan's ID3 and C4.5 [2], Murthy et al.'s OC1 [3,4], and Breiman et al.'s older CART [5] system. This tutorial section describes some of the features shared by all these algorithms, and a few specific properties of OC1, which is used in the experiments later in the chapter. The interested reader is directed to the books by Quinlan and Breiman et al. for more details.

All decision trees operate using the same basic algorithm. The input is a set S of examples; in the current study, S consists of a set of non-overlapping DNA subsequences. An example is nothing more than a collection of feature values and a class label – note that a raw DNA sequence is *not* an example per se, but must be processed to extract features (such as GC content). The label depends on the classification task, so if we are trying to distinguish exons and introns, our labels might simply be "exon" and "intron." In that case, the examples used to build the tree would comprise only whole exons and introns, rather than arbitrary subsequences of DNA.

The basic tree-induction algorithm is:

Algorithm Build-Tree (S)

(1) Find a test that splits the set S into two (or more) subsets. The test should group examples with similar feature values together.

(2) Score the two subsets (see below for scoring rules).

(3) If a subset is pure (all examples belong to the same class), then make a leaf node for the subset and stop.

(4) Else, call Build-Tree recursively to split any subsets that are not pure.

The result of this algorithm is a decision tree with test nodes and leaf nodes. The leaf nodes contain class labels.

Decision trees vary primarily in how they choose a "split" of the data, and in how they prune the trees they produce. The data should be formatted as a set of features and a class label, and it is up to the programmer to choose the features. An excellent review of features used to distinguish coding and non-coding DNA was written by Fickett and Tung [6], who described 21 different coding measures including hexamer frequencies, open reading frame length, codon frequencies, and periodicity measures. Each of these features is a number that can be computed for any subsequence of DNA.

1.2. Splitting rules

Given this numeric input, the most common approach to choosing a test or "split" is to consider each feature in turn, and to choose a threshold for the test; e.g., $x_1 \geqslant 34.5$. Because only one feature is used, we call this a univariate test. In geometric terms, this test is equivalent to a plane in feature space, where the plane is parallel to every axis except x_1. (In 2D, with just features x and y, the test would be simply a line parallel to the y-axis.) Therefore we also call this an axis-parallel test. With a univariate test, there are only $N - 1$ different tests for any set of N examples. These tests would occur at the midpoints between successive feature values; geometrically, we are just splitting the examples using lines. Here is a simple graphical explanation:

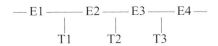

In this illustration, we have four examples, E1–E4, and just a single feature, which we will call x_1. The natural places to split the examples are T1, T2, and T3, which are at the midpoints between successive examples. So for four points (examples) we have three tests.

With D features, we have $D(N - 1)$ possible tests to consider. A decision tree algorithm will try all of these, and choose the best. The best must be selected based on a score, known as an *impurity* measure. The simplest way to compute impurity is to count the number of examples that would be misclassified by the test [7]. A more sophisticated measure, *information gain*, measures the entropy of the initial set and the subsets produced after splitting [2]. Other measures use statistical properties such as the probabilities of each class in each subset, and balance these with the size of each subset [5] (it is generally preferable to split the data roughly in half with each test, since this will build the smallest trees). For mathematical definitions of half a dozen of the most common impurity measures, see Murthy et al. [4].

Another, more complicated type of test is the linear discriminant, also called an *oblique* split. Instead of $x_1 \geqslant 34.5$, the test uses all the features in combination, such as

$$a_1 x_1 + a_2 x_2 + \cdots + a_d x_d \geqslant a_{d+1}.$$

This equation defines a plane that is not parallel to the axes of feature space, and therefore we call this an *oblique* test. While oblique tests are much more powerful than the simpler univariate tests, they are also more expensive to compute. Even using strictly linear combinations, such as the one shown here, requires much more computation. (A non-linear combination would include terms such as x_1^2, and would correspond to a curved surface rather than a plane.) Fortunately, recent work on randomized decision tree algorithms has resulted in efficient methods for finding these oblique tests [4]. Our

OC1 system was the first such system to use a practical, efficient algorithm for finding oblique tests as part of decision tree construction. OC1 also finds the best axis-parallel split (since this is easy to do, not to mention fast), and it only uses oblique tests if they work better than the simpler univariate tests. For extra speed and simplicity, the OC1 system can be set to use axis-parallel splits only, in which case it behaves just like a more traditional decision tree system.

One interesting result that came out of this study was that virtually all impurity measures are designed so as to maximize, in one form or another, the overall accuracy of the tree. Thus if one class is much larger than another, decision trees tend to optimize accuracy on the larger class. Special steps have to be taken to correct this problem if the classes are skewed. Some ideas on how to handle skewed classes are described by Breiman et al. [5], and Salzberg [8] describes some practical adjustments to make for decision trees on a DNA sequence classification problem. The MORGAN system has to handle severely skewed classes, and we had to develop special training procedures to handle them. MORGAN uses decision trees to determine when a subsequence is a true exon versus a "pseudo-exon", and it also uses trees to distinguish true introns from pseudo-introns. A pseudo-exon has a potential acceptor site on the 5′ end and a potential donor site on the 3′ end, with an open reading frame spanning the sequence. A pseudo-intron is defined similarly, but with a donor site on the 5′ end and an acceptor site on the 3′ end – and there is no requirement for an orf to span the sequence. In a long stretch of DNA, there are many more pseudo-exons and pseudo-introns than true exons and introns, typically tens or hundreds of times more. If we use all the pseudo-exons from a sequence to form a training set, the decision tree will have to discriminate between classes with vastly different numbers of examples.

1.3. Pruning rules

After building a decision tree, it will correctly classify every example in the training set. This "complete" tree usually overfits the training data; it may contain nodes that contain only one or two examples, and these nodes are typically making distinctions that are really just artifactual. Such a tree will perform poorly on additional data. A smaller tree often performs better, and offers the additional advantages of being simpler and faster to use. To create a smaller tree from a complete tree, we use a procedure called pruning.

The OC1 decision tree system uses a pruning technique called cost complexity pruning [5], which operates as follows. Before building the tree, OC1 separates the data randomly into two sets, the training set (T) and the pruning set (P). The system first builds a complete tree that classifies every example in T correctly. It then looks at every non-leaf node in the tree, and measures the cost complexity of that node. The cost complexity is a simple function of two numbers: (1) the number of examples that would be mis-classified if that node were made into a leaf, and (2) the size of the subtree rooted at that node. These two numbers are combined into the cost complexity measure. The node whose cost complexity is greatest is then "pruned", meaning it is converted into a leaf node. This pruning now gives us a new, smaller tree, and we can repeat the whole process. Pruning thus produces a succession of smaller trees, stopping when the tree is pruned down to a single node.

OC1 then examines the series of increasingly smaller trees that have been generated by the pruning procedure. For each of these trees, it computes their accuracies on the pruning set P, which gives an independent estimate of the accuracy of these trees on new data. The tree with the highest accuracy on the pruning set then becomes the output of the system. An alternative strategy chooses the smallest tree whose accuracy is within a certain amount from the highest accuracy. By choosing a tree whose accuracy is not quite the highest on P, this strategy usually picks even smaller trees.

1.4. Internet resources for decision tree software

The decision tree system used for all our experiments is called OC1, or Oblique Classifier 1. It has the ability to generate both oblique and univariate tests at every node, and it can use any one of seven different goodness measures, including information gain [9], the twoing rule [5], and others. The code for OC1, including sources and documentation, is available via ftp or the World Wide Web. For those using ftp, the address is ftp.cs.jhu.edu. Connect to the directory pub/oc1 and transfer the files there. For those on the World Wide Web, connect to

`http://www.cs.jhu.edu/labs/compbio/home.html`

and select the OC1 decision tree system listed there. There is also a pointer from `http://www.cs.jhu.edu/~salzberg`. These directories also contain a copy of the journal paper describing OC1 [4].

Another program, C4.5, is one of the most popular and well-known decision tree systems in the machine learning community. C4.5 was developed by J.R. Quinlan at the University of Sydney. The software includes routines for converting decision trees to sets of rules. Information about the software, including technical papers, can be found at Quinlan's ftp site at

`ftp://ftp.cs.su.oz.au/pub/ml/`

The code is included on a disk with the book *C4.5: Programs for Machine Learning* [2]. The newest version of this system is C5.0, which can be found on the World Wide Web at `http://www.rulequest.com`.

The IND package is another nice decision tree system. This package includes modules that emulate CART [5], C4.5, Wray Buntine's smoothing and option trees, and Wallace and Patrick's minimum message length method. The system was developed by Wray Buntine, and a variety of technical papers, as well as information about obtaining his software, can be found on his home page, at

`http://www.ultimode.com/~wray/`

Although it is not a decision tree learning system, the CN2 system learns rules in a straightforward if–then format that is intuitively simple, and that works well for many domains. To get this system and Peter Clark's other machine learning software from the University of Texas, go to

`http://www.cs.utexas.edu/users/pclark/software.html`

CN2 inductively learns a set of propositional if–then rules from a set of training examples. To do this, it performs a general-to-specific beam search through rule-space for the "best" rule, removes training examples covered by that rule, then repeats until no more "good" rules can be found. The original algorithm defined "best" using a combination of entropy and a significance test. The algorithm was later improved to replace this evaluation function and also to induce ordered rule lists ("decision lists"), which are a simplified type of decision tree. The software implements the latest version, but has flags which can be set to return it to the original version. The algorithm was designed by Tim Niblett and Peter Clark.

2. Decision trees to classify sequences

From the description above, it may not be immediately obvious how to use decision trees for sequence analysis. This requires that the sequences first be converted into an appropriate form. In the tutorial section above, we described how trees may be used to classify *examples*, not sequences. It turns out that any sequence or subsequence can be treated as an example; all that is required is to first convert the sequence into a set of features. Fortunately, there has been a significant body of work focusing on exactly what features are most useful in determining if a DNA sequence is a coding region or not. This work has used the term "coding measure" to describe what is called a feature in the decision tree context.

Briefly, a coding measure is any sort of statistic computed on a DNA sequence. A set of these numbers together with a class label (e.g., "exon" or "intron") is exactly what we need to represent a sequence for the purposes of decision tree creation. Ideally we will choose coding measures that are most relevant to the classification problem we wish to solve.

2.1. Coding measures

To build the MORGAN gene-finding system, we considered using most of the 21 coding measures reviewed and summarized in Fickett and Tung [6]. These measures fall into several distinct groups, and within a group some measures are just variations of each other. After quite a bit of experimentation, we settled on a small subset of features which seemed to perform as well as any other set we tried.

For distinguishing coding and non-coding DNA, probably the most useful of all the coding measures is the in-frame hexamer statistic, for which we used the definition given by Snyder and Stormo [10]:

$$IF_6(i, j) = \max \begin{cases} \sum_{k = 0,3,6,\dots,j-6} \log(f_k/F_k) \\ \sum_{k = 1,4,7,\dots,j-6} \log(f_k/F_k) \\ \sum_{k = 2,5,8,\dots,j-6} \log(f_k/F_k) \end{cases}$$

where f_k is the frequency of the hexamer from a table of in-frame hexamers computed over the coding regions in the training set, and F_k is the frequency of the hexamer among all hexamers (in all frames) in the training set. This statistic is defined so that a value greater than zero indicates that the sequence looks more like coding DNA than non-coding DNA.

We also used the position asymmetry statistic defined by Fickett and Tung [6], which counts the frequency of each base in each of the three codon positions. If we let $f(b, i)$ be the frequency of base b in position i, where $i \in (1, 2, 3)$, then $\mu(b) = \sum_i f(b, i)/3$, and the asymmetry A is $A(b) = \sum_i (f(b, i) - \mu(b))^2$. Thus position asymmetry is actually four separate feature values, one for each base.

In addition, we developed a method for scoring the donor and acceptor sites using a second-order Markov chain. (Markov models and chains are described in more detail elsewhere in this volume.) This gave us two additional features that measured the signals on either end of each intron. We used a first-order Markov chain to score the sequence around potential starts for translation.

All of these features were then given to OC1 to generate trees that could distinguish exons, introns, and intergenic DNA. We used many additional features in our early experiments, and the decision tree software automatically selected those (using the information gain impurity measure) that were most useful for distinguishing coding and non-coding sequences. Ultimately, the features described above were the only ones needed.

3. Decision trees as probability estimators

Now that we have a method for using decision trees to classify sequences, we need to add one more capability. Rather than simply classifying a sequence as exon, intron, intergenic, or some other class, we would like to be able to measure our confidence in this decision. This is best reported as a probability, which should be interpreted as the probability that the prediction is correct. So instead of reporting that a sequence is an exon, we would like a decision tree to report that the probability of exon is p, where $0 \leqslant p \leqslant 1$.

This is a very simple thing to do, as long as we use pruned decision trees (which we do). Recall that in a pruned tree, each leaf node contains a set of examples that may belong to different classes. For example, a leaf node may contain 20 examples, of which 16 might be exon sequences and 4 might be non-coding or "pseudo-exon" sequences. Then we can report that a new example that ends up at this node has an 80% chance of being an exon. An example of one of these probability trees is shown in Fig. 2.

In order to make these probability estimates more robust, we have adopted a strategy of generating multiple trees for the same training data. The need for this arises from the fact that some leaf nodes might have very few examples, and thus our probability estimates could be too coarse. For instance, a node with just 3 examples might estimate the probability of one class at 67% or 100%, although the true value might lie somewhere in between. Thus we would like to use a larger set of examples to estimate the probability. We get this by averaging together the distributions from ten different trees.

Unlike other decision tree software, OC1 is a randomized system. This unique feature allows OC1 to build different trees each time it is run, even though the input is identical each time. (It is also possible to turn off randomization so that OC1 produces the same output each time.) So to get ten different trees, all that is required is to take the training data and run it through OC1 ten times, saving each of the pruned trees. When a new example is processed, it is given to all ten trees, each of which reports a probability

194

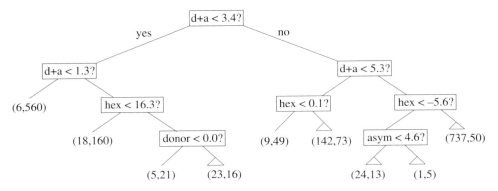

Fig. 2. A probability tree, which has probability distributions instead of class labels at the leaf nodes. Each internal node tests a variable or a linear combination of variables, and if the test is satisfied, the program proceeds down the left branch until it reaches the bottom of the tree. Each leaf node contains the distribution of classes in the set of examples that match that node. This tree, which is derived from one of the actual trees used by the MORGAN system, classifies DNA sequences into exons and pseudoexons. The features tested in this tree include the donor site score (donor), the sum of the donor and acceptor site scores (d+a), the in-frame hexamer frequency (hex), and Fickett's position asymmetry statistic (asym). The leaf nodes contain class distributions for the two classes "exon" and "pseudoexon." For example, the leftmost leaf node contains the distribution (6,560), which means that in the training set, 6 examples that reached this point in the tree were exons and 560 were pseudo-exons. The nodes shown as triangles represent un-expanded subtrees.

distribution. The average of these distributions is then used to estimate the probability that the new example belongs to each class.

4. MORGAN, a decision tree system for gene finding

MORGAN is an integrated system for finding genes in vertebrate DNA sequences. MORGAN uses a combination of decision trees, Markov chains, and dynamic programming to parse a genomic DNA sequence into coding and non-coding regions. By using dynamic programming, MORGAN guarantees that the parse will be optimal with respect to the scoring methods it uses. Of course, the scoring methods (decision trees and Markov chains) are themselves fallible, which means that MORGAN is not guaranteed to be correct. The design of MORGAN and experiments demonstrating its accuracy are explained in detail in [1]. This section is only intended as a summary; the interested reader should read the technical paper. The gene-finding problem is one of the central problems of computational biology, in large part because it is crucial to understanding the large volumes of sequence data now being generated by genome projects around the world. The systems developed to find genes in eukaryotic sequences include GRAIL [11,12], GeneID [13], GeneParser [14,10], VEIL [15], Genie [16], FGENEH [17], SorFind [18], and GENSCAN [19]. In addition, there are two separate systems for finding genes in prokaryotes, GeneMark [20] and Glimmer [21]. MORGAN, which stands for Multi-frame Optimal Rule-based Gene Analyzer, is the only system that uses decision trees, and below we explain more about how this works.

4.1. The MORGAN framework

The framework of MORGAN is a dynamic programming (DP) algorithm that looks at all possibly optimal parses of a DNA sequence. The DP algorithm is built around the *signals*: start codons, stop codons, donor sites, and acceptor sites. As mentioned above, MORGAN uses Markov chains to identify each of these signals. The more accurately it can find the signals, the better (and faster) the whole system will work. Consider the limiting case where the Markov chains do a perfect job: only one start site and one stop site are identified (the true ones), and likewise only the true donor and acceptor sites pass the Markov filters. In this case, parsing is trivial, because there is only one possible way to parse the sequence. A "parse," by the way, has a specific syntax for eukaryotic genes: first comes the non-coding sequence, then the first coding region beginning with the start codon ATG, then any number of alternating introns and exons, and finally another non-coding region.

The first coding region may occur in the first exon or it may occur later, and likewise the last coding region may occur in the last exon or earlier. MORGAN only attempts to identify coding regions, and therefore its annotation begins with the location of the start of translation and ends with a stop codon. Currently, neither MORGAN nor any other gene finder can effectively locate the start of transcription (as opposed to translation). This remains an important open problem for further investigation.

4.2. Markov chains to find splice sites

All gene finders look for splice sites, but not all of them use separate splice site recognition programs. Sometimes the splice site recognition is integrated into the system, as in HMM-based approaches [15]. In MORGAN, a separate program scans the sequence looking for potential donor and acceptor sites. Good candidates are marked for use later in the dynamic programming algorithm.

The most common method for characterizing splice sites is the position weight matrix (PWM), which tallies for each location in a site what the probabilities of the four bases are. For example, in the donor site, the bases G and T occur with 100% probability in locations 0 and 1 of the intron; i.e., the first two positions in the intron. These positions alone are not sufficient to distinguish true from false donor sites, however, since the dinucleotide GT occurs in roughly 1/16 of all positions. The bases near the GT, especially on the intron side of the site (where there is no constraint for the bases to be coding), also have strongly preferred bases. For example, the base in position 3 is adenine over 70% of the time in vertebrate donor sites. The PWM method uses these preferences to create a table of base probabilities, and from this table a scoring scheme. (One can easily use dinucleotide probabilities instead, as in ref. [22], or conditional probabilities, as in ref. [23].)

MORGAN originally used the PWM method, but after much experimentation we found that a second-order Markov chain was the best method for the data available. (An introduction to Markov chains appears in chapter 2.) To construct this model, we need the probability of each base in each position given the two previous bases. Thus we need to compute 64 probabilities for each position in a splice site. The vertebrate data set used to create MORGAN had over 2000 splice sites, but this was still not enough to give reliable estimates of all 64 probabilities for every position. Therefore we made a slight modification: in the first position, only purines (A and G) and pyrimidines (C and T) were counted. Thus we computed probabilities such as $P(A_i|C_{i-1}, R_{i-2})$, i.e., the probability of A in position i given C in position $i-1$ and a purine in position $i-2$. This modification meant that only 32 probabilities were necessary to build the model. Because there are many fewer start sites than donor and acceptor sites, MORGAN uses a conditional probability matrix [23] to score start sites. This is equivalent to a first-order Markov chain.

The latest version of MORGAN goes one step beyond these Markov chains, and incorporates another type of decision tree at donor sites. This is the *maximal dependence decomposition* (MDD) tree, explained in detail in the chapter by Burge (and also in ref. [19]). We found that the use of MDD trees, with second-order Markov chains at the leaf nodes, gave a small but noticeable improvement over the Markov chains alone, at least for donor sites. More details on site selection are given in [1], which describes the technical design of MORGAN.

For further background on efforts to characterize start sites and splice junctions, the reader should look at Kozak [24,25] for translational start sites and Senapathy and Mount [26–28] for splice sites. Thus far, no one has yet come up with an accurate model of stop sites; the sequences surrounding known stop codons do not seem to have much in common.

After filtering the input sequence to find all candidate start sites, donor sites, acceptor sites, and stop codons, the sequence plus this list of candidates are passed on to the dynamic programming module, where the bulk of the computation in MORGAN takes place.

4.3. Parsing DNA with dynamic programming

Dynamic programming (DP) is at the heart of the MORGAN system. Dynamic programming is a very general name given to a wide range of different algorithms, all of which use a common strategy: in order to find the optimal solution to a problem, break the problem into smaller problems, compute their optimal solution, and then glue those solutions together. (See chapter 2 for some simple examples.) As it happens, the gene-finding problem fits into this framework very comfortably, as is reflected in the fact that DP is the basis of quite a few gene-finding systems, including GeneParser [14,10] and GRAIL [11], both of which use DP in combination with a feedforward neural network. Genie [16], VEIL [15], and GENSCAN [19] use variations on a hidden Markov model algorithm that is a special type of dynamic programming. The DP formulation of Wu [29] is the most similar to that of MORGAN, in that both algorithms explicitly consider all possible reading frames.

MORGAN's DP algorithm finds an optimal segmentation of a DNA sequence into alternating exons and introns. Even though there are an exponential number of ways to parse a sequence into exons and introns, DP is able to prune away large numbers of these possibilities, considering only those alternatives that are candidates for the optimal solution. Currently, MORGAN assumes that its input is a complete genomic DNA sequence containing a single gene (as do most other gene finders), but this assumption will be removed in the future. Its output is a parse of the sequence into an initial non-coding region containing (possibly) intergenic and 5′ untranslated sequence (UTR), followed by alternating exons and introns, following by the 3′ UTR and any additional non-coding DNA. This parse will always be optimal with respect to the scores produced by the scoring algorithm, which in MORGAN is a set of decision trees and some hand-coded rules. Thus the "exons" found by MORGAN are only the coding portions of exons, spanning the translated portion of the sequence.

The following is just a high-level summary of how the DP algorithm works in MORGAN. For more details, see [1]. The algorithm goes through a sequence S from left to right, pausing to consider each of the signals identified in the previous (filtering) phase. Recall that there are four signal types:

(1) start signals;
(2) donor sites (end of exon/beginning of intron);
(3) acceptor sites (end of intron/beginning of exon);
(4) stop codons (end of the coding portion of the gene).

At each signal location, MORGAN keeps track of the best (optimal) parse of the sequence *up to that point*. This means that the system must score the preceding sequence and find the best parse. It actually keeps track of the best parse in all three frames, so three different scores must be stored at every signal.

For start sites, MORGAN simply marks the site and gives it a constant score. This is because the system does not currently score the upstream region – the region prior to the start site. The filtering phase, in which start sites were identified, looked carefully at the bases immediately before the start site in order to find the start site itself. However, upstream promoters and other transcription factors are not currently identified by MORGAN. This is an important area for future improvement of the system.

For donor sites, processing is more complicated. Suppose the system is considering a possible donor site at location i. It must consider two possibilities.

• The donor site at i is the end of the first coding exon. In this case, MORGAN searches back through the sequence for all matching start codons. A start codon at location j matches if there is an open reading frame (the frame is defined by the location of the start codon) stretching from j to i.

• The donor site at i is the end of an internal exon. In this case, MORGAN looks back for all matching acceptor sites. An acceptor site at location j matches if there is any open reading frame spanning the interval from j to i.

For each of the sites that matches the donor site at i, MORGAN must score the intervening region from j to i. For this purpose it uses the decision trees described above. If the matching sites are a start site and a donor site, MORGAN uses trees that are trained to score initial exons. If the two sites are an acceptor site at j and a donor site at i, then MORGAN uses the trees that score internal exons.

After scoring a subsequence (j, i) as above, MORGAN multiplies that score by the length of the subsequence. This is a means of normalizing the scores so that different parses can be compared. For each j, the score of the newly added subsequence (j, i) is added to whatever score was already stored at j. Only the highest scoring parse is kept at each location; so from all the j, MORGAN saves only the best score to store at the new site i.

At each acceptor site, MORGAN must scan back in the sequence looking for a matching donor site. The process is similar to the scanning just described except that no open reading frame constraints apply. Introns can also be much longer, so there is a very large number of possible introns to consider. This is why it is important that the filtering phase eliminate as many of the false positive introns as possible.

At the stop sites, MORGAN scans back to find the previous acceptor sites, and scores the intervening sequence as a final coding exon. As above, this score is first normalized by multiplying it by the length of the subsequence, and then it is added to whatever score was stored at the previous acceptor site.

When MORGAN reaches the end of the sequence, it simply looks at the scores stored at all the stop codons, and the best among them is the optimal score. It then traces back through a set of pointers to figure out the parse represented by that score.

4.4. Frame consistent dynamic programming

To guarantee that the parse is optimal, MORGAN must keep track of the best parse at every signal in all three frames. This requires roughly three times as much work, but it is necessary. Figure 3, which is taken from ref. [1], illustrates the problem. In the situation shown in the figure, we have two different possible initial exons, in different frames, followed by introns that could end at the same location i. If only one value could be stored at i, it would be the higher-scoring parse. Thus, if the parse with *Exon 1* scored higher, the parse with *Exon 1'*, which we are assuming is in a different reading frame, would be lost. It is quite possible, however, that *Exon 2* is in a reading frame that is compatible only with *Exon 1'*. And the parse using *Exon 1'* followed by *Exon 2* might be the highest scoring parse. Clearly, then, we need to keep a separate score for each of the three reading frames at each candidate start, donor, acceptor, and stop site. The importance of keeping separate optimal scores for each frame has also been emphasized by Wu [29].

non-coding →|←— Exon 1 →|←— Intron 1 ——→|←— Exon 2 —→
non-coding ——→|←— Exon 1'—→|←— Intron 1'—→|←— Exon 2 —→
$$i$$

Fig. 3. An illustration of why MORGAN keeps track of separate parses for all three frames

4.5. Downstream sequence

MORGAN does not score the region downstream (on the 3' side) of the stop codon, because currently we do not have good methods of distinguishing that sequence from other types of sequence. Instead, the system includes a custom-coded rule that looks for the polyadenylation binding site, which has the consensus sequence AATAAA. This site is where the enzyme poly-A polymerase binds before attaching the poly-A tail to the pre-mRNA molecule. In addition to this poly-A binding site, one could add additional rules to look for terminators, enhancers, and other signals that might appear both downstream and upstream of genes. This is an important area for future work.

5. Data and experiments

MORGAN was tested on a benchmark database of 570 vertebrate sequences. This database was created by Burset and Guigo to compare seven major gene-finding systems [30]. It has now been used as a benchmark in several additional studies, and it provides a good basis for comparison among gene finders. Of course, the number and variety of genes available in GenBank is increasing daily, so we expect that soon the Burset and Guigo database will be supplanted by a larger database. In this data, every sequence contains exactly one gene, and every gene contains at least one intron. All of the introns use standard splicing machinery; i.e., they begin with the dinucleotide GT and end with AG . Burset and Guigo edited the data to remove other problematic sequences such as pseudo-genes, but they did not remove homologous sequences.

To train MORGAN, 80% of the data was randomly chosen to form the training set and the rest was the test set. The 454 training sequences contained 2.3 million bases and 2146 exons. The test set, which had 114 sequences, contained 607 924 bases and 499 exons. Two sequences were discarded because they contained no bases upstream of the start codon. Of course, by randomly dividing the data into training and test sets, the two sets may contain sequences that are closely homologous, which might bias the experiment. (This point is debatable, however; for genes sequenced in the future, there will be a good chance that some of these will be homologous to one of the sequences used to train MORGAN. In that case, a test set with some homology to the training set gives a better estimate of the expected future accuracy of the system.) We therefore also created a second test set by removing from the original test set all sequences that had at least 80% identity to any sequence in the training set. The resulting non-homologous test set has 97 sequences and 566 962 bases.

Several measures of MORGAN's accuracy for the vertebrate sequences are given in Table 1. Overall, MORGAN did very well on the benchmark data. There are several ways to measure the accuracy of a gene-finding program. The most useful numbers reflect what percentage of the exons a system predicted correctly. (Because human genes are

Table 1
Prediction accuracies for finding genes using MORGAN[a]

Data Set	TEx	PEx	OvEx	TPE	1Edge	Sn	Sp	CC	P(I)	P(All)
Train	2146	2363	0.88	0.62	0.82	0.86	0.80	0.80	0.96	0.94
Test	499	508	0.83	0.59	0.78	0.81	0.83	0.79	0.97	0.95
N-H Test	453	461	0.83	0.57	0.77	0.81	0.82	0.78	0.96	0.95

[a] The three data sets are the training set, the test set, and the non-homologous (N-H) test set. TEx is the total number of true exons, PEx is the total number of exons predicted by MORGAN, OvEx is the fraction of true exons that overlap a predicted exon, TPE is the fraction of perfectly predicted exons (both boundaries are correct), and 1Edge is the fraction of exons for which at least one edge (either the $5'$ or $3'$ end) is predicted correctly. Sn is the sensitivity for coding exon bases; *i.e.*, the fraction of true exon bases that were correctly predicted as coding. Sp is the specificity for coding bases: the fraction of bases predicted to be in coding regions that actually were coding. CC is the correlation coefficient, $P(I)$ is the probability that if a given base is truly an intron we will mark it correctly, and $P(\text{All})$ is the probability that the system will mark any base correctly.

Table 2
A comparison of several leading gene-finding systems[a]

Gene finder	Coding bases			Exact exons			ME
	Sn	Sp	AC	Sn	Sp	Avg	
MORGAN	0.81	0.83	0.79	0.59	0.59	0.59	0.17
GENSCAN	0.83	0.93	0.91	0.78	0.81	0.80	0.09
VEIL	0.83	0.72	0.73	0.53	0.49	0.51	0.19
Genie	0.78	0.84	0.77	0.61	0.64	0.62	0.15
FGENEH	0.77	0.85	0.78	0.61	0.61	0.61	0.15
GRAIL 2	0.72	0.87	0.75	0.36	0.43	0.40	0.25
GeneID	0.63	0.81	0.67	0.44	0.46	0.45	0.28
GeneParser2	0.66	0.79	0.67	0.35	0.40	0.37	0.29
GenLang	0.72	0.79	0.69	0.51	0.52	0.52	0.21
SorFind	0.71	0.85	0.73	0.42	0.47	0.45	0.24
Xpound	0.61	0.87	0.68	0.15	0.18	0.17	0.33

[a] AC is the approximate correlation proposed by Burset and Guigo (1996) [30] as a replacement for the correlation coefficient. Sensitivity (Sn) is the fraction of true coding bases that were correctly predicted as coding, and specificity (Sp) is the number of bases predicted to be in coding regions that actually were coding; their average is given in the Avg column. The "exact exon" columns show the corresponding results for prediction of whole exons. ME (missing exons) is the fraction of whole coding exons that are missed completely.

typically about 80% intron, the overall percentage of bases predicted correctly is not necessarily a good indicator of performance. Simply predicting "intron" everywhere gives 80% accuracy.) The *sensitivity* for coding bases is the percentage of true coding bases that the system correctly predicted as coding. MORGAN obtained 81% sensitivity on the test set, and 81% on the non-homologous test set. (Apparently the presence of homology did not make a big difference to the system.) The *specificity* for coding bases is the percentage of the system's predicted coding bases that were actually coding. MORGAN's specificity here was 83% on the test set and 82% on the non-homologous test set.

An even more stringent test is how many of the exons are predicted exactly; i.e., how many times does the system predict exactly the right locations for both ends of an exon? This tests the system's ability to locate signals accurately. All gene finders do much worse here than they do on the measures that count bases. For the test data, MORGAN predicted both ends of 293 exons (out of 499 total) exactly right. Overall it predicted 508 exons, so its specificity was slightly lower. However, note that 416 of the true exons overlapped MORGAN's predicted exons, so only 83 out of 499 were missed completely.

Finally, Table 2 shows how MORGAN fares in comparison with other gene finders on the same data. Most of the results in the table are taken from Burset and Guigo's study [30]. The exceptions are the results from three new systems that appeared since that study: Genie [31], VEIL [15], and GENSCAN [19]. Interestingly, all three of these new systems are based on hidden Markov models (HMMs), which none of the older systems use. VEIL is a "pure" HMM, while Genie is a hybrid of neural network classifiers and HMMs.

GENSCAN uses a semi-Markov model, which is a more general HMM that allows the system to use the lengths of exons and introns as part of the model (e.g., the system can prefer exons that are around 150 bp if it likes).

A few points are worth noting about the comparison table. First, until the GENSCAN results appeared, MORGAN was among the top systems, along with Genie and FGENEH. Second, most of the systems were not trained and tested on separate data, because Burset and Guigo did not have access to source code and could not re-train the systems. So the numbers may be biased in some cases. The Genie system was trained on a non-homologous database of 353 human sequences, and GENSCAN used the same training data. We found that 122 of these sequences were contained in the 570 sequences of Burset and Guigo's database, so there is some overlap between training and test sets for these cases. VEIL was trained and tested using cross validation, so that the training and test sets do not overlap in that result. Despite these caveats, it seems that currently GENSCAN is the leading contender among gene finders. Clearly, though, there are many areas of improvement for MORGAN, and several of these are currently being pursued.

6. Next steps: interpolated Markov models

Recently, our group has developed a new gene finder for microbial DNA called Glimmer [21], which is based on interpolated Markov models (IMMs). Glimmer performs extremely well thus far; it finds 97.4% of the annotated genes in the bacterium *H. pylori* and 97.9% of the genes in the bacterium *H. influenzae*. These high figures were accomplished using a very simple training procedure, in which the system was simply given all long open reading frames (greater than 500 bp) that did not overlap each other. Subsequently, Glimmer has been used as the primary gene-finding method for the more recently sequenced genomes of *B. burgdorferi*, the bacterium that causes Lyme disease (Fraser et al., 1997, Nature 390, 580–586 [32]), and *T. pallidum*, the bacterium that causes syphilis (Fraser et al., in preparation). The secret to Glimmer's success is the IMM technique, which combines together 0th-order, 1st-order, 2nd-order, 3rd-order, and so on up through 8th-order Markov models.

In order to combine the nine different Markov models, Glimmer determines how much weight to give to the predictions coming from each one. One possibility is to use a single weight for each model; this strategy, however, ignores the fact that some oligomers are much more abundant than others. Thus it makes sense to give each oligomer its own weight, based on how frequently is has been observed. Although this requires the computation of a very large number of weights, the performance of Glimmer demonstrates that values for these weights can be determined efficiently and accurately.

We have already described how 1st- and 2nd-order Markov chains are used in MORGAN to create excellent start site and splice site recognizers. Based on the success of Glimmer, and obvious next step is to apply IMMs to the task of exon and intron recognition. We anticipate that MORGAN would use a lower-order IMM, perhaps only 1st- through 5th-order Markov chains, but this still represents a more sophisticated model than is presently employed in the system. This will be one of the most important areas for further testing and development of MORGAN.

7. Summary

MORGAN's strong performance on the benchmark data is probably due to a combination of factors. The Markov chain models do an excellent job at filtering out false donor, acceptor, and start sites, and as the amount of available data increases, these models should get better. A natural improvement is to use higher-order models, which will only be possible with more data. We have recently developed a new splice site recognition module that combines the maximal dependence decomposition (MDD) trees described by Burge in chapter 8 with the 2nd-order Markov chains described here. This combined Markov–MDD method outperforms either of the two methods alone. The decision tree classifiers are central to MORGAN, and the use of four separate tree types seems to be an effective design. However, decision trees are not limited to just two-class problems, so an alternative is to use one set of trees but allow four or five classes to be output (initial exon, internal exon, intron, final exon, non-coding).

MORGAN does not yet incorporate database lookup, but several gene finders have recently added this capability, and obtained 15–20% improvements as a result [10,30]. In addition, the rapidly growing EST databases should provide an excellent resource for identifying exons, since ESTs should come exclusively from exonic regions (though there are some introns known to be included in the EST databases). Clearly this is a promising direction for improvement. Finally, MORGAN does not yet look explicitly for any of the many important regulatory signals that occur in DNA sequences: promoters, enhancers, repressors, and other signals. Effective methods for finding these signals should significantly improve performance.

Acknowledgments

Thanks to the co-developers of the MORGAN system: Kenneth Fasman, Arthur Delcher, and John Henderson. This material is based upon work supported by the National Center for Human Genome Research at NIH under Grant No. K01-HG00022-1 and by the National Science foundation under Grant No. IRI-9530462.

References

[1] Salzberg, S., Chen, X., Delcher, A., Henderson, J. and Fasman, K. (1997) MORGAN: A decision tree system for finding genes in DNA. Technical Report JHU-97/02 Dept. of Computer Science, Johns Hopkins University, Baltimore, MD 21218.
[2] Quinlan, J.R. (1993) C4.5: Programs for Machine Learning. Morgan Kaufmann Publishers, San Mateo, CA.
[3] Murthy, S.K., Kasif, S., Salzberg, S. and Beigel, R. (1993) OC1: Randomized induction of oblique decision trees. In: Proc. 11th Natl. Conf. on Artificial Intelligence, Washington, D.C. MIT Press, Cambridge, MA, pp. 322–327
[4] Murthy, S.K., Kasif, S. and Salzberg, S. (1994) J. Artif. Intell. Res. 2, 1–33.
[5] Breiman, L., Friedman, J., Olshen, R. and Stone, C. (1984) Classification and Regression Trees. Wadsworth Intl. Group.
[6] Fickett, J. and C.-S. Tung (1992) Nucleic Acids Res. 20(24), 6441–6450.

[7] Heath, D., Kasif, S. and Salzberg, S. (1993) Learning oblique decision trees. In: Proc. 13th Int. Joint Conf. on Artificial Intelligence, Chambery, France. Morgan Kaufmann, pp. 1002–1007

[8] Salzberg, S. (1995) J. Comput. Biol. 2(3), 473–485.

[9] Quinlan, J.R. (1986) Machine Learning 1, 81–106.

[10] Snyder, E.E. and Stormo, G.D. (1995) J. Mol. Biol. 248, 1–18.

[11] Xu, Y., Mural, R. and Uberbacher, E. (1994) Comput. Appl. Biosci. (CABIOS) 10(6), 613–623.

[12] Xu, Y., Mural, R., J.R. Einstein, Shah, M. and Uberbacher, E. (1996) Proc. IEEE 84(10), 1544–1552.

[13] Guigo, R., Knudsen, S., Drake, N. and Smith, T. (1992) J. Mol. Biol. 226, 141–157.

[14] Snyder, E.E. and Stormo, G.D. (1993) Nucleic Acids Res. 21(3), 607–613.

[15] Henderson, J., Salzberg, S. and Fasman, K. (1997) J. Comput. Biol. 4(2), 127–141.

[16] Kulp, D., Haussler, D., Reese, M. and Eeckman, F. (1996) A generalized hidden Markov model for the recognition of human genes in DNA. In: ISMB-96: Proc. 4th Int. Conf. on Intelligent Systems for Molecular Biology. AAAI Press, Menlo Park, CA, pp. 134–141.

[17] Solovyev, V., Salamov, A. and Lawrence, C. (1994) Nucleic Acids Res. 22, 5156–5163.

[18] Hutchinson, G. and Hayden, M. (1992) Nucleic Acids Res. 20, 3453–3462.

[19] Burge, C. and Karlin, S. (1997) J. Mol. Biol. 268, 78–94.

[20] Borodovsky, M., McIninch, J., Koonin, E., Rudd, K., Medigue, C. and Danchin, A. (1995) Nucleic Acids Res. 23, 3554–3562.

[21] Salzberg, S., Delcher, A., Kasif, S. and White, O. (1997) Microbial gene identification using interpolated Markov models. Nucleic Acids Res. 26, 544–548.

[22] Zhang, M.Q. and Marr, T.G. (1993) Comput. Appl. Biosci. (CABIOS) 9(5), 499–509.

[23] Salzberg, S. (1997) Comput. Appl. Biosci. (CABIOS) 13(4), 365–376.

[24] Kozak, M. (1987) Nucleic Acids Res. 15(20), 8125–8148.

[25] Kozak, M. (1992) Crit. Rev. Biochem. Mol. Biol. 27, 385–402.

[26] Senapathy, P., M.B. Shapiro, and N.L. Harris (1990) Methods Enzymol. 183, 252–278.

[27] Mount, S., Burks, C., Hertz, G., Stormo, G., White, O. and Fields, C. (1992) Nucleic Acids Res. 20, 4255–4262.

[28] Mount, S., Peng, X. and Meier, E. (1995) Some nasty little facts to bear in mind when predicting splice sites. In: Gene-Finding and Gene Structure Prediction Workshop, Philadelphia, PA.

[29] Wu, T. (1996) J. Comput. Biol. 3(3), 375–394.

[30] Burset, M. and Guigo, R. (1996) Genomics 34(3), 353–367.

[31] Reese, M., Eeckman, F., Kulp, D. and Haussler, D. (1997) J. Comput. Biol. 4(3), 311–323.

[32] Fraser, C.M., Casjens, S., Huang, W., Sutton, G., Clayton, R., Lathigra, R., White, O., Ketchum, K., Dodson, R., Hickey, E., Gwinn, M., Dougherty, B., Tomb, J.-F., Fleischmann, R., Richardson, D., Peterson, J., Kerlavage, A., Quackenbush, J., Salzberg, S., Hanson, M., van Vugt, R., Palmer, N., Adams, M., Gocayne, J., Weidman, J., Utterback, T., Watthey, L., McDonald, L., Artiach, P., Bowman, C., Garland, S., Fujii, C., Cotton, M., Horst, K., Roberts, K., Hatch, B., Smith, H.O. and Venter, J.C. (1997) Genomic sequence of a Lyme disease spirochaete, *Borrelia burgdorferi*. Nature 390, 580–586.

PART III

Protein Structure Modeling and Prediction

S.L. Salzberg, D.B. Searls, S. Kasif (Eds.), *Computational Methods in Molecular Biology*
© 1998 Elsevier Science B.V. All rights reserved

CHAPTER 11

Statistical analysis of protein structures

Using environmental features for multiple purposes

Liping Wei, Jeffrey T. Chang and Russ B. Altman

*Section on Medical Informatics, Department of Medicine, Stanford University School of Medicine,
Stanford, CA 94305-5479, USA. Phone: +1 650-723-6979;
Fax: +1 650-725-3394; Email: {wei, jchang, altman}@smi.stanford.edu*

List of Abbreviations

3D	three-dimensional	PDB	Protein Data Bank
ATP	adenosine triphosphate	SCWRL	Side Chain placement With a Rotamer
LPFC	Library of Protein Family Cores		Library
NMR	nuclear magnetic resonance		

1. Protein structures in three dimensions

In its native state, a protein folds into a beautiful, three-dimensional (3D) structure [1]. Generally, the amino acid sequence of a protein can determine its three-dimensional conformation, which in turn determines the protein's functions. However, the precise manner in which sequence determines structure and structure determines function is not always clear. The 3D structures of proteins are, for the most part, determined experimentally by X-ray crystallography or nuclear magnetic resonance (NMR). In recent years, the number of protein structures determined experimentally and deposited in structural databases such as the Protein Data Bank (PDB) [2] has been increasing exponentially. However, the acquisition of this structural information is still a slow and expensive process, and the structural information is extremely valuable. For these reasons, careful scrutiny of the available structures to extract a maximum of useful information is critical. In the PDB, many of the known protein structures belong to the same structural or functional families. This offers us the opportunity to statistically study multiple structures with the same general three-dimensional shape (also called the "fold") or the same function. The advantage of statistical analysis on a large set of structures, rather than case-by-case analysis, is that it becomes possible to distinguish general, conserved features from individual idiosyncratic ones, to discover patterns that could only be revealed by pooled data, and to use these patterns for analysis of new sequences and structures.

This chapter is the first on topics related to protein 3D structures. It outlines a statistical approach to collectively analyzing protein structures, which are represented with physical and chemical properties defined at multiple levels of detail. As we will show, the statistical descriptions of protein families contain rich biochemical information and can be used to characterize the molecular environments within protein structures, recognize functional sites in new proteins, compare the environments around amino acids, and predict some protein 3D structures from their sequences.

In particular, we have implemented a system called FEATURE that can be used to describe 3D motifs within protein families. FEATURE has been used to detect sites within protein structures – such as binding sites for cations (calcium, magnesium, sodium) and small molecules such as ATP. FEATURE has also been used to compare the average environments around the twenty amino acids, in order both to understand the different chemical milieu in which they exist, and to create a similarity matrix for performing protein sequence alignments.

There are many ways to represent a protein structure. The most basic description includes the set of atoms in the structure, their three-dimensional coordinates, and their connectivity. Other useful representations can be a specification of the secondary structural segments (helices, β-strands, and loops), a plot of all the distances between amino acids, or a computation of the electrostatic field generated by the atoms in the protein. The system we will discuss in this chapter, called FEATURE, represents protein structures using the spatial distributions of a variety of physical and chemical properties [3]. When multiple structures from the same structural family or with the same functional site are aligned, one can apply statistical analysis to the ensemble of all structures and generate a statistical description of the family or site. These statistical descriptions are useful for a variety of applications discussed later in this chapter.

Due to the rapidity of gene sequencing and the increasingly successful efforts of the genome projects, thousands of DNA and protein sequences are determined every year. Each of the protein sequences must be associated with a function and, ideally, a 3D structure in order to fully understand the basic molecular biology of the organisms from which they are obtained. However, only a small number of these protein 3D structures can be determined by the traditional experimental methods which are both time- and labor-intensive. The full potential of the genome projects to catalyze the biotechnology, pharmaceutical, and chemical engineering industries depend on methods for rapid and accurate analysis of gene sequences. Thus, as the gene sequences come out of the automated sequencing machines and are stored in large databases, the problem of assigning structure and function becomes a fundamentally computational one. The most basic approach relies on aligning a new sequence with sequences whose structure or function is already known. In addition to seeking alignment of identical amino acids, these methods allow a substitution matrix to be used which gives partial match credit to amino acids that are in some way similar. This approach works well when the sequence identity (the number of exact matches in a high-quality alignment) between the two sequences is reasonably high (for example, greater than 25%). However, two proteins with sequence identity less than 20% can adopt the same fold, and simple sequence comparison approaches may fail to detect similarity. One way to address this problem is to test the compatibility of a new protein sequence with a set of previously determined folds. This is often called the "threading", or "fold recognition", approach to protein structure prediction [4–6]. It is based on the assumption that the side chains of an amino acid chain, folded into their native conformations, form energetically favorable interactions. Thus, when a sequence is mounted upon a backbone similar to its own, its side chains will form favorable local stabilizing interactions. Threading may be more sensitive than sequence alignment because it models interactions among amino acids, and not just sequential similarity.

Statistical descriptions of protein structures are useful in both of these structure prediction approaches. They can be used to construct amino acid similarity matrices for sequence comparison [7]. When sequence similarity is low, the same statistical descriptions can be used as the basis for threading methods.

Section 2 is a tutorial section that introduces the concepts underlying statistical comparison of two distributions, classification methods and evaluation, conditional probability, and Bayes' Rule. We provide the basic review in order to make the chapter relatively self-contained. More detailed information about statistical inference techniques can be found in refs. [8,9]. Readers familiar with the material can go directly to section 3, which discusses a method for representing and statistically analyzing protein structures. Section 4 includes four applications of the statistical approach. The chapter concludes with a brief summary.

2. Statistics, probability, and machine learning concepts

Determining the key features that distinguish two populations is a common statistical problem. Given all the properties that describe members from each population, which are most important for distinguishing between the populations? Given the set of distinguishing features, how should we classify a new individual into one of the populations, based on the attributes of that individual? The first task is called the inference problem, and the second is known as the classification problem. This section first introduces statistical inference methods, with emphasis on the Mann–Whitney Rank-sum test. It then discusses the classification problem and the evaluation of classification algorithms. There is an important class of algorithms that are based on Bayes' Rule. Section 2.3 will introduce Bayes' Rule and the notion of conditional probability.

2.1. Statistical inference from two sets of data

Quite often we are faced with inference problems involving comparison of two probability distributions based on two sets of data samples, with each set being randomly sampled from one of the distributions. For example, we often need to determine whether the mean values of two distributions are the same, based on our data samples. Many statistical methods have been developed to solve this problem, and they can be roughly divided into two categories: parametric and nonparametric [9]. Parametric methods, such as the z-test or t-test, should be used when both distributions are, or are close to, Gaussian normal. As long as the assumption of Gaussian normality is valid, the parametric methods are more powerful in that they are more likely to detect small differences when they exists. However, if either of the distributions is non-Gaussian, nonparametric methods must be used.

The Mann–Whitney Rank-sum test is a nonparametric test that compares two distributions with roughly the same shape, which need not be Gaussian. All the data points in the two sets are ordered (or ranked) by their value. If the two distributions are roughly equivalent, then the average rank of members originally from the first set should be about the same as the average rank of data points from the second set. Moreover, the data points from the two data sets should be mixed together evenly throughout the ordered set of combined data. If one distribution has a higher mean value than the other, then the average rank of its members will be lower than that of the members of the other distribution. The average rank for data from each set can be computed. Let T be the sum of the ranks for a data set with n_1 members when compared to a data set with n_2 members. When both data sets are drawn from the same distribution, and the larger of the data sets contains more than 8 members, the statistic T approaches the normal distribution with mean (μ_T) and variance (σ_T^2) given by

$$\mu_T = \frac{n_1(n_1 + n_2 + 1)}{2}, \qquad \sigma_T^2 = \frac{n_1 n_2}{12}(n_1 + n_2 + 1).$$

Thus, the z-statistic,

$$z = \frac{T - \mu_T}{\sigma_T},$$

measures the difference between the two distributions, and also has a standardized normal distribution. The statistical significance of z, the p-level (ranging from 0 to 1), can be looked up in any statistics books or widely available software. The p-level is the probability that the observed difference is due to chance alone. Thus, the lower the p-level, the more likely it is that the observed difference is significant.

The implementation of the Rank-sum test is somewhat more complicated when there are ties in the values of the data set (hence multiple data points with the same rank) and is beyond the scope of this discussion. For more details on the Rank-sum test and other statistical methods, see ref. [9]. Nevertheless, this test (also known as the Wilcoxon Rank test) is good at detecting difference in nonparametric distributions and can work with a fairly low sample size. We show one application of this test in section 4.

2.2. Classification problems and evaluation of classification algorithms

The problem of classification is that of assigning a new individual into one of a set of possible classes or populations on the basis of a set of features. Classification methods can be divided into two types: supervised classification and unsupervised classification. Unsupervised classification (also referred to as clustering methods) is used when the classes are not known ahead of time, and so the classes are defined by looking at individuals and grouping individuals together into clusters. Supervised classification is used when the classes have already been specified, and the task is to determine which features of the classes should be used to evaluate and assign a new unlabeled individual to a class. One example of a supervised classification problem is the task of distinguishing protein structures that bind calcium ions from those that cannot bind calcium. Proteins that bind calcium ions will have different structural and biochemical features than those that do not. Now, if given a new 3D protein structure, how can we predict if it will bind calcium? We look at its structural and biochemical features, compare them to those in the examples we have seen before, and decide whether to "classify" the region in the new structure as a calcium binding site or not.

There are many classification algorithms that address the classification problem by statistics, probability, or machine learning methods. Most of them infer a model (or mapping function) that relates the features with the categories based on the training data. The model will be used to classify new, previously unseen instances. A perfectly accurate algorithm should be able to classify each instance into its proper category. The accuracy of a classification algorithm can be defined by two criteria: sensitivity and specificity, defined by

$$\text{sensitivity} = \frac{\text{TP}}{\text{TP} + \text{FP}}, \qquad \text{specificity} = \frac{\text{TN}}{\text{TN} + \text{FN}},$$

where TP is the number of true positives, FP is the number of false positives, TN is the number of true negatives, and FN is the number of false negatives. For our calcium example, "sensitivity" measures the ability to detect when there really is a calcium binding site, and "specificity" measures the ability to reject regions that are not calcium sites.

When a model is built from a set of training data, it is not unusual for the model to "match" or "explain" the training data better than it explains other previously unseen sets of data. Very often models contain subtle sources of bias that make them perform better on the training data sets than they do on test data sets. Therefore, it is important to calculate sensitivity and specificity on a test data set not used in constructing the classification model. Otherwise the model may only memorize and recognize examples it was given, but may not have extracted the more general features necessary for classification of previously unseen examples. When a model is built that performs very well on the data used to construct it, but poorly on equivalent data not used in the construction, we say that the model "overfits" the data. Instead, a better approach is to calculate the sensitivity and specificity of the algorithm on a separate, previously unseen, test data set. It is common to divide the whole data set into a training set and a test set. Then, hiding the test set, construct a classification model using only the training set. Finally, calculate the sensitivity and specificity of the classification model on the test set. Although this train-and-test approach avoids the overfitting problem, it has other potential problems. First, the estimated accuracy may be influenced by the different ways of dividing the whole data set into training and test set if this division is in some way biased. Second, in some cases we have so little available data that we cannot afford to divide them. One solution to building and evaluating models with sparse data sets is to estimate accuracy using cross-validation. The complete data set is divided into n subgroups. For each subgroup, a classification model is built using the remaining $n-1$ subgroups. The model is then tested on the subgroup that was left out in order to measure its accuracy. This procedure is repeated for every subgroup. The overall accuracy will be the averaged accuracy from the n steps. Cross-validation has the advantages of being unbiased and making full use of the available data.

2.3. Conditional probability and Bayes' Rule

Some classification algorithms are based on Bayes' Rule. Before we introduce Bayes' Rule, we need to understand the principles of basic probability. Let the probability of an event S be $P(S)$, and the probability of another event E be $P(E)$. Denote the probability that S and E both occur, which is often called the joint probability, by $P(SE)$. The complement of event S is the event that S *does not* occur, and is often written as $\overline{S}$. We have

$$P(\overline{S}) = 1 - P(S).$$

The probability of event S, given the fact that event E has already occurred, is called the conditional probability of S given E, and is written as $P(S|E)$. For example, suppose we are interested in finding out if a site in a protein structure is a calcium binding site. Let S be the event of the site being a calcium binding site. Let E be the event of the site having five oxygen atoms in the shell about 2 Å away from the site center. Now $P(S|E)$ is the conditional probability of seeing a calcium binding site *given* that we have already seen five oxygen atoms in the 2 Å shell. Knowing that there are many oxygen atoms around the site makes the probability of its being a calcium binding site much greater. Given new evidence about the abundance or deficit of oxygen atoms, our estimate of the possibility of seeing a calcium binding site changes accordingly.

Bayes' Rule can be used to update conditional probabilities of events when given new evidence [8]. The "updated" conditional probabilities are useful in decision making. For one event and one piece of evidence, Bayes' Rule gives

$$P(S|E) = \frac{P(E|S)\,P(S)}{P(E|S)\,P(S) + P(E|\overline{S})\,P(\overline{S})} = \frac{P(E|S)\,P(S)}{P(E|S)\,P(S) + P(E|\overline{S})\,(1 - P(S))}.$$

The goal is to find the posterior probability $P(S|E)$. Unconditioned probability $P(S)$, often referred to as prior probability, needs to be specified from experience or previous data. The conditional probabilities of the evidence given different states of the event, $P(E|S)$ and $P(E|\overline{S})$, also need to be provided. Prior and posterior refer to probabilities before and after observing the evidence E. Bayes' Rule gives a simple method for computing the new probabilities of an event based on our previous best estimate and the information contained in the new evidence.

3. Statistical description of protein structures

In the FEATURE system, we represent proteins using the spatial distributions of critical properties. A comprehensive set of chemical and physical properties are used to describe structures at multiple levels of detail. The properties can be grouped roughly into categories based on whether they depend on the identity of the atom, the chemical groups to which they belong, the amino acids to which they belong, the secondary structures to which they belong, or some other properties that can be computed knowing the protein structure. A summary list of one set of properties is shown in Table 1. This list is representative, and the detailed list of the properties and their definitions can be found in ref. [3]. Other properties of interest may exist, and they may certainly be added.

Our FEATURE method starts with a 3D protein structure as reported in the Protein Data Bank (PDB [2]). A three-dimensional grid keeps track of the locations of the properties

Table 1
A summary list of chemical and physical properties used by FEATURE

Atom-based properties	Atom types; hydrophobicity; charge; positive charge; negative charge
Residue-based properties	Residue types; hydrophobicity classification
Chemical group-based properties	Hydroxyl; amide; amine; carbonyl; ring system; peptide
Secondary structure-based properties	Secondary structure classification
Other properties	VDW-volume; B-factor; mobility; solvent accessibility

of the protein's atoms. The grid is cubic, with a unit-cell diagonal chosen to be the length of a carbon–oxygen single bond, so that two nearby atoms rarely occupy the same cell. The axes of the grid are determined by the coordinate system specified in the PDB file. Each grid box stores the values of the properties for the atom(s) it contains.

Very often, a small region of the protein structure (sometimes called a site) is of particular interest. A site is a microenvironment within a structure, distinguished by a structural or functional role. A site includes the whole local neighborhood around the center of interest within a certain distance (10 Å, for example). (The dimensions of medium-sized proteins are 30 to 100 Å, on average.) In order to study the spatial distribution of the properties within a site, the FEATURE system combines the grid cells into bigger spatial volumes and sums the property values in the grid cells that fall within the volume. (The individual grid cells are too small to analyze for significant occurrence of properties.) For example, FEATURE can combine grid cells into radial shells around the central point of interest. It can also collect grid cells into oriented volumes to preserve detailed spatial relationships [10]. Whatever the strategy for dividing space into smaller volumes, the final product from FEATURE is a set of summed property values, one value for every property at every volume.

To take advantage of the increasing number of protein structures available and, more importantly, to avoid characterizing the idiosyncrasies of a single protein, one can analyze multiple, aligned sites with the same structural or functional characteristic, and compare them to a set of background "control" nonsites – microenvironments that do not have the specific structural or functional characteristic. For each property, at each spatial volume, there are a list of values from the sites, and a list of values from the nonsites. From these, we have the basic lists of numbers for sites and nonsites that form distributions that can be analyzed with various statistical approaches for different purposes.

4. Applications of statistical analysis of protein structures

What is the value in collecting these statistics on the microenvironments of protein structures? First, one can characterize the key features of protein sites with important structural and functional roles. Then, the characterizations can be used to recognize sites in new protein structures that have not been analyzed. The description is also useful in protein structure prediction, first by constructing substitution matrices based on a comparison of amino acid environments, and second by recognizing the compatibility of protein structure environments with sequences using threading methods.

4.1. Characterizing the microenvironment surrounding protein sites

The first use of FEATURE [3] is simply to assist in the analysis of protein structure and function, in order to identify the special features of a structural or functional site, by comparing examples of sites with examples of control nonsites. For this purpose, FEATURE looks at corresponding volumes in sites and nonsites and reports those volumes in which there is a significant difference in the abundance of particular properties. Structures with common shape or function should share certain features in their physical layout, and these should be discerned by comparison with other structures which lack these shared features.

As discussed in section 3, for each site and nonsite, the property values are computed and stored in the grid. A common strategy is to divide the space of interest into radial shells of thickness 1 Å and sum the property values within each shell. In order to determine if a property within a shell has significantly different distributions for sites and nonsites, FEATURE uses the Rank-sum test (discussed in section 2.1) to compare the list of values for that property in the sites with the list of values from nonsites. The significance level of the observed difference is given by the p-level returned by the Rank-sum test. If the p-level is lower than a specified cutoff value (for example, 0.01), that property at that shell is reported to be significantly different for the sites compared to nonsites.

It may be surprising that looking at the environment surrounding a site with concentric radial shells is sufficient to pick out many key features. Indeed, the analysis of sites can sometimes benefit from a more localized division of space, so that the spatial relationships are preserved and not radially averaged. FEATURE can also analyze properties in oriented blocks [10]. These analyses often require more data since the total volume being compared are smaller and so need better sampling to reliably detect differences.

FEATURE has been used to characterize a number of microenvironments, including calcium binding sites, disulfide bonding cysteines, serine protease active sites, ATP binding sites, and sodium binding sites. It has found many of the previously observed features of the sites as well as some novel findings. FEATURE summarizes the findings in a graphic plot, an example of which – the description of the microenvironment of ATP binding sites – is shown in Fig. 1. Some of the significant findings are summarized below:
(1) There is a deficit of atoms from 0 to 4 Å, reflecting the empty binding pocket.
(2) There are significantly more sulfur atoms at radius 10 Å than observed in the random control nonsites.
(3) There is an abundance of positive charges at radii 3 to 10 Å, and an abundance of negative charges at radii 10 to 12 Å. The ATP molecules are surrounded by positive charges, and the positive charges are, in turn, surrounded by an outer shell of negative charges.
(4) There are significantly more ASP, LYS, GLY, and MET residues in ATP binding sites observed in controls.
(5) There is an abundance of charged residues at radii around 7 to 12 Å.
(6) There is an abundance of both helices and β-strands.
  It is well known that the charged ASP and LYS residues are important in binding ATPs.

215

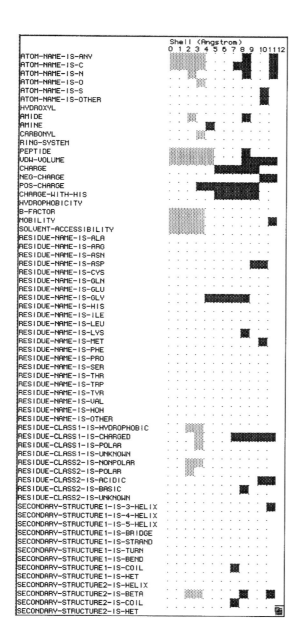

Fig. 1. Description of the microenvironment of an ATP binding site. Along the vertical axis are properties, and along the horizontal axis are shell volumes (for example, shell 1 is the shell from 1 to 2 Å around the origin). Dark-gray cells mark property–volume pairs for which the ATP binding sites have significantly high values; the light-gray are cells for which the ATP binding sites have significantly low values. Empty cells show no significant difference.

Thus, it is not surprising that they are over-represented in the ATP binding sites (as shown in Fig. 1). Their presence also causes the abundance of charges in the environment. Glycine is a conserved residue, because any larger residue causes steric hindrance to the binding of the ribose ring [11]. Our observation that helices and β-strands are abundant in ATP binding sites is consistent with the secondary structural motif found in nucleotide-binding proteins – the Rossmann Fold [11]. We currently have no feature corresponding to supersecondary structural motifs. Adding such a feature to the system would allow it to recognize the nucleotide binding fold directly, instead of inferring it as a collection of low level features – helices and β-strands.

4.2. Recognizing protein sites using environmental statistics

A second use of FEATURE is to recognize microenvironments using the description model built by comparing sites with nonsites. Given a new, unannotated protein structure, one of the first things we want to know is whether it has any functional sites. Although we can perform biochemical experiments to elucidate functional sites, they are often expensive. Therefore, computational methods offer a useful screening strategy. Given a protein structure, we would like to have an automated method for recognizing a functional site of interest. If we use known examples of sites and nonsites, the problem becomes one of supervised classification. In the last section, we established that FEATURE statistically characterizes the properties of sites by comparing them with nonsites. In this section, we will outline a general purpose, automated recognition system that uses the distinguishing features of a site and a description of a structure to decide if the structure contains the site [12]. In order to explain its decision, the system generates a prose report of those features which contribute most to its classification decision.

As described, FEATURE calculates the spatial distributions of properties in the example sites and nonsites. The values of these properties in the region of interest can also be computed. Conceivably, if the region is a site, then the property values for the region should be similar to the property values of the example sites. How can we assess similarity, since the ranges of property values for the sites and nonsites may overlap? How can we decide whether the observed values from the new structure are more compatible with the statistical distribution of site values or the nonsite values? We use a Bayesian scoring function to score the likelihood that a property value is drawn from the site distribution. For one property within one spatial volume, the posterior probability that the property value, v, of the new structure, comes from the statistical distribution of site values can be calculated using Bayes' Rule:

$$P(\text{site}|v) = \frac{P(v|\text{site})\,P(\text{site})}{P(v|\text{site})\,P(\text{site}) + P(v|\text{nonsite})\,P(\text{nonsite})}, \qquad (1)$$

where $P(\text{site})$ and $P(\text{nonsite})$ are the prior probabilities of site and nonsites, and need to be provided to the system. They are based on the overall probabilities of seeing a site of interest, and they must be estimated. $P(v|\text{site})$ is the probability of seeing the observed value in a site, while $P(v|\text{nonsite})$ is the probability of seeing the observed value in a nonsite. These probabilities can be evaluated by dividing the total range of site and nonsite values into bins, and comparing the counts in each bin for sites and nonsites. Given the observed value, v, in a new structure, we identify the bin in which v falls and assign $P(v|\text{site})$ to be the ratio of counts in that bin for sites over the total number of counts.

Protein: 1ARC
Center of site of interest: (-2.284 -4.002 10.663)

Likelihood score: -86.063
It is most probably not an ATP-binding site.

Strongest evidence SUPPORTING that it is a site:
  VDW-VOLUME at shell 0, 1, 2 is low.
  ATOM-NAME-IS-ANY at shell 0, 1 is low.
  ATOM-NAME-IS-C at shell 1, 2 is low.
  RESIDUE-CLASS2-IS-ACIDIC at shell 10 is high.

Strongest evidence AGAINST that it is a site:
  RESIDUE-CLASS1-IS-CHARGED at shell 7, 8, 9 is low.
  RESIDUE-NAME-IS-ASP at shell 10 is low.
  POS-CHARGE at shell 5, 6 is low.
  NEG-CHARGE at shell 10, 11 is low.
  CHARGE at shell 5, 6, 7, 8 is low.
  VDW-VOLUME at shell 8, 10, 11 is low.
  VDW-VOLUME at shell 4 is high.
  MOBILITY at shell 11 is low.
  MOBILITY at shell 3 is high.

Fig. 2. An example of prose explanation: why the binding pocket in serine protease 1ARC is identified as not an ATP binding site. The key properties are presented in order of strength of evidence, grouped by whether they are supporting or refuting evidence.

The score for the overall likelihood of the query region being a site is the sum of the logarithm of the probability update ratios:

$$\sum_{\substack{\text{properties at} \\ \text{associated volumes}}} \log \left(\frac{P(\text{site}|v)}{P(\text{site})} \right). \tag{2}$$

A detailed explanation of the evidence contributing to the classification of a site is a useful adjunct to the actual classification result. FEATURE generates a report of the strongest individual pieces of evidence contributing to the total score in eq. (2). This report can be used to understand how different features contribute to the classification and to provide information about which properties would need to be modified in order to change the degree of fit between the new structure and the site/nonsite models. Figure 2 shows an example of a report explaining why the binding pocket in a serine protease structure (PDB identification number 1ARC) is not an ATP binding site. 1ARC binds N-P-tosyl-L-Lys chloromethyl ketone hydrochloride (TLCK), a molecule similar to ATP in both shape and size [13]. Thus, the binding pocket in 1ARC has similar geometric properties to ATP binding sites, shown in "strongest evidence SUPPORTING that it is a site" in Fig. 2. However, the "strongest evidence AGAINST that it is a site" suggest that the binding pocket does not possess other important biochemical and biophysical properties necessary for binding ATP. Since there is more evidence suggesting that the pocket does not have the right microenvironment to bind ATP, the pocket should be classified as not being an ATP binding site.

Fig. 3. Scanning the structure of phosphotransferase (PDB id 1CSN) for ATP binding sites. The top 20 highest scoring points are marked with dots. Only the backbone of the protein structure is shown. An ATP molecule is depicted at the actual binding sites, with the actual orientation, for comparison with the computational results.

FEATURE has been successfully applied to recognizing several ionic binding sites and ligand active sites [12]. The accuracy of this method has been evaluated using sensitivity and specificity, computed both on independent test sets and in cross-validation. The sensitivity and specificity typically range from 86% to 100%. Extensive sensitivity analyses have been performed on parameters of the program such as the prior probability of site, the number of bins, the size of the environment, and significance cutoff. These studies have shown that the method is robust to a wide range of values for each parameter. In particular, the method performs consistently over a wide range of prior probabilities – the most difficult parameter to assess accurately.

FEATURE has also been successfully used to scan whole protein structures for the occurrence and location of sites, such as calcium binding sites and ATP binding sites. First, a search grid is defined on top of a new structure. Then, the recognition function is applied at each grid point to compute the likelihood that the region around that point is a site. Finally, high-scoring grid points are labeled and can be visualized graphically. Figure 3 shows a typical result of scanning the structure of phosphotransferase (PDB identification number 1CSN) for ATP binding sites. The actual ATP binding site in phosphotransferase, experimentally determined by X-ray crystallography, is clustered and well localized by high-scoring points. There are a few high-scoring points that are far away from the actual binding sites, and they should be considered false positives according to PDB documentation which we use as gold standard.

4.3. Comparing the environments of amino acids and constructing a substitution matrix

The techniques developed in the previous sections provide a general framework for describing and recognizing the patterns in the biochemical environments of a set of protein sites. An interesting application of this technology is to study the environments found around each of the amino acids. We have created statistical descriptions of the biochemical features found around each of the amino acids [7]. These descriptions provide insights into the roles that the 20 amino acids play within protein structure. Comparing the similarities and differences in the environments between pairs of amino acids provides information on their likelihood to mutate into one another.

For our analysis, ~100 instances of each amino acid were randomly selected from a set of 20 nonhomologous proteins [14]. The environment around each of the amino acid instances was described by FEATURE as a spatial distribution of physical and chemical properties. FEATURE compared the environments of each amino acid (considered as sites) with random control environments (considered as nonsites), then compared each amino acid environment (site) with each of the other amino acid environments (nonsites). All significantly different features between each amino acid environment and the random control environments, and different features between any two amino acid environments, were reported. Figure 4 shows two examples of comparisons of amino acid environments. Graphical descriptions of the all comparison results are available on the WWW at

```
http://www.smi.stanford.edu/projects/helix/pubs/wacpsb/
```

During the process of evolution of a protein sequence, one amino acid may be substituted by another at some location. The tolerance of a microenvironment for one amino acid or another is not only a function of the properties of the lost amino acid, but also a function of the properties of the environment surrounding the lost amino acid. We used the statistical descriptions of amino acid environments to construct a substitution matrix (the WAC matrix) to illustrate the validity and utility of comparing amino acid environments.

When one compares two amino acid environments using FEATURE, the Rank-sum test yields a z-value that describes the normalized difference between the two amino acid environments for a property at a particular volume. The overall difference between the two amino acid environments can be estimated by taking the sum of the squared z-values for each property and volume. This yields a single number that represents comparatively the difference between the features found in the environments of the amino acid pair. If

220

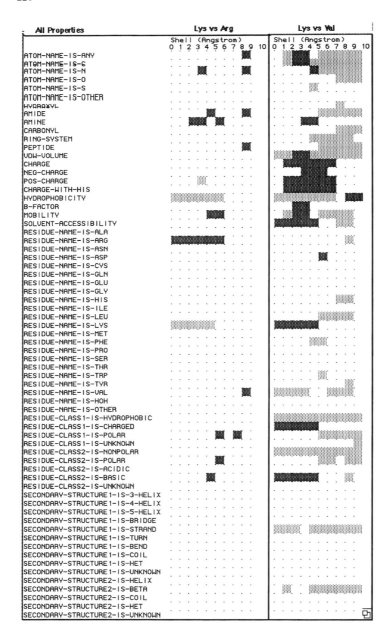

Fig. 4. Example comparison results of environments around amino acids. The left-hand column lists all the properties. All properties are analyzed at 1 Å volumes up to 10 Å from the C$_\beta$ carbon. The volumes for which the properties were significantly more prevalent for the first sites are marked in light gray, and dark gray for the second. "LYS versus ARG" shows much fewer significantly different properties than "LYS versus VAL" indicating that LYS is much more similar to ARG than it is to VAL.

C	4																			
S	0	4																		
T	0	1	4																	
P	-1	0	0	4																
A	0	0	0	0	4															
G	0	0	0	0	-1	4														
N	0	1	1	0	0	0	4													
D	-2	-1	-1	-1	-2	-2	0	4												
E	-2	-2	-1	-1	-3	-2	0	0	4											
Q	0	0	1	0	0	0	1	-1	0	4										
H	0	-1	-1	0	0	-1	0	-1	-1	0	4									
R	-1	-1	0	0	-1	-1	0	-1	-1	0	0	4								
K	-2	0	0	0	-1	-1	0	-1	-1	0	0	2	4							
M	1	0	0	1	1	0	0	-1	-1	1	0	0	0	4						
I	0	-1	0	0	0	-2	-1	-4	-4	0	-1	-2	-2	2	4					
L	0	-1	0	0	0	-1	-1	-3	-3	0	-1	-1	-2	2	1	4				
V	0	-1	0	0	1	-2	-1	-3	-4	-1	-1	-2	-3	1	2	1	4			
F	0	-2	-2	0	0	-2	-1	-4	-4	0	0	-2	-2	2	0	1	0	4		
Y	0	-1	-1	-1	0	-2	0	-3	-4	1	0	-1	-2	0	0	0	0	1	4	
W	0	-1	-1	0	0	-2	-1	-3	-2	0	0	0	-2	2	0	1	0	2	1	4
	C	S	T	P	A	G	N	D	E	Q	H	R	K	M	I	L	V	F	Y	W

Fig. 5. The WAC substitution matrix computed by comparing the amino acid environments. The other half can be generated by symmetry.

the two environments are similar, then there are not many distinguishing features with high z-values. If they are different, then there are many distinguishing features. The summed difference of the amino acid pairs were used to create a 20×20 substitution matrix, shown in Fig. 5. To evaluate the performance of WAC, the sensitivities of three substitution matrices, WAC, BLOSUM62, and PAM250 [15] were compared in detecting related protein sequences in the same family (as defined in PROSITE [16]). Although BLOSUM62 was the most sensitive matrix overall, WAC was more sensitive for some families and showed overall performance similar to PAM250.

4.4. Threading a protein sequence onto a fold

The final use of FEATURE's statistical description that we will discuss is in the area of protein threading or fold recognition. Given a query sequence, we want to determine if it can adopt any of the known 3D folds in the PDB. In fold recognition, the query sequence is aligned to each structure in a library of folds. Using a fitness scoring function, an alignment algorithm searches for the sequence-fold alignment with the highest score. Then, the scores for the alignment of the sequence to each fold in the library are ranked to find the most probable fold for the sequence. For a step-by-step review of threading methods, see ref. [17] and chapter 12 of the present volume. Here, we will describe the set of choices we have made for our fold library, alignment methodology, and fitness function.

4.4.1. Library of folds

The Library of Protein Family Cores (LPFC) is a resource with information about the folds adopted by protein families. It can be found at

`http://www.smi.stanford.edu/project/helix/LPFC`

LPFC is based on the FSSP database, which clusters protein structures based on structural similarity. For each fold family, LPFC provides the multiple structural alignment of the protein structures within the family and a consensus backbone structure that includes all common aligned backbone positions that are found in each family member – called the core positions. Positions that are occupied by some – but not all – fold family members are not included in the set of core positions. If there is a sequence of consecutive core positions in the multiple alignment that all belong to the core, it is called a "core segment". The core backbone positions provide the scaffold upon which a query sequence is mounted in order to assess its compatibility with the fold family.

When all the members of a fold family have been structurally aligned, then each core position (centered at the α-carbon of the core amino acid) can be described statistically using FEATURE. The core positions are the sites, and randomly selected α-carbon environments are the nonsites. Thus, the fold family can be considered a string of distinctive microenvironments which can be used to recognize structure. Query sequences can be mounted (or threaded) onto the structural backbone of the core, and then evaluated to determine if their side chain interactions are similar to those observed in the natural structures.

The description of the α-carbon environments can be used in two ways to identify matches between query sequences and family fold core models. In the first method, based on environmental profiles, the statistical description of each core position is compared to the statistical description of each amino acid (as described in previous sections) [4]. A score is determined to summarize the match between the two descriptions. For each core position, the environment of each structure in the fold family is extracted as sites. In addition, the environment of ~100 instances of each amino acid is also collected. Then, the difference in the environments of each amino acid to the core position is scored in a manner similar to that used to construct the substitution matrix. The squared z-scores representing the statistical differences for each property of the two environments are summed to yield a score representing the total difference in the two environments. Repeating this for each core position yields a matrix that provides the score comparing each amino acid with each backbone environment. A new sequence can be aligned to the core backbone optimally using dynamic programming sequence alignment algorithms and the matrix provided by this analysis as a score matrix. If the score of a query sequence resembles the scores of sequences which are known to adopt the fold, then it is likely that the query sequence also adopts that fold.

This algorithm is relatively simple to implement, but it loses information because it compares core backbone environments with average side chain environments, and not with the detailed side chain environments created when the query sequence is mounted onto the core backbone.

The second fold recognition method is based on detailed scoring of a threaded structure with environmental descriptions. The statistical description of each core backbone position is used to score a query sequence that has been aligned with the core backbone. The side chains of the query sequence are instantiated upon the backbone to produce a detailed atomic model. The compatibility of the atomic model and the statistical description at each core backbone position can be scored using our site recognition algorithm.

One can view a protein structure as a string of microenvironments surrounding the α-carbons within the sequence. Given a set of aligned structures from a fold family, we can generate a statistical description of the environments surrounding each α-carbon. If we take a template backbone and mount the side chains of a new sequence upon the template, then we can compute the score of each α-carbon environment (created by the side chains as they hang off the template backbone), and sum these to get an overall measure of the compatibility of this sequence to the template backbone. This is the essence of how we use FEATURE for sensitive threading.

After obtaining a structural alignment file from the LPFC database, we select structures with less than 25% sequence identity. Using the average backbone fold of the proteins in the multiple alignment, the structures within the fold family are superimposed, thus creating a set of environments around each α-carbon position. To extract the statistical environmental descriptions, FEATURE is applied to the region surrounding each of the core α-carbons within the training proteins. For the nonsite controls, α-carbons are randomly selected from twenty nonhomologous proteins. FEATURE collects information about the key biochemical and biophysical features of the microenvironments around the aligned α-carbons, as compared with the randomly chosen α-carbons from these proteins as previously described [3,18]. The analysis is centered at the α-carbon of the sites and nonsites, and a threshold of $p=0.01$ is used to determine if the differences in the environments for given properties at specific radii were statistically significant.

4.4.2. Alignments

Since the core positions typically occur within a set of ungapped core segments, the task of generating an alignment is equivalent to generating loop lengths between the core segments. For each query sequence (that sequence not used in training the model), we generate possible alignments with the core. We generate random sample alignments using information about the average loop lengths gathered from the multiple alignment file.

Using the multiple sequence alignment from the LPFC, we can compute the mean length of each known loop, its variance, and its covariance with other loop lengths. Assuming a multivariate, correlated normal distribution of loop lengths, we can generate sample gap lengths that follow a similar distribution as the known protein fold family members. Alternatively, we can use standard sequence alignments and then systematically vary them by shifting segments to the left and right to generate a good sample of plausible alignments of the query sequence with the known family members. With these alignments, we can search for high-scoring alignments, and determine whether any alignment scores high enough to reliably classify the query sequence as belonging to the fold family.

4.4.3. Generating and scoring actual and sample structures

Given a test alignment, we can not score the environment around the α-carbon using FEATURE until the side chains of the query sequence are mounted upon the template backbone. The SCWRL (Side Chain placement With a Rotamer Library) program uses a backbone-dependent rotamer library to place side chains onto a protein backbone [19]. We therefore supply SCWRL with the core backbone positions and the alignment of the query sequence with the backbone. SCWRL positions the side chains onto the template backbone and then performs some optimizations to minimize steric clashes. The SCWRL program for placing side chains is able, in our tests, to reproduce crystal structure side chains to about 1.5–2 Å rmsd. After attaching the side chains onto the core backbone, we use the scoring method to determine the compatibility of the newly created environments with those observed in real family members.

It is important to note that the quality of the threading results is dependent on the quality of the structural alignment in the FSSP resource. Any inaccuracies in this alignment will affect the results. In fact, improvements in the alignment would most likely lead to a more pronounced difference in the score between the correct alignment and incorrect alignments, since the environmental descriptions would be more precisely defined.

5. Summary

This chapter has described an approach for statistical analysis of the three-dimensional features within protein structures and has shown that the method can be used to: (1) characterize protein sites, (2) recognize protein sites in new structures, (3) compare amino acid environments, (4) generate an amino acid substitution matrix, (5) thread query sequence with environment profiles, and (6) thread query sequence with detailed environments. Until recently, much effort has focused on statistical analysis of sequence information. Although powerful methods have been developed, they are always limited by the reality that sequence is often a proxy for structure and function. Statistical analysis of aligned 3D structures has now become feasible as the number of experimentally determined structures increases. As the database of structures continues to grow, our ability to describe the conserved structural and functional features in a sensitive and specific manner will improve. The availability of large data sets will increase the power of the methods described in this chapter, as they are applied to problems of protein structure analysis and engineering. We also expect that analysis of structure will create new insights about function.

Acknowledgments

This work has been supported in part under grants from the Culpeper Foundation, the National Institutes of Health (NIH-LM05652, LM-06244, LM-05305), the National Science Foundation (NSF-BIR9600637), and the IBM corporation. Steve Bagley, Allison Waugh and Michelle Whirl have participated in the construction and/or testing of the methods described in this chapter. Kevin Lauderdale and Michelle Whirl have read and commented on the text. We also thank Dr. Simon Kasif and Dr. Steven Salzberg for their editorial comments.

References

[1] Creighton, T.E. (1993) Proteins: Structures and Molecular Properties, W.H. Freeman and Company, New York.

[2] Bernstein, F.C., Koetzle, T.F., Williams, G.J.B., Meyer, E.F.J., Brice, M.C., Rodgers, J.R., Kennard, O., Shimanouchi, T. and Tasumi, M. (1977) J. Mol. Biol. 112, 535–542.

[3] Bagley, S.C. and Altman, R.B. (1995) Protein Sci. 4, 622–635.

[4] Bowie, J.U., Lüthy, R. and Eisenberg, D. (1991) Science 253, 164–170.

[5] Bryant, S.H. and Lawrence, C.E. (1993) Proteins: Struct. Funct. Genet. 16, 92–112.

[6] Sippl, M.J. and Weitckus, S. (1992) Proteins: Struct. Funct. and Genet. 13, 258–271.

[7] Wei, L., Altman, R.B. and Chang, J.T. (1997) In: R.B. Altman, A.K. Dunker, L. Hunter and T.E. Klein (Eds.), Proc. Pacific Symp. on Biocomputing. World Scientific, Singapore, pp. 465–476.

[8] Ott, L.R. (1993) An Introduction to Statistical Methods and Data Analysis, Wadsworth Publishing Company, Belmont, CA 94002.

[9] Glantz, S.A. (1987) Primer of Biostatistics, McGraw-Hill, New York.

[10] Bagley, S.C., Wei, L., Cheng, C. and Altman, R.B. (1995) In: C. Rawlings, D. Clark, R. Altman, L. Hunter, T. Lengauer and T. Wodak (Eds.), 3rd Int. Conf. on Intelligent Systems for Molecular Biology. AAAI Press, Cambridge, pp. 12–20.

[11] Rossmann, M.G., Moras, D. and Olsen, K.W. (1974) Nature 250, 194–199.

[12] Wei, L. and Altman, R.B. (1998) In: R.B. Altman, A.K. Dunker, L. Hunter and T.E. Klein (Eds.), Proc. Pacific Symp. on Biocomputing. World Scientific, Singapore, pp. 497–508.

[13] Tsunasawa, S., Masaki, T., Hirose, M., Soejima, M. and Sakiyama, F. (1989) J. Biol. Chem. 264, 3832–3839.

[14] Holm, L. and Sander, C. (1994) Nucleic Acids Res. 22, 3600–3609.

[15] Dayhoff, M.O., Schwartz, R.M. and Orcutt, B.C. (1978) Atlas Protein Sequence Struct. 5(Suppl. 3), 345–352.

[16] Bairoch, A. (1991) Nucleic Acids Res. 19(Suppl.), 2241–2245.

[17] Fischer, D., Rice, D., Bowie, J.U. and Eisenberg, D. (1996) Faseb J. 10, 126–136.

[18] Altman, R.B., Whirl, M., Waugh, A., Wei, L. and Chang, J.T. (1997) Stanford Medical Informatics Technical Report SMI-97–0682.

[19] Bower, M.J., Cohen, F.E. and Dunbrack, R.L.J. (1997) J. Mol. Biol. 267, 1268–1282.

S.L. Salzberg, D.B. Searls, S. Kasif (Eds.), *Computational Methods in Molecular Biology*

CHAPTER 12

Analysis and algorithms for protein sequence–structure alignment

Richard H. Lathrop[1], Robert G. Rogers Jr.[2], Jadwiga Bienkowska[2],
Barbara K.M. Bryant[3], Ljubomir J. Buturović[4], Chrysanthe Gaitatzes[2],
Raman Nambudripad[5], James V. White[2,6] and Temple F. Smith[2]

[1] *Department of Information and Computer Science, 444 Computer Science Building,
University of California, Irvine, Irvine, CA 92697-3425, USA;* [2] *BioMolecular Engineering Research Center,
Boston University, 36 Cummington Street, Boston, MA 02215, USA;* [3] *Millennium Pharmaceuticals, Inc.,
640 Memorial Drive, Cambridge, MA 02139, USA;* [4] *Incyte Pharmaceuticals, Inc., 3174 Porter Drive,
Palo Alto, CA 94304, USA;* [5] *Molecular Computing Facility, Beth Israel Hospital, 330 Brookline Avenue,
Boston, MA 02215, USA;* [6] *TASC, Inc., 55 Walkers Brook Drive, Reading, MA 01867, USA*

Chapter overview

This chapter discusses analytic and algorithmic results for computational protein structure prediction by protein sequence–structure alignment, an approach also known as protein threading. Biological results are beyond the scope of this chapter, but may be found in refs. [1–5]. See also chapter 13 by David Jones in this volume, which discusses another approach to protein threading.

The chapter visits in turn: motivation, intuition, formalization, analysis, complexity, algorithm, computational cost, discussion, and conclusions. The early sections are tutorial in nature; the rest represent original research results. The overall conclusions are that: (1) computational techniques can render vast search spaces approachable by exploiting natural constraints; and (2) advances in knowledge-based objective functions and protein structural environment definitions represent an important opportunity for future progress.

A long-range goal of this work is to integrate structural and functional pattern recognition. The reader will notice that gapped block alignment is conceptually similar to block patterns, consensus patterns, weight matrices, profile patterns, and hierarchical patterns, among many other gapped block pattern methods (reviewed in ref. [6]). Combined structural and functional pattern recognition is likely to prove more powerful than either one alone.

1. Introduction

Simply stated, the protein folding problem is to transform information. The input is a string of characters drawn from an alphabet of 20 letters. In the simplest case, the desired output annotates each character with three numbers, giving its *XYZ* coordinates in the protein's three-dimensional folded shape. Surprisingly, in many cases these coordinates are unique and depend only on the input string. There, protein structure prediction from sequence simply transforms implicit information into an explicit final form.

The protein folding problem is also the premiere computational problem confronting molecular biology today: it has been called the "holy grail of molecular biology" and "the second half of the genetic code" [7]. It is important because the biological function of proteins (enzymes) underlies all life, their function is determined by their three-dimensional shape, and their shape is determined by their one-dimensional sequence. The importance of computational solutions is escalating rapidly due to the explosion of sequences and genomes becoming available, compared to the slow growth in the number of experimentally determined three-dimensional protein structures.

The problem is unusually accessible to computer scientists because it is (in its essence) a pure information processing transformation, from implicit to explicit. No single computer program would so transform the face of experimental molecular biology practice today as one that correctly, reliably, and rapidly computed this function. It is a Grand Challenge problem for computer science [8].

1.1. Why is it hard?

The problem, although simply stated, is quite difficult. The process by which nature folds the string is complicated, poorly understood, and most likely the global sum of a large number of weak, local, interacting effects. Quantum mechanics provides a solution in principle, but the computation becomes intractable when confronted with the many thousands of atoms comprising a protein.

The direct approach to protein folding, based on modeled atomic force fields [9,10] and approximations from classical mechanics, seeks to find the folded conformation having minimum free energy. This is difficult because a folded protein results from the delicate energetic balance of powerful atomic forces, and because the vast number of possible conformations poses a formidable computational barrier. The forces involved are often difficult to model accurately, and include stabilizing and destabilizing terms making large contributions of opposite sign summed over a very large number of atoms [11]. Thus, small cumulative approximation errors may dominate the smaller net stabilization. For technical reasons it is difficult to model surrounding water properly [10,12], yet hydrophobic collapse is believed to be the main effect driving protein folding. Classical macroscopic parameters such as the dielectric constant become problematic at the atomic level. We may not know the protein's cellular folding context, which may include chaperon proteins, post-translational modifications, and hydrophobic interfaces to which the protein conforms. The search space may exceed 10^{50} plausible folded conformations even for medium-sized proteins. Simulation time-steps are measured in femtoseconds while folding time scales of interest are measured in milliseconds, a ratio of 10^{12}. Unless sophisticated methods are used, the basic time-step computation is $O(N^2)$ where N may approach 10^6 atoms with surrounding water. The simulation time may exceed 10^{12} CPU-years at current supercomputer speeds. The direct approach has been applied successfully to smaller molecules, but as yet faces stiff challenges for large proteins [12,13], though recent versions using cruder force fields are promising [14,15].

One important alternative approach is to use the wealth of information contained in already known protein structures. The structures can serve as spatial folding templates, impose constraints on possible folds, and provide geometrical and chemical information.

This is an attractive strategy because proteins exhibit recurring patterns of organization; there are estimated to be only around 1000 to 10 000 different protein structural families [16,17]. In this approach, the known structure establishes a set of possible amino acid positions in three-dimensional space. These template spatial positions generally include only the backbone atoms, though sometimes the implied beta-carbon is used as well. The highly variable surface loops are not included in the template positions. Based on topological and physicochemical criteria, an alignment of an amino acid sequence to the set of positions in one such core template is chosen. Each amino acid of the sequence is given the three-dimensional coordinates of the template position to which it is aligned. Estimation of the complete structure still requires some means of assigning positions to the amino acids in the loop regions [18,19], of assigning amino acid side-chain orientations and packing [20,21], and of searching the immediate structural neighborhood for a free energy minimum [9,10]. In this chapter we focus on the choice of core template and the method of identifying the optimal alignment of the sequence to that core template.

Initially, such methods employed primary sequence string similarity between the candidate sequence and the structure's native sequence in order to perform the alignment ("homology modeling" or "homological extension"). Computing the sequence similarity yields a direct alignment of amino acids in the candidate's and structure's sequences [18, 22]. In cases where the sequence similarity is high this is still the most successful protein structure prediction method known. Unfortunately, it is of limited generality because novel sequences rarely have sufficiently high primary sequence similarity to another whose structure is known. Indeed, of the genomic sequences known at present, fully 40% have no homologs to any sequence of known function, let alone known structure [23].

1.2. Why threading?

Many evolutionarily unrelated sequences (non-homologs) contain similar domain folds or structural cores [16,17,24–26]. Recently, approaches have been devised which exploit this fact by aligning a sequence directly to a structure or structural model. The process of aligning a sequence to a structure and thereby guiding the spatial placement of sequence amino acids is referred to as "threading" the sequence into the structure [27,28], and "a threading" means a specific alignment between sequence and structure (chosen from the large number of possible alignments). In this way "threading" specializes the more general term "alignment" to refer specifically to a structure (considered as a template) and a sequence (considered as being arranged on the template).

These new approaches exploit the fact that amino acid types have different preferences for occupying different structural environments (for example, preferences for being in alpha-helices or beta-sheets, or for being more or less buried in the protein interior). Additionally, some of the new approaches also exploit the fact that there appear to be distinct preferences for side-chain contact (e.g., contact potentials [29]), or more generally for spatial proximity (e.g., potentials of mean force [30]), as a function of those environments. For example, a buried charged residue may be more likely to be adjacent to another buried residue of opposite charge. These interaction preferences have been quantified statistically and used to produce a score function reflecting the extent to which amino acids from the sequence are located in preferred environments and adjacent to preferred neighbors. The known protein structures can be represented in a way that makes explicit the structural environment at each position, as well as the spatially adjacent structural positions. This done, the sequence can be aligned or threaded into the structure by searching for an alignment or threading that optimizes the score function. The optimal threading(s) maximize(s) the degree to which environment and adjacency preferences are satisfied. This has been a very active area of recent research, in part because it has been somewhat successful, and numerous scoring schemes and threading algorithms have been proposed. For reviews see refs. [31–39], while for cautionary notes see refs. [1,4,13,40–43].

2. Protein threading – motivating intuitions

The logic behind the threading approach to the prediction of an amino acid sequence's expected three-dimensional fold is almost seductively simple, if not obvious. Given the extreme difficulty of any direct, de novo, or quantum level approach to protein structure prediction, combined with our esthetic sense as expressed in Ockham's principle of parsimony [44], this seduction is understandable. In practice, however, as of this writing the threading approach has not yet lived up to our expectations [37]. The threading research challenge, only partially met at present, is to devise a theory of protein structure sufficiently information-rich to allow accurate prediction yet sufficiently concise to retain the simplicity of discrete alignment.

2.1. Basic ideas

Threading rests on two basic ideas: first, that there is a limited and rather small number of basic "protein domain core" folds or architectures found in nature; and, second, that some average over an entire sequence of amino acid propensities to prefer particular structural/solvent environments is a sufficient indicator for the recognition of native-like versus non-native-like folds. The first of these ideas is supported by our understanding of polymer chemistry, which suggests that there is a limited number of ways to fold a repetitive polymer of two basic unit types (amino acids are to a first approximation either hydrophilic or hydrophobic) in an aqueous environment. These are helical structures in which the helical repeat is synchronous with the polymer unit repeat, and the extended sheet structures in which neighboring polymer strands place the adjacent polymer repeat in synchrony. This results in six basic protein fold building blocks: helices and sheets, either of which can be all hydrophobic, all hydrophilic, or amphipathic (or two-sided). Now, given the requirement that a protein's interior is hydrophobic and its surface is hydrophilic, there appears to be a limited number of ways to pack these six building blocks together. Indeed the vast majority of currently determined structures fall into only a few core architectures, or arrangements of helices and sheets [17], as perhaps first clearly stated by Jane Richardson in her "taxonomy of proteins" [26]. Nonetheless, a major limitation of the threading approach is that if an appropriate core is not already present in the structure library, correct prediction is obviously impossible. Currently most of the estimated thousands of fold families remain unseen, and entirely novel folds appear frequently. Attempts to assemble structure fragments into a novel core include refs. [45–47].

The second threading concept, that the various preferences of the different amino acids for different structural environments can provide sufficient information to choose among alternate basic fold architectures, is less obvious. Protein structures, and even their functions, are known to be very robust to most single amino acid substitutions. This is a requirement of an evolving system. As has been noted many times, the thousands of distinct hemoglobin amino acid sequences all fold to nearly identical three-dimensional structures [48]. In addition, protein structures must be stable enough to tolerate some amino acids in very unfavorable positions, such as hydrophobic ones on the surface in a site designed to bind a hydrophobic substrate. It thus seems very unlikely that particular atomic details of particular amino acids determine the overall architecture of the fold. Given that many different amino acid pairs can produce nearly equivalent interactions, in terms of contact energies and packing density, it generally has been assumed that some average over these pairwise contact interactions determines relative positioning of the helices and sheets with respect to one another.

2.2. Common assumptions of current threading work

All current threading proposals replace the three-dimensional coordinates of the known structure by an abstract template description in terms of core elements and segments, neighbor relationships, distances, environments, and the like. This avoids the computational cost of atomic-detail molecular mechanics or dynamics [9,10] in favor of a much less detailed, hence much faster, discrete alignment between sequence and structure. However, important aspects of protein structure, such as intercalated side-chain packing, excluded volume, constraints linking different environments, higher-order interactions, and so forth, also may be abstracted away. This depends on the theory employed.

The common assumptions underlying all threading proposals are as follows:
(1) the known structures provide sets of template positions in three-dimensional space;
(2) a sequence–structure alignment, or threading, specifies the placement of sequence residues into one of the sets of template positions;
(3) different candidate threadings arise from different structural templates, different sequences, and different possible alignments of template positions and sequence amino acids; and
(4) a score function (often statistical) can distinguish good from bad threadings.

2.3. Principal requirements for structure prediction

To predict accurately the structure of a novel protein sequence using the threading approach, it is necessary both to select the proper core template from a library of known examples ("fold recognition", Fig. 1), and to align the sequence to it correctly ("sequence–structure alignment", Fig. 2), simultaneously (Fig. 3). There are four components to any practical application of the threading approach to the prediction of the three-dimensional fold for an amino acid sequence:

232

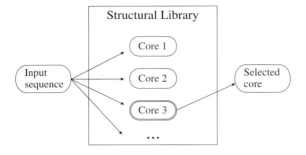

Fig. 1. Selecting a core from a structural library. Cores in the structural library are rank ordered by their probability according to eq. (28). The globally most probable core is shown selected. It is also possible to enumerate all cores in order of probability, or to sample most-probable cores.

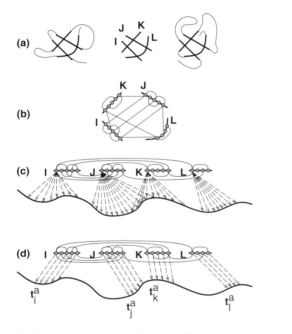

Fig. 2. A schematic view of the gapped block alignment approach to protein threading (adapted from ref. [1]). (a) Conceptual drawing of two structurally similar proteins and a common core of four secondary structure segments (dark lines, I–L). Note that there is no restriction on core segment length, which may range from a single residue position upwards. To form the core templates used here, side-chains were replaced by a methyl group resulting in polyalanine, and loops or variable regions were removed resulting in discrete core segments. (b) Abstract core template showing spatial adjacencies (interactions). Small circles represent amino acid positions (core elements), and thin lines connect neighbors that interact in the objective function. The structural environments and interacting positions will be recorded for later use by the objective function. (c) Illustration of the combinatorically large number of threadings (sequence–structure alignments) possible with a novel sequence. t_x^a indexes the sequence amino acid placed into the first element of segment X. Sequence regions between core segments become connecting turns or loops, which are constrained to be physically realizable. All alignment gaps are confined to turn or loop regions. (d) A sequence is threaded through the core template by placing successive sequence amino acids into adjacent core elements. Alignments are rank-ordered by their probability according to eq. (24). The globally most probable alignment is shown selected. It is also possible to enumerate all alignments in order of probability, or to sample the near-optimal alignments.

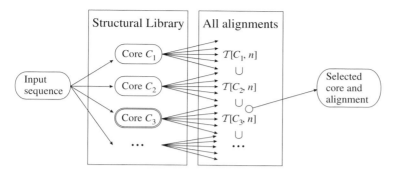

Fig. 3. Selecting a core and alignment jointly. Conceptually, every core template of the structural library (C_i) is used to generate a pool of all possible sequence–structure alignments (All alignments, $\cup_i T[C_i,n]$) with the input sequence. The pooled (core, alignment) pairs are rank ordered by probability according to eq. (34). The globally most probable core template and alignment pair is shown selected. It is also possible to enumerate all (core, alignment) pairs in order of probability, or to sample near-optimal pairs.

(i)   Construction of as complete as possible a library of potential core folds or structural templates.
(ii) An objective function (i.e. a score function) to evaluate any particular placement or alignment of a sequence into each core template.
(iii) Some means of searching over the vast space of possible alignments between a given sequence and each core template in the library for those that give the best total score.
(iv) A means of choosing the best template from among the best scoring alignments of a sequence to all possible templates.

This chapter focuses on the theory of selecting the best core templates and alignments (requirements iii and iv). It analyzes closely the consequences of choosing "best = globally highest conditional probability". For this case, it provides a probabilistic Bayesian theory that unifies core recognition and sequence–structure alignment (requirements iii and iv). The theory is Bayes-optimal because it rigorously selects cores and alignments that are globally most probable, regardless of the particular theory of protein structure adopted.

In contrast, the theory of protein structure does in fact determine the particular forms of the core templates and the score or objective function (requirements i and ii). We assume that the objective function may be interpreted as encoding the probability of observing a given sequence in a given alignment to a given core structure, and otherwise consider requirements (i) and (ii) to be arbitrary and fixed in advance. For approaches to requirements (i) and (ii) see refs. [28,29,38,41,49–58]. A rigorous probabilistic derivation of the objective function from Markov random field (MRF) theory may be found in White et al. [59] and Stultz et al. [60].

2.3.1. Library members (requirement i)
Library members (requirement i) variously are termed structures, folds, cores, folding motifs, folding patterns, topology fingerprints, three-dimensional profiles, contact profiles, adjacency graphs, spatial patterns, domains, structural models, tertiary templates, structural templates, core templates, and so on. Here we refer to them as core templates

to emphasize the loop region variability; but the name has no special meaning and the analysis applies to many structure representations currently in the literature. Library members usually consist of abstractions of known structures, constructed by erasing atomic detail of specific native side-chains and variable or loop regions while retaining core backbone structural features. The core template is annotated with environmental features that describe local structural neighborhoods, such as spatial adjacencies, distances, angles, secondary structure, solvent accessibility or exposure, backbone parameters, and so on. The environmental features chosen depend upon the particular theory of protein structure adopted. Sometimes idealized coordinates replace database coordinates, allowing variable-length segments.

The core template corresponds to an annotated backbone trace of the secondary structure segments in the conserved core fold. Core segments are connected by variable loop or coil regions. Loops are not considered as part of the conserved fold, and are modeled by an arbitrary loop score function that is "sequence-specific", because it may depend upon the specific sequence residues forming the loop. In contrast, the gap penalty used in dynamic programming alignment usually is a function of gap length only [22], and usually does not depend on the identity of sequence residues in the gap. Core template spatial "positions", implying a specific three-dimensional location, are abstracted to become spatially neutral core "elements", implying only a discrete place-holder that may be occupied by a residue from the sequence. Depending upon the requirements of the particular theory of protein structure adopted, the core template may record various local structural environments. In this way, the annotated core template organizes its core elements: each is embedded in an implied structural environment and interacts with structurally implied neighbors.

Pairs of elements are "neighbors" if they interact within the given score function. It is necessary to record on the core template whatever information the score function will use to assign scores to candidate threadings (e.g., Bryant and Lawrence [28] and Sippl [61] both record discretized distances). Such information comprises the abstract "structural environment" as seen by the threaded residues. If the score function quantifies individual residue preferences (e.g., for being more or less buried, or for certain secondary structures), the (singleton) structural environment at each element must be recorded. This is easily done by labeling the element. If the score function quantifies neighbor preferences (e.g., for being a certain distance apart), the (pairwise) structural environment between neighbors must be recorded. Several equivalent data structures have been used; the common theme is that certain pairs of elements are distinguished as neighbors, and the pair is labeled in some manner. One common data structure constructs a matrix having one row and one column for each element. Each cell contains the corresponding pairwise environment (e.g., a distance matrix [61] results when all elements are neighbors and the pairwise environment is Euclidean distance). An equivalent approach, the adjacency graph used here, constructs a graph with one vertex for each element. Neighbor elements are connected by a (directed) edge, and the edge is labeled with the pairwise environment. The edge in the graph corresponds to the cell in the matrix, and the edge label corresponds to the label contained in the cell. Related representations include adjacency matrix, contact graph, and so on. For pairwise interactions this is fully general, because each label could (in principle) specify a different 20×20 table with an arbitrary score for any possible pair.

2.3.2. Objective function (requirement ii)

Each distinct threading is assigned a score by a specified objective function or score function (requirement ii). This chapter generally restricts attention to score functions that can be computed by considering no more than two core segments at a time.

The objective function usually describes the degree of sequence–structure compatibility between sequence amino acid residues and their corresponding positions in the core template as indicated by the alignment (e.g., contact potentials, knowledge-based potentials, potentials of mean force, etc.). In the general case, the objective function may reflect the sequence residue types placed elsewhere in the structure. For example, a polar residue in a buried structural environment may be more likely to interact with a complementary polar residue in a neighboring buried position than with a hydrophobic residue there. In contrast, what we refer to below as the "singleton-only" objective function ignores interactions between pairs (or higher) of sequence residues. It is restricted to reflect only individual sequence residue preferences for the structural environments that annotate their core template position. For example, such a function may utilize the fact that a hydrophobic sequence residue may be more likely to occur in a buried structural environment than in an exposed one.

Most scoring schemes proposed thus far utilize pseudo-energies associated with each amino acid type for its placement in any given structural environment and neighbor relationships. Such pseudo-energies are statistics derived from the observed occurrence frequencies for amino acid types in each structural environment among a set of determined structures. Frequencies are converted to a pseudo-energy using Gibbs' or Boltzmann's relationship between system state energy and state probability, or simply by taking negative logarithms of occurrence frequency-derived probabilities or odds ratios. The structural environments used have included the degree of solvent exposure, the type of secondary structure, the "distance" between spatial neighboring amino acid positions, and so on, across the types of amino acid residues that are observed as typical neighbors.

2.3.3. Alignment (requirement iii)

The alignment of the sequence to a given core template (requirement iii) usually is selected by searching for the best alignment under the objective function. An alignment is optimal if its score is the global minimum score, and near-optimal if it approximates this. A search method is said to be exact if it guarantees to find the optimal alignment under the objective function, and otherwise is said to be near-optimal or approximate.

A given sequence is threaded through a given structure by searching for a sequence–structure alignment that places sequence amino acids into preferred structural environments and near other preferred amino acid types. The two key conditions that determine the complexity of this search [62] are whether or not (1) variable-length gaps are admitted into the alignment, and (2) interactions between neighboring amino acids from the sequence being threaded are admitted into the score function. Finding the optimal alignment is NP-hard in the general case where both conditions are allowed [62,63], and low-order polynomial in the singleton-only case allowing variable-length gaps.

If variable-length gaps are not permitted, then alignments are restricted to substructures and subsequences of equal length that are extracted from a database. However, in a predictive setting the alignment method must allow alignment gaps to account for the loop length variability that is observed across structural families. Ignoring variable-length gaps means that the structure and a novel sequence almost invariably will be partially out of hydrophobic registration [12]. Consequently, this alternative is of little use for prediction.

If variable-length gaps are permitted but pairwise or higher-order interactions between sequence amino acids are not, then the global optimum alignment can be found using dynamic programming alignment [22]. Dynamic programming alignment may employ a gap penalty which biases the search to prefer loop lengths present in the core template's original sequence, and so would make distant structural homologs more difficult to recognize if their loop lengths differed substantially [42]. In addition, ignoring amino acid interactions means giving up a potentially rich source of structural information.

Alternatively, if both variable-length gaps and interactions between neighboring amino acids are allowed, then finding the global optimum threading is NP-hard [62]. This means that in order to find an optimal solution, any known algorithm must require an amount of time that is exponential in protein structure size. Consequently, approximate search algorithms are relatively fast and capable of finding a good but not necessarily the optimal solution, while exact search algorithms are guaranteed to find the optimal solution but sometimes require exponential time. Approximate alignment search methods include double dynamic programming, which employs a secondary level of dynamic programming to fix the neighbors for the first level [64,65]; the "frozen approximation", a dynamic programming adaptation which substitutes the original motif residues initially and previous aligned sequence residues in subsequent steps [52]; and the Gibbs sampler, an iterative improvement method not based on dynamic programming which estimates a probability distribution and samples accordingly [66]. Exact methods include the branch and bound search described below [1]; exhaustive search [28]; and a cut set method [67] that may be faster for "modest" numbers of pair interactions because it is exponential in the maximum cut size of pair interactions rather than structure size.

2.3.4. Core template selection or fold recognition (requirement iv)

Selecting a core template from the library (requirement iv) usually is approached by aligning the sequence to each member of the library and selecting the one yielding the best alignment score, usually after normalizing or correcting the raw scores in some way. Aligning a given sequence to several cores produces raw scores that usually are not directly comparable. Normalizing terms attempt to correct for biases due to search space size (larger search spaces have larger extreme tail values), core size (larger cores have more terms in the objective function), sequence composition, structural environment frequency, and so on. Score normalizations have included z-scores measured relative to near-optimal threadings of library structures, shuffled sequences of the same composition, random structures, or sequences of random composition; sequence composition corrections; accounting for the variable size of the core templates, or of the alignment search space; and reference states based on assumptions from statistical mechanics, mathematical statistics, or sampling theory.

2.4. Gapped block alignment

The gapped block alignment definition of threading used here follows Greer [18], Jones et al. [68], and Bryant and Lawrence [28]: (1) alignment gaps are prohibited within modeled secondary structure segments; (2) specific pairwise amino acid interactions are confined to the core template; and (3) loops are scored by an arbitrary sequence-specific function.

When a sequence is threaded through the core template, successive core elements of each segment are occupied by adjacent amino acids from the sequence. Note that conserved super-secondary structures such as beta-hairpins or tight turns could be included as additional types of core segments. No alignment gaps are permitted within segments, as this would correspond to an unphysical break in the sequence mainchain. Alignment gaps are confined to the connecting non-core loop regions, which undergo evolutionary insertions and deletions over time and are not usually considered part of the conserved core motif. Because the loops are viewed as variable, they do not participate in specific pairwise amino acid interactions under the objective function. The loop score function depends only on how the sequence is threaded through pairs of core segments; threading two adjacent core segments fixes the subsequence in the intervening loop region. Since loop endpoints are known, the sequence residues occupying a single loop in principle could be modeled in three-dimensional detail, but not modeled in general with other loops. More commonly, loop regions are treated simply as one or more additional structural environment types (e.g., tight turn, short, medium, and long), and objective function parameters are generated for them just as for any other singleton-only score term.

In contrast, with the use of an alignment method based on gap penalties, some threadings delete portions of the sequence or the structure. In analytical terms, this causes the global sum in eq. (22) below to be over different effective structures and different effective sequence lengths. See Flöckner et al. [69] or Maiorov and Crippen [70] for cogent criticism of allowing parts of the sequence or structure to "vanish" in this way. In probabilistic terms, the usual linear or affine gap penalty forces loops to become exponentially unlikely in the length of the insertion or deletion. See Benner et al. [71] for empirical data showing that an exponential distribution does not provide an adequate fit to observed gap lengths. See Lemer et al. [37] for a discussion of inappropriate gap penalties, leading to gaps that are obviously far too small, as one aspect of threading algorithms that contributes to error.

Alternative methods that do use amino acid pair interaction terms directly in the loop score function require additional information to determine the loop placement in three-dimensional space. This information generally is not available in a predictive setting because the loop backbone coordinates often shift substantially to accommodate insertions or deletions. Maiorov and Crippen [70] propose an elegant approach that would include the needed three-dimensional information for pairwise terms. An open computational problem is to provide a practical algorithm for their scheme.

3. Formalization

The formulation below deliberately isolates the computational methods as much as possible from any particular theory of protein structure, from the way structural

Table 1
Notational usage of this chapter

Notation	Usage
$\mathbf{a}$	A sequence or string over A of length n
A	An alphabet of 20 characters (amino acid types)
A^n	The set of all strings over A of length n
$\mathbf{b}$ (or $\mathbf{d}$)	A vector of m integers; segment lower (or upper) bounds
c_i	$\|C_i\|$, the length of the ith core segment C_i
C	A core structure; its ith segment is C_i, whose jth element is $C_{i,j}$
f	An objective function or score function
f^1	A sequence singleton-only version of f
f^A	A per-residue version of f^1
f_a (or f_v, or f_e)	f restricted to amino acid residue types (or vertices, or edges)
f_l	f restricted to loops or variable regions
f_s	f restricted to core segments
$\bar{f}$	A mean or expected value of f
g	A per-segment encapsulation of f
h	$\exp(-f)$
h_λ	The loop length prior probability
H	$\sum_{\mathbf{w} \in A^k} h$
l_i (or l_i^{min}, or l_i^{max})	$\|\lambda_i\|$, the variable (or minimum, or maximum) length of the ith loop λ_i
$\tilde{l}_i$	$l_i^{max} - l_i^{min}$, the variability of l_i
lb	A function returning a lower bound on scores achievable within a set
$\mathcal{L}$	A library of core structures
m	$\|C\|$, the number of core segments in C
n	$\|\mathbf{a}\|$, the length of the sequence $\mathbf{a}$
$\tilde{n}$	$n + 1 - \sum_i (c_i + l_i^{min})$, the relative sequence length
$P(A\|B)$	The conditional probability of A given B
$P_1(i, t_i)$ (or $P_2(i, j, t_i, t_j)$)	The probability that a random threading places C_i at t_i (and C_j at t_j)
q, r	The inactive and active components of lb
s	A function returning a structural environment label
S, $S[\mathbf{b}, \mathbf{d}]$, $S\langle i, t_i \rangle$	The sizes of $\mathcal{T}[C, n]$, $\mathcal{T}[\mathbf{b}, \mathbf{d}]$, $\mathcal{T}\langle i, t_i \rangle$
$\mathbf{t}$ (or $\mathbf{t}^a$)	A vector of m integers; t_i (or t_i^a) is the ith relative (or absolute) coordinate
$\mathcal{T}$	A set of alignments
$\mathcal{T}[C, n]$	The set of all alignments between core C and any sequence of length n
$\mathcal{T}[\mathbf{b}, \mathbf{d}]$	The set of alignments satisfying $b_i \leq t_i \leq d_i$
$\mathcal{T}\langle i, t_i \rangle$	The set of alignments that place C_i at t_i
V	The variance of a search space score distribution
$\mathbf{w}$ (or $\mathbf{x}$)	A summation variable over A^n (or over $\mathcal{T}[C, n]$)
Z, $Z_{\langle x, y, z \rangle}$	A partition function; a global sum specified by x, y, and z
$\alpha(i)$ (or $\beta(i, j)$)	An indicator of whether axis i (or either axis i or j) is active

continued on next page

Table 1, *continued*

Notation	Usage
$\mathcal{E}$	A set of adjacency graph edges, e or $\{u,v\}$; a subset of $\mathcal{V} \times \mathcal{V}$
λ	A set of loops; the ith loop is λ_i, whose j^{th} element is $\lambda_{i,j}$
$\mu_{\langle x,y,z \rangle}$	A global mean specified by x, y, and z
σ	The standard deviation of a search space score distribution
$\mathcal{V}$	A set of adjacency graph vertices, v, corresponding bijectively to $\{C_{i,j}\}$
$B, J, J^*, K, K^*, Q, Q_j, R$	Recurrence functions

environments are defined, and from the score function employed. Consequently the methods apply to a wide variety of score functions that utilize pairwise amino acid interactions. Additional background on the problem formalization and notational conventions may be found in refs. [1,3,5,59,60,62], which this chapter follows where possible but revises where necessary. Table 1 summarizes the notation used.

3.1. Sequence

The sequence **a** is a string of length n over an alphabet A of twenty characters (amino acid residue types). The set A^n consists of all strings over A of length n. The sequence **w** is a summation variable over A^n.

3.2. Core templates and library

The core template C is drawn from a library $\mathcal{L}$ of cores. Core template C is composed of m core segments C_i, each of length $c_i = |C_i|$ and representing a set of contiguous amino acid positions. Core segments are usually the pieces of conserved structure comprising the tightly packed internal protein core, and may correspond to the backbone trace of conserved secondary structure segments. Each segment C_i is composed of primitive core elements $C_{i,j}$, for $1 \leqslant j \leqslant c_i$. Each element $C_{i,j}$ corresponds to a spatial position that may be occupied by a residue from the sequence. For generality we make no restriction on segment length, and when $c_i = 1$ the segments may correspond to single amino acid residue positions.

For simplicity in the presentation we assume in this chapter that core segment length is fixed, even though Lathrop and Smith[1] showed how this can lead to biological threading errors. Some important approaches [72,73] treat core segment length as variable by adding residue positions to, or deleting them from, core segment endpoints. This would be modeled according to section 4.8 with additional parameters specifying each segment endpoint adjustment relative to the core template. Similarly, in this chapter we assume fixed core topology (i.e. segment rank order and direction). Variable topology, e.g. alternate arrangement of beta-strands in a beta-sheet, arises easily from core segment permutations or reversals. This would be modeled according to section 4.8 with additional parameters specifying the rank order and direction of each segment. Except for section 4.8, however, this chapter assumes fixed segment length and topology.

3.3. Loops

Core segments are connected by a set λ of loop regions. The loop regions might equally well be the "gaps" (in a dynamic programming alignment sense) used by some formulations. Loop λ_i connects segment C_i to C_{i+1}, N-terminal leader λ_0 precedes C_1, and C-terminal trailer λ_m follows C_m. Knowing the endpoints of λ_i is equivalent to knowing the threadings of C_i and C_{i+1}.

The length of loop λ_i is the variable l_i and its maximum (respectively minimum) length is $l_i^{\max}$ (respectively $l_i^{\min}$). Unless stated otherwise, $l_i^{\max} = +\infty$ and $l_i^{\min} =$ the minimum geometric spanning loop length (i.e. the minimum loop length capable of spanning the distance between the end of C_i and the beginning of C_{i+1}). Other values may reflect knowledge of additional constraints. For example, loops assigned length zero or one by the crystallographer usually reflect constrained "kinks" in the secondary structure which should be retained ($l_i^{\max} = l_i^{\min} = l_i^{\text{native}}$) or restricted ($l_i^{\max} = 1$ and $l_i^{\min} = 0$). As another example, Bryant and Lawrence [28] set $l_i^{\max}$ and $l_i^{\min}$ based on the maximum and minimum loop lengths in an aligned homologous family. To simplify notation we assume $l_0^{\min} = l_m^{\min} = 0$, i.e. the leader and trailer loops have zero minimum length.

3.4. Adjacency graph

An adjacency graph (also called interaction matrix, interaction graph, neighbor matrix, contact map, etc.) describes core element positions that are "neighbors". Positions are defined to be neighbors if they interact in the score function. The adjacency graph consists of a set V of vertices and a set $\mathcal{E}$ of edges. Each core element $C_{i,j}$ corresponds one-to-one to a graph vertex $v \in V$. Consequently, the adjacency graph vertices merely relabel the core elements. Pairs of vertices u and v which interact in the score function (neighbors) are connected by a graph edge $e \in \mathcal{E}$, sometimes written $e = \{u,v\}$. Each vertex v and each edge e is labeled by an environment function s. The vertex (residue) environment labels, $s(v)$, may describe local factors such as degree of solvent exposure, local secondary structure type, and so forth. The edge environment labels, $s(e)$, may encode distance or contact between amino acids, the local environments at each end of the edge, and so forth. The edges are directed because the local environments at the edge head and tail may differ. The unaligned loop regions do not participate directly in the adjacency graph of pairwise relations.

3.5. A threading of a sequence into a core

A given alignment ("threading") of sequence $\mathbf{a}$ to core C associates each core element $C_{i,j}$ with exactly one amino acid from $\mathbf{a}$ (i.e. the core segments may not overlap). A legal threading is subject to the further constraints that successive amino acids in the sequence necessarily fall into successive core elements within each segment C_i (i.e. the core segments may not have internal gaps), that the core segments retain their original topological ordering (i.e. the threading does not permute or reverse segments), and that loop sizes are within the legal ranges (i.e. every loop region is long enough to span between its flanking segments). A sequence–structure alignment ("threading") of $\mathbf{a}$ into C is completely described by the primary sequence indices of the amino acids placed into the first element of each core segment. This results in a vector of m integers, denoted by $\mathbf{t}^a$ in absolute coordinates and $\mathbf{t}$ in relative coordinates. Each absolute coordinate t_i^a specifies the index in the sequence $\mathbf{a}$ that is aligned to the first element of the ith core segment. That is, sequence residue $\mathbf{a}[t_i^a]$ occupies core element $C_{i,1}$.

For simpler notation [1] we generally replace absolute sequence coordinates $\mathbf{t}^a$ by relative coordinates $\mathbf{t}$, defined by $t_i = t_i^a - \sum_{j < i}(c_j + l_j^{\min})$. Let $\tilde{n} = n + 1 - \sum_i(c_i + l_i^{\min})$ and $\tilde{l}_i = l_i^{\max} - l_i^{\min}$. Then $t_i = 1$ corresponds to the lowest legal value of t_i^a and $t_i = \tilde{n}$ to the highest. Below, the absence of the superscript a will indicate relative coordinates.

The ith loop length l_i and segment length c_i are related to $\mathbf{t}^a$ and $\mathbf{t}$ by $l_i = t_{i+1}^a - t_i^a - c_i = t_{i+1} - t_i + l_i^{\min}$. Due to the minimum spanning loop length constraints, $1 + \sum_{j < i}(c_j + l_j^{\min}) \leqslant t_i^a \leqslant n + 1 - \sum_{j \geqslant i}(c_j + l_j^{\min})$. Due to core segment topological ordering constraints, $t_i^a + c_i + l_i^{\min} \leqslant t_{i+1}^a \leqslant t_i^a + c_i + l_i^{\max}$. In relative coordinates, the minimum loop length constraints simplify to $1 \leqslant t_i \leqslant \tilde{n}$ and the ordering constraints simplify to $t_i \leqslant t_{i+1} \leqslant t_i + \tilde{l}_i$.

Fictitious segments C_0 (respectively C_{m+1}) are fixed at the beginning (respectively end) of the sequence whenever it is convenient for indicated summations or recurrence limits. By convention, $c_0 = c_{m+1} = 0$, i.e. fictitious segments have zero length; and $t_0 = 1$ and $t_{m+1} = \tilde{n}$, i.e. they are fixed.

3.6. Sets of threadings

The set of threadings $T[C, n, \mathbf{b}, \mathbf{d}]$ consists of all legal alignments of any sequence of length n to the core template C such that $b_i \leqslant t_i \leqslant d_i$ is satisfied. Where the entire search space is the intended set, we simplify the notation by writing $T[C, n]$. The vector $\mathbf{x}$ is a summation variable over $T[C, n]$.

Where the sequence and core are clear from context, we simplify the notation by writing $T[\mathbf{b}, \mathbf{d}]$. The integers b_i and d_i define an interval, $[b_i, d_i]$, made up of the allowed sequence coordinates for core segment C_i. These m intervals may be represented compactly by two m-length vectors, $\mathbf{b}$ and $\mathbf{d}$ (mnemonic for "**B**egin" and "en**D**"). This allows us to represent all sets $T[\mathbf{b}, \mathbf{d}]$ that have the particularly simple form of an m-dimensional axis-parallel hyper-rectangle whose two opposite corners are the vectors $\mathbf{b}$ and $\mathbf{d}$. The entire search space is represented by the hyper-rectangle $1 \leqslant t_i \leqslant \tilde{n}$. Thus $T[C, n] = T[\mathbf{1}, \tilde{\mathbf{n}}] = \{\mathbf{t} | 1 \leqslant t_i \leqslant \tilde{n}\}$.

The ability to represent and manipulate the search space directly allows for controlling the search. A list of several hyper-rectangles corresponds to the union of the sets they represent. If a particular list of hyper-rectangles is used to initialize the search in section 8.2, the subsequent search will examine only the corresponding threadings.

Each hyper-rectangle also contains a large number of illegal threadings that violate spacing or ordering constraints. Illegal threadings are always ignored. By convention, if $\mathbf{t}^{\text{ill}}$ is an illegal threading then $f(\mathbf{t}^{\text{ill}}) = +\infty$. Whenever we speak of a set of threadings we mean only the legal ones. Whenever search space sizes are computed, only legal threadings are counted.

3.7. Objective function

For a specific core motif C and protein sequence $\mathbf{a}$, the score of a candidate threading is defined to be the sum of the scores of the vertices and edges in the adjacency graph and the bulk composition of the loop regions. By analogy to energy minimization, lower scores are considered better. We assume the availability of an objective function f satisfying

$$P(\mathbf{a}|n, C, \mathbf{t}) \propto \exp\left[-f(\mathbf{a}, C, \mathbf{t})\right], \tag{1}$$

where $\mathbf{a}$ is a sequence of length n; C is a core template; $\mathbf{t}$ is a vector that specifies a sequence–structure alignment (a threading) and whose ith component t_i specifies the alignment of core segment i; and $P(A|B)$ is the conditional probability of A given B. Where arguments are clear from context, we simplify the notation by omitting them, writing $f(i, t_i)$ to abbreviate $f(\mathbf{a}, C, i, t_i)$, and so on. The function f is the negative logarithm of an unnormalized conditional probability. It encodes the probability of observing sequence $\mathbf{a}$, given the alignment $\mathbf{t}$ to core C. For example, White et al. [59] and Stultz et al. [60] describe how to construct an objective function based on Markov random field (MRF) theory that satisfies eq. (1).

Many published threading approaches are grounded in an underlying probabilistic objective function of this general nature. In practice, they may convert the underlying probabilistic objective function from a strict conditional probability to an odds ratio relative to some assumed reference state, say $P(\mathbf{a}|n, C, \mathbf{t})/P_{\text{ref}}(\mathbf{a}|n, C, \mathbf{t})$. In contrast, the Bayesian analysis uses only the strict conditional probability, $P(\mathbf{a}|n, C, \mathbf{t})$. This is equivalent to setting $P_{\text{ref}}(\mathbf{a}|n, C, \mathbf{t}) = \text{constant}$ in some odds ratio approaches. The reference state is derived from first principles of probability theory.

3.7.1. The fully general objective function
Given a specific protein sequence and a core template of m core segments, the fully general form of the score function is

$$f(\mathbf{t}) = \sum_i g_1(i, t_i) + \sum_i \sum_{j>i} g_2(i, j, t_i, t_j) + \cdots$$
$$+ \sum_i \sum_{j>i} \cdots \sum_{l>k} g_m(i, j \ldots k, l, t_i, t_j \ldots t_k, t_l), \tag{2}$$

where $i, j \ldots k, l$ index core segments and $t_i, t_j \ldots t_k, t_l$ give their relative positions in the sequence. The final sum is repeated over m indices representing all m core segments, and reflects amino acid interactions among all m core segments simultaneously.

In practice there is insufficient data to specify all the free parameters such a function would imply. Approaches differ in where they terminate the expansion. Profile-based methods employ only the g_1 term. Dynamic programming methods that do not allow pairwise amino acid interactions employ the g_1 term plus an affine gap penalty g_2 term of the form $g_2(i, i+1, t_i, t_{i+1}) = a + b|l_i - l_i^{\text{native}}|$. Methods that permit pairwise interactions, as here, employ a full g_2 term. Triplet interactions [52,54] would require a g_3 term. A score function requiring g_4 or higher terms [74–77] would arise in treatment of steric packing among multiple core segments, linked constraint equations on structural environments, detailed geometric or environment modeling, and so forth.

3.7.2. General pairwise interaction objective function

In the pairwise case, the score of a candidate threading is defined to be a function only of the sum of (1) a series of g_1 terms, each of which depends only on the threading of a single core segment, plus (2) a series of g_2 terms, each of which depends only on the threading of a pair of core segments:

$$f(\mathbf{t}) = \sum_i g_1(i, t_i) + \sum_i \sum_{j>i} g_2(i, j, t_i, t_j). \tag{3}$$

The functions g_1 and g_2 are the essential point of contact between the search algorithm and any particular choice of scoring function, neighbor relationships, or structural environments. The search algorithm is driven only by g_1 and g_2, regardless of how the score function assigns values to them.

In principle, every core element and every possible pair of elements could be assigned a unique structural environment encoding a different score table, and each loop region could assign a different score to every possible spanning subsequence. Consequently eq. (3) is fully general for pair interactions. In most threading schemes, the score of a candidate threading is built up from the scores of the vertices and edges in the adjacency graph, and the sequence composition of the loop regions. Score functions that depend on separation in the sequence, proximity to the N- or C-terminal end of the sequence, or specialized identities of particular segments (e.g., including a regular expression pattern match based on known enzymatic function) are accommodated easily because the segment numbers (i, j) and segment indices (t_i, t_j) appear explicitly in the g_1 and g_2 argument lists. Other score components may be included provided they depend only on singleton or pairwise core segment threadings as shown.

3.7.3. A typical pairwise score function

Here we give an example of one way that a score function might be constructed. Details will vary with the particular score function and environment definitions chosen.

For any threading $\mathbf{t}$, let $f_v(v, \mathbf{t})$ be the score assigned to core element or vertex v, $f_e(\{u, v\}, \mathbf{t})$ the score assigned to interaction or edge $\{u, v\}$, and $f_l(\lambda_i, \mathbf{t})$ the score assigned to loop region λ_i. Then the total score of the threading is

$$f(\mathbf{t}) = \sum_{v \in V} f_v(v, \mathbf{t}) + \sum_{\{u,v\} \in \mathcal{E}} f_e(\{u, v\}, \mathbf{t}) + \sum_{\lambda_i \in \lambda} f_l(\lambda_i, \mathbf{t}). \tag{4}$$

We can rewrite this as a function of threadings of pairs of core segments as follows:

$$
f(\mathbf{t}) = \sum_i \sum_{v \in C_i} f_v(v, \mathbf{t}) + \sum_i f_l(\lambda_i, \mathbf{t})
$$

$$
+ \sum_i \sum_j \sum_{\substack{\{u,v\} \in \mathcal{E} \\ u \in C_i \\ v \in C_j}} f_e(\{u, v\}, \mathbf{t}) \tag{5}
$$

$$
= \sum_i \left[\sum_{v \in C_i} f_v(v, \mathbf{t}) + \sum_{\substack{\{u,v\} \in \mathcal{E} \\ u,v \in C_i}} f_e(\{u, v\}, \mathbf{t}) \right] + f_l(\lambda_0, \mathbf{t}) + f_l(\lambda_m, \mathbf{t})
$$

$$
+ \sum_i \left[f_l(\lambda_i, \mathbf{t}; 0 < i < m) + \sum_{j \neq i} \sum_{\substack{\{u,v\} \in \mathcal{E} \\ u \in C_i \\ v \in C_j}} f_e(\{u, v\}, \mathbf{t}) \right] \tag{6}
$$

$$
= \sum_i g_1(i, t_i) + \sum_i \sum_{j > i} g_2(i, j, t_i, t_j). \tag{7}
$$

The singleton terms, in g_1, include contributions from sources such as (in order of eq. 6) individual core elements assigned to particular structural environments, pairwise interactions within a single core segment, and loop scores of the N- and C-terminal loop regions. The pairwise terms, in g_2, include contributions from sources such as (in order of eq. 6) interior loop scores, and pairwise interactions between different core segments.

3.7.4. Computing g_1 and g_2 efficiently

Precomputing g_1 and g_2 and storing them in arrays permits rapid evaluation of individual threadings as in eq. (7), compared to their time-consuming ab initio evaluation as implied by eq. (4). Storing g_1 requires $O(m)$ arrays of size $\tilde{n}$, and storing g_2 requires $O[m(m-1)/2]$ arrays of size $\tilde{n}(\tilde{n}+1)/2$, though in practice less storage is required because some core segment pairs do not interact.

3.7.5. Singleton-only objective function

In this section we define an objective function that allows an arbitrary sequence-specific score function for each segment or loop, but ignores all pairwise or higher-order interactions between non-adjacent segments. We use f^1 to distinguish this singleton-only objective function from the general case, and Z^1 and μ^1 to distinguish corresponding global sums and means.

Let $f_s(i, t_i)$ be the score for occupying segment C_i by the substring of length c_i beginning at $\mathbf{a}[t_i]$, and let $f_l(i, t_i, t_{i+1})$ be the score for occupying loop λ_i by the substring of length $l_i = t_{i+1} - t_i + l_i^{\min}$ beginning at $\mathbf{a}[t_i + c_i]$. If desired, pair interactions between segments C_i and C_{i+1} may be encoded in $f_l(i, t_i, t_{i+1})$ as well.

Assume that the threading score is the sum of the segment and loop scores separately:

$$f^1(\mathbf{a}, C, \mathbf{t}) = \sum_{i=1}^{m} f_s(i, t_i) + \sum_{i=0}^{m} f_l(i, t_i, t_{i+1}). \tag{8}$$

Functions h_s and h_l are the unnormalized probability functions corresponding to f_s and f_l.

$$h_s(i, t_i) \quad = \exp[-f_s(i, t_i)], \tag{9}$$
$$h_l(i, t_i, t_{i+1}) = \exp[-f_l(i, t_i, t_{i+1})]. \tag{10}$$

By convention, all illegal or out-of-range indices imply score $+\infty$ (infinitely bad) and probability zero; and $h_s(0, x) = h_s(m + 1, x) = 1$, i.e. fictitious segments have zero score and unit probability; and $h_l(0, x, x) = h_l(m + 1, x, x) = 1$, i.e. they have zero length.

Function H_s (respectively H_l) is the sum of h_s (respectively h_l) over all strings over A of length c_i (respectively l_i):

$$H_s(i, t_i) \quad = \sum_{\mathbf{w} \in A^{c_i}} h_s(\mathbf{w}, C, i, t_i), \tag{11}$$

$$H_l(i, t_i, t_{i+1}) = \sum_{\mathbf{w} \in A^{l_i}} h_l(\mathbf{w}, C, i, t_i, t_{i+1}). \tag{12}$$

Function $h_\lambda(i, t_i, t_{i+1})$ is the sequence-independent probability of observing loop length $l_i = t_{i+1} - t_i + l_i^{\min}$ at loop λ_i. The assumption that loop lengths are independent yields

$$P(\mathbf{t}|n, C) = \prod_{i=0}^{m} h_\lambda(i, t_i, t_{i+1}). \tag{13}$$

If uninformative priors are used, then $h_\lambda(i, t_i, t_{i+1}) = |\mathcal{T}[C, n]|^{-1/(m+1)}$ and the equation is exact. If an empirical loop length distribution [71] is used, then h_λ is taken from empirical tables; in this case the equation is approximate because $\sum_i \tilde{l}_i = \tilde{n} - 1$ so the assumption of loop length independence is violated, but it may yield a biologically more plausible result in some cases.

Recall that the relative coordinates shown must be converted to absolute coordinates to obtain an actual index into $\mathbf{a}$; specifically, add $\sum_{j < i}(c_j + l_j^{\min})$ (respectively $\sum_{j < i+1}(c_j + l_j^{\min})$) to the second argument t_i (respectively the third argument t_{i+1}) of $f_s, f_l, h_s, h_l, h_\lambda, H_s$, and H_l; and add $\sum_{j < i}(c_j + l_j^{\min})$ to the relative coordinate y in $\mathbf{a}[y]$.

3.7.6. Per-residue singleton-only objective function

In many current proposals, the singleton-only f_s (respectively f_l) is specialized further to be the sum of the individual sequence residue scores at each element of the segment (respectively loop). Here we give a simple way to derive f_s, f_l, H_s, and H_l, in such proposals.

246

Let $s(C_{i,j})$ be the structural environment assigned to core element $C_{i,j}$ and $s(\lambda_i)$ be the structural environment assigned to loop λ_i. $s(C_{i,j})$ potentially reflects a different structural environment for each core element, as annotated by the theory of protein structure used. The loop structural environment $s(\lambda_i)$ might be used to divide loops into categories, e.g., tight, short, medium, and long; or all loops might be assigned to a single generic loop environment. Let $f_a^A(a',s)$ be the score assigned to amino acid residue type $a' \in A$ in environment s, and let $h_a^A(a',s) = \exp[-f_a^A(a',s)]$.

$$f^A(\mathbf{a}, C, \mathbf{t}) = \sum_{i=1}^{m} f_s^A(i, t_i) + \sum_{i=0}^{m} f_l^A(i, t_i, t_{i+1}), \tag{14}$$

$$f_s^A(i, t_i) = \sum_{j=1}^{c_i} f_a^A(\mathbf{a}[t_i + j - 1], s(C_{i,j})), \tag{15}$$

$$f_l^A(i, t_i, t_{i+1}) = \sum_{j=1}^{l_i} f_a^A(\mathbf{a}[t_i + c_i + j - 1], s(\lambda_i)), \tag{16}$$

$$h_s^A(i, t_i) = \exp[-f_s^A(i, t_i)] = \prod_{j=1}^{c_i} h_a^A(\mathbf{a}[t_i + j - 1], s(C_{i,j})), \tag{17}$$

$$h_l^A(i, t_i, t_{i+1}) = \exp[-f_l^A(i, t_i, t_{i+1})] = \prod_{j=1}^{l_i} h_a^A(\mathbf{a}[t_i + c_i + j - 1], s(\lambda_i)), \tag{18}$$

$$H_s^A(i, t_i) = \prod_{j=1}^{c_i} \sum_{a' \in A} h_a^A(a', s(C_{i,j})), \tag{19}$$

$$H_l^A(i, t_i, t_{i+1}) = \prod_{j=1}^{l_i} \sum_{a' \in A} h_a^A(a', s(\lambda_i)). \tag{20}$$

4. Analysis – selection tasks

For a given sequence, this section develops formulae for selecting
(1) the most probable alignment $\mathbf{t}$ to a given core template, which maximizes $P(\mathbf{t}|\mathbf{a}, n, C)$ (eq. 24 in section 4.2);
(2) the most probable core template C across the entire library, which maximizes $P(C|\mathbf{a}, n)$ (eq. 28 in section 4.3);
(3) the most probable joint core template and alignment $\langle C, \mathbf{t} \rangle$ across the entire library, which maximizes $P(C, \mathbf{t}|\mathbf{a}, n)$ (eq. 34 in section 4.4), and which need not be the most probable alignment of the most probable core template; and
(4) the most probable core template segment alignments across the entire library, which maximize $P(C, i, t_i|\mathbf{a}, n)$ (eq. 36 in section 4.5), and which may potentially allow for the construction of a core template for a sequence whose core template is not in the

library by selecting piecewise the most probable segment alignments from different core templates.

Item 1 above corresponds to requirement (iii) in section 2.3, item 2 corresponds to requirement (iv), item 3 corresponds to (iii) and (iv) simultaneously, and item 4 corresponds to (iii) and (iv) for individual core segments. See Figs. 1–3.

4.1. Bayesian analysis

Bayes [78] provided the first exact treatment of inference based on inverting conditional probabilities. His interpretation of the formula, $P(A|B) = P(B|A)P(A)/P(B)$, is well known. Today the well-understood mathematics of Bayesian methods is a central component of optimal statistical inference [79,80]. Conditional probability and Bayesian methods have been applied to protein secondary structure [60,81], side-chain packing [82], fragment assembly [47], multiple sequence alignment [66], solvent exposure prediction [83], and structure classification [60,84,85], all with good results.

Here, the Bayesian analysis provides a compact account of the globally most probable cores and alignments. The Bayesian analysis presented here is closely related to other threading approaches as shown in Fig. 4. All approaches compute a transform from observed database frequencies to predicted probable structures. The Bayesian analysis is somewhat more direct and compact, but somewhat less biologically intuitive, than pseudo-energy potential approaches.

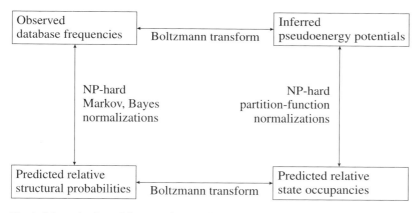

Fig. 4. Schematic view of the general-case information transformation pathways.

4.1.1. Prior probabilities

Three prior probabilities are necessary for this analysis: $P(a|n)$, $P(C|n)$, and $P(t|n, C)$. $P(a|n)$ is constant for a given sequence and may be ignored. $P(C|n)$ corresponds to the sequence-independent part of the core template probability. It reflects at least two influences: the relative frequencies of different core templates, and the way these shift with sequence length. $P(t|n, C)$ corresponds to the sequence-independent part of the loop probability. It reflects the loop length probability distribution for C and n, independent of the specific amino acid residue types that actually occupy the loops. Assuming uninformative priors and fixed n, $P(C|n) = |\mathcal{L}|^{-1}$ and $P(\mathbf{t}|n, C) = |\mathcal{T}[C, n]|^{-1}$.

There are several plausible biological reasons why the assumption of uninformative priors might be relaxed. For example, $P(C|n)$ might instead reflect the observation that some folds are more probable than others [17,25,86]; or that fold-space attractors have unequal population densities [87]; or that proteins are roughly half secondary structure and half coil; or that longer sequences are more likely to fold into larger structures. $P(\mathbf{t}|n, C)$ might instead reflect an empirical loop length distribution [71] constructed by tabulating the loop lengths observed to connect loop endpoints in various geometries across a structural database; or a linear or affine gap penalty, in which case it becomes exponentially improbable in the number and length of insertions and deletions. This is not to argue for or against any particular set of priors. Rather, different informative priors might be plausible or not under different circumstances and assumptions.

4.1.2. Global sums
Only four global sums are sufficient to accomplish all of the probabilistic selections described above. It is easy to see that global sums and global means are equivalent, because if we know one we easily can produce the other. Below, Z denotes global sums and μ denotes global means. As with many biological problems where large search spaces and non-local influences co-occur, their computation is NP-hard if specific pair interactions and gaps are both permitted. Consequently either pruned search, approximations, or long computation must be employed. Section 7.3 provides brief proof sketches of NP-hardness for the general pair interaction case.

4.2. Selecting an alignment given a core template

For fixed sequence and core template, the task of sequence–structure alignment is to select an alignment of the sequence to the core template (see Fig. 2). White et al. [59] show that

$$P(\mathbf{a}|n, C, \mathbf{t}) = \frac{\exp(-f(\mathbf{a}, C, \mathbf{t}))}{Z_{\langle n, C, \mathbf{t}\rangle}}, \tag{21}$$

$$Z_{\langle n, C, \mathbf{t}\rangle} = \sum_{\mathbf{w} \in A^n} \exp(-f(\mathbf{w}, C, \mathbf{t})), \tag{22}$$

$$P(\mathbf{t}|\mathbf{a}, n, C) = P(\mathbf{a}|n, C, \mathbf{t})\frac{P(\mathbf{t}|n, C)}{P(\mathbf{a}|n, C)} \tag{23}$$

$$= \frac{P(\mathbf{t}|n, C)}{P(\mathbf{a}|n, C)}\frac{\exp(-f(\mathbf{a}, C, \mathbf{t}))}{Z_{\langle n, C, \mathbf{t}\rangle}}, \tag{24}$$

where $Z_{\langle n, C, \mathbf{t}\rangle}$ is the global sum over all possible sequences of length n aligned by $\mathbf{t}$ to core template C. When f is modeled as a Markov random field, as in White et al. [59] and Stultz et al. [60], $Z_{\langle n, C, \mathbf{t}\rangle}$ is the same for every $\mathbf{t} \in \mathcal{T}[C, n]$. This case is treated here. In this case, assuming uninformative priors, the globally most probable alignment also is the one of globally lowest alignment score. Section 4.8 treats the case where $Z_{\langle n, C, \mathbf{t}\rangle}$

is allowed to vary with **t**. In this case, even assuming uninformative priors, the globally most probable alignment may not have the globally lowest alignment score; the variability of $Z_{\langle n,C,\mathbf{t}\rangle}$ also must be accounted for.

4.3. Selecting a core template

For a fixed sequence, the task of fold recognition is to select a core template from the structure library (see Fig. 1). There is general agreement that one would like to select the core that has the highest conditional probability given the sequence.

$$P(C|\mathbf{a},n) = \sum_{\mathbf{x} \in T[C,n]} \frac{P(\mathbf{a},n,C,\mathbf{x})}{P(\mathbf{a},n)} \tag{25}$$

$$= \sum_{\mathbf{x} \in T[C,n]} \frac{P(\mathbf{a}|n,C,\mathbf{x})P(\mathbf{x}|n,C)P(C|n)P(n)}{P(\mathbf{a}|n)P(n)} \tag{26}$$

$$= \frac{P(C|n)}{P(\mathbf{a}|n)} \sum_{\mathbf{x} \in T[C,n]} \frac{\exp(-f(\mathbf{a},C,\mathbf{x}))P(\mathbf{x}|n,C)}{Z_{\langle n,C,\mathbf{t}\rangle}} \tag{27}$$

$$= \frac{P(C|n)}{P(\mathbf{a}|n)} \frac{\mu_{\langle \mathbf{a},n,C\rangle}}{Z_{\langle n,C,\mathbf{t}\rangle}}, \tag{28}$$

$$\mu_{\langle \mathbf{a},n,C\rangle} = \sum_{\mathbf{x} \in T[C,n]} \exp(-f(\mathbf{a},C,\mathbf{x}))P(\mathbf{x}|n,C), \tag{29}$$

$$\mu_{\langle n,C,\mathbf{t}\rangle} = \sum_{\mathbf{w} \in A^n} \exp(-f(\mathbf{w},C,\mathbf{t}))P(\mathbf{w}|n,C). \tag{30}$$

Equation (28) orders all cores by conditional probability across the library.

Normalizing by $\sum_{C' \in \mathcal{L}} P(C'|\mathbf{a},n)$ imposes the fundamental threading assumption that the proper core is indeed in the library, and converts the ordering into a conditional probability reflecting that assumption. Similar normalizations reflecting the fundamental threading assumption apply throughout.

Assuming uninformative priors, the most probable core template maximizes the ratio $\mu_{\langle \mathbf{a},n,C\rangle}/Z_{\langle n,C,\mathbf{t}\rangle}$. Uninformative priors implies that $\mu_{\langle n,C,\mathbf{t}\rangle} = P(\mathbf{a}|n)Z_{\langle n,C,\mathbf{t}\rangle}$, in which case it also maximizes the ratio $\mu_{\langle \mathbf{a},n,C\rangle}/\mu_{\langle n,C,\mathbf{t}\rangle}$. This ratio is the mean probability across all possible alignments holding the sequence fixed, divided by the mean probability across all possible sequences holding the alignment fixed.

4.4. Selecting structure and alignment jointly

The central problem of inverse structure prediction is to select simultaneously both a core template and an alignment, given a sequence (see Fig. 3). To predict accurately, both the core template and the alignment must be selected correctly. However, it is evident by comparing eqs. (24) and (28) to eqs. (31) and (34) that the most probable alignment to the most probable core template is not necessarily equivalent to the most probable structure–alignment pair considered jointly:

$$P(C, \mathbf{t} | \mathbf{a}, n) = P(\mathbf{t} | \mathbf{a}, n, C) P(C | \mathbf{a}, n) \tag{31}$$

$$= \frac{P(\mathbf{a} | n, C, \mathbf{t}) P(\mathbf{t} | n, C)}{P(\mathbf{a} | n, C)} \frac{P(\mathbf{a} | n, C) P(C | n)}{P(\mathbf{a} | n)} \tag{32}$$

$$= \frac{P(\mathbf{a} | n, C, \mathbf{t}) P(\mathbf{t} | n, C) P(C | n)}{P(\mathbf{a} | n)} \tag{33}$$

$$= \frac{P(C | n) P(\mathbf{t} | n, C)}{P(\mathbf{a} | n)} \frac{\exp \big(-f(\mathbf{a}, C, \mathbf{t}) \big)}{Z_{\langle n, C, \mathbf{t} \rangle}}. \tag{34}$$

This orders all $\langle$ structure, alignment $\rangle$ pairs by conditional probability jointly across the entire structure library.

4.5. Selecting individual core segment alignments

By selecting alignments to the most probable segments across the entire library, it might in principle be possible to construct a new core template piecewise out of the selected segments even though the constructed core template does not yet appear in the library. In this way it might in principle be possible to work around a current limitation of protein threading, namely that only known core structures may be predicted:

$$P(C, i, t_i | \mathbf{a}, n) = \sum_{\{\mathbf{x} \in \mathcal{T}[C, n] \,|\, x_i = t_i\}} P(C, \mathbf{x} | \mathbf{a}) \tag{35}$$

$$= \frac{P(C | n)}{P(\mathbf{a} | n)} \frac{\mu_{\langle \mathbf{a}, n, C, i, t_i \rangle}}{Z_{\langle n, C, \mathbf{t} \rangle}}, \tag{36}$$

$$\mu_{\langle \mathbf{a}, n, C, i, t_i \rangle} = \sum_{\{\mathbf{x} \in \mathcal{T}[C, n] \,|\, x_i = t_i\}} \exp \big(-f(\mathbf{a}, C, \mathbf{x}) \big) P(\mathbf{x} | n, C). \tag{37}$$

This orders all $\langle$ structure, segment number, sequence index $\rangle$ triples by conditional probability across the entire library. The triples so generated (a) potentially arise from multiple different cores in the library, (b) have no overlap constraints between them, and (c) are selected from the set of legal threadings for each core, and so reflect its mean-field intra-template preferences and constraints. Furthermore, arranging for consistent pairwise or higher interactions between selected triples would result in a challenging constraint satisfaction problem. The problem of actually assembling such triples in a consistent manner into a novel "meta-core" is left open.

4.6. Super-secondary structures, or core template subsets

In many cases a core template may fit only partially to a structural analog. Some secondary structure segments may correspond, while others may not. This might be the case, for example, when a common super-secondary structure motif is shared but the rest of the protein diverges; or when part of the core superposes but another part does not. Suppose that k of the m segments correspond, that the corresponding segments are $I = \{i_1, i_2, \ldots, i_k\}$, and that the corresponding indices are $T = \{t_{i_1}, t_{i_2}, \ldots, t_{i_k}\}$. The previous section gave the special case when $k = 1$, and the caveats there apply here too:

$$P(C, I, T | \mathbf{a}, n) = \sum_{\{\mathbf{x} \in T[C,n] \,|\, j \in I \Rightarrow x_j = t_j\}} P(C, \mathbf{x} | \mathbf{a}) \tag{38}$$

$$= \frac{P(C|n)}{P(\mathbf{a}|n)} \frac{\mu_{\langle \mathbf{a}, n, C, I, T \rangle}}{Z_{\langle n, C, \mathbf{t} \rangle}}, \tag{39}$$

$$\mu_{\langle \mathbf{a}, n, C, I, T \rangle} = \sum_{\{\mathbf{x} \in T[C,n] \,|\, j \in I \Rightarrow x_j = t_j\}} \exp\left(-f(\mathbf{a}, C, \mathbf{x})\right) P(\mathbf{x} | n, C). \tag{40}$$

This orders, by conditional probability across the entire library, all super-secondary structures or core template subsets that consist of k segments all taken from the same core template.

4.7. Secondary structure prediction

Let helix(j) denote the event that the jth sequence residue $\mathbf{a}[j]$ is found in a helical conformation, and $\Phi_{\mathrm{helix}}(\mathbf{a}, j, C, i) = \{t_i \,|\; C_i \text{ is helix and } t_i \text{ places } C_i \text{ over } \mathbf{a}[j]\}$; that is, the set of all threading indices for C_i that thread $\mathbf{a}[j]$ to a helix position. Then

$$P(\mathrm{helix}(j) | \mathbf{a}, n) = \sum_{C \in \mathcal{L}} P(\mathrm{helix}(j) | \mathbf{a}, n, C) P(C | \mathbf{a}, n) \tag{41}$$

$$= \sum_{C \in \mathcal{L}} P(C | \mathbf{a}, n) \sum_{\mathbf{t} \in T[C,n]} P(\mathrm{helix}(j) | \mathbf{a}, n, C, \mathbf{t}) P(\mathbf{t} | \mathbf{a}, n, C) \tag{42}$$

$$= \sum_{C \in \mathcal{L}} P(C | \mathbf{a}, n) \sum_{i=1}^{m} \sum_{t_i \in \Phi_{\mathrm{helix}}(\mathbf{a}, j, C, i)} P(C, i, t_i | \mathbf{a}, n). \tag{43}$$

The final equation follows because for physical reasons $\mathbf{a}[j]$ cannot simultaneously be in two different helices or two different positions of the same helix. $P(\mathrm{extended}(j) | \mathbf{a}, n)$ is defined similarly (extended $= \beta$-sheet). For 3-state prediction, coil is defined as anything that is not helix or extended. Let $\Phi_{\mathrm{coil}}(\mathbf{a}, j, C, i) = \{\langle t_i, t_{i+1} \rangle \,|\; t_i \text{ places } C_i \text{ before } \mathbf{a}[j] \text{ and } t_{i+1} \text{ places } C_{i+1} \text{ after it}\}$, where by convention $t_0 = $ the beginning and $t_{m+1} = $ the end of the sequence accounts for the leader and trailer loop regions. Then

$$P(\mathrm{coil}(j) | \mathbf{a}, n)$$

$$= \sum_{C \in \mathcal{L}} P(C | \mathbf{a}, n) \sum_{i=0}^{m} \sum_{\langle t_i, t_{i+1} \rangle \in \Phi_{\mathrm{coil}}(\mathbf{a}, j, C, i)} P(C, \{i, i+1\}, \{t_i, t_{i+1}\} | \mathbf{a}, n). \tag{44}$$

The terms are given by eqs. (28), (36), and (39) with $k = 2$, adjusted for boundary cases at sequence endpoints. As elsewhere, the values correspond to unnormalized probabilities.

4.8. Variable $Z_{\langle n,C,\mathbf{t}\rangle}$

The equations above treat the case where $Z_{\langle n,C,\mathbf{t}\rangle}$ is the same for every $\mathbf{t} \in \mathcal{T}[C,n]$. Here we treat the case where $Z_{\langle n,C,\mathbf{t}\rangle}$ varies with $\mathbf{t}$. This would occur with non-physical loop functions such as an affine gap penalty, or with variable segment length, rank order, connectedness, or topology. In this case, f induces a partition on $\mathcal{T}[C,n]$ such that two threadings $\mathbf{t}$ and $\mathbf{u}$ are in the same partition element if and only if $Z_{\langle n,C,\mathbf{t}\rangle} = Z_{\langle n,C,\mathbf{u}\rangle}$. Let the induced partition be $\mathcal{T}^*[C,n] = \{\mathcal{T}_i^*[C,n]\}$ where $\mathcal{T}_i^*[C,n] \subseteq \mathcal{T}[C,n]$ is the ith partition element and the $\mathcal{T}_i^*[C,n]$ are disjoint and cover $\mathcal{T}[C,n]$. Let $_iZ$ (respectively $_i\mu$) represent global sums (respectively global means) over threadings in partition element $\mathcal{T}_i^*[C,n]$. Define

$$_iZ_{\langle n,C,\mathbf{t}\rangle} \;=\; \sum_{\mathbf{w} \in A^n} \exp\big(-f(\mathbf{w},C,\mathbf{t})\big), \quad \text{where } \mathbf{t} \in \mathcal{T}_i^*[C,n], \tag{45}$$

$$_i\mu_{\langle \mathbf{a},n,C\rangle} \;=\; \sum_{\mathbf{x} \in \mathcal{T}_i^*[C,n]} \exp\big(-f(\mathbf{a},C,\mathbf{x})\big)P(\mathbf{x}|n,C), \tag{46}$$

$$_i\mu_{\langle \mathbf{a},n,C,I,T\rangle} \;=\; \sum_{\{\mathbf{x} \in \mathcal{T}_i^*[C,n]\,|\,j \in I \Rightarrow x_j = t_j\}} \exp\big(-f(\mathbf{a},C,\mathbf{x})\big)P(\mathbf{x}|n,C). \tag{47}$$

Equation (28) must be generalized to

$$P(C|\mathbf{a},n) = \frac{P(C|n)}{P(\mathbf{a}|n)} \sum_{\mathcal{T}_i^*[C,n]} \frac{_i\mu_{\langle \mathbf{a},n,C\rangle}}{_iZ_{\langle n,C,\mathbf{t}\rangle}}. \tag{48}$$

Equation (39) must be generalized to

$$P(I,T,C|\mathbf{a},n) = \frac{P(C|n)}{P(\mathbf{a}|n)} \sum_{\mathcal{T}_i^*[C,n]} \frac{_i\mu_{\langle \mathbf{a},n,C,I,T\rangle}}{_iZ_{\langle n,C,\mathbf{t}\rangle}}. \tag{49}$$

Recall that eq. (36) was the special case of eq. (39) when $k = 1$, and must be generalized accordingly. Equations (24) and (29) are unchanged, but must be interpreted with variable $Z_{\langle n,C,\mathbf{t}\rangle}$.

4.9. Recurrence equations for singleton-only objective functions

Here we describe recursive formulae for the singleton-only objective function. This disallows pairwise interactions between sequence residues, but otherwise allows arbitrary sequence-specific score functions for the segments and loops and an arbitrary function for h_λ. We have implemented the relations below in Common Lisp [88]. In practice we actually compute the logarithm of the quantities shown, then exponentiate only as needed, to avoid floating point problems.

4.9.1. Equations for $Z^1_{\langle n,C,\mathbf{t}\rangle}$

In the singleton-only case, if h_s and h_l are normalized probabilities, then $Z^1_{\langle n,C,\mathbf{t}\rangle} = H_s = H_l = 1$; otherwise, $Z^1_{\langle n,C,\mathbf{t}\rangle}$ corresponds to a normalizing constant for $\exp(-f^1)$:

$$Z^1_{\langle n,C,\mathbf{t}\rangle} = \sum_{\mathbf{w}\in A^n} \exp\bigl(-f^1(\mathbf{w},C,\mathbf{t})\bigr) \tag{50}$$

$$= \prod_{i=0}^{m} H_l(i,t_i,t_{i+1})\, H_s(i,t_i). \tag{51}$$

4.9.2. Recurrence equations for $\mu^1_{\langle \mathbf{a},n,C\rangle}$

$$\mu^1_{\langle \mathbf{a},n,C\rangle} = \sum_{\mathbf{x}\in T[C,n]} \exp\bigl(-f^1(\mathbf{a},C,\mathbf{x})\bigr) P(\mathbf{x}|n,C) \tag{52}$$

$$= \sum_{\mathbf{x}\in T[C,n]} \prod_{i=0}^{m} h_l(i,x_i,x_{i+1})\, h_\lambda(i,x_i,x_{i+1})\, h_s(i+1,x_i). \tag{53}$$

Define an intermediate function R by the recurrence

$$R(m,x) = h_l(m,x,\tilde{n})\, h_\lambda(m,x,\tilde{n}), \tag{54}$$

$$R(i,x) = \sum_{y=x}^{\tilde{n}} h_l(i,x,y)\, h_\lambda(i,x,y)\, h_s(i+1,y)\, R(i+1,y), \quad 0 \leqslant i < m. \tag{55}$$

$R(i,x)$ is the unnormalized probability corresponding to placing segment i at relative coordinate x but assigning it zero score, together with all following segments and loops, summed over all possible placements of the following segments. That is, $R(i,x)$ is $\mu^1_{\langle \mathbf{a},n,C\rangle}$ restricted to segments $i+1$ and above and the substring $\mathbf{a}[x]$ and beyond. Consequently,

$$\mu^1_{\langle \mathbf{a},n,C\rangle} = R(0,1). \tag{56}$$

4.9.3. Recurrence equations for $\mu^1_{\langle \mathbf{a},n,C,i,t_i\rangle}$

$$\mu^1_{\langle \mathbf{a},n,C,i,t_i\rangle} = \sum_{\{\mathbf{x}\in T[C,n]\,|\,x_i = t_i\}} \exp\bigl(-f^1(\mathbf{a},C,\mathbf{x})\bigr) P(\mathbf{x}|n,C) \tag{57}$$

$$= \sum_{\{\mathbf{x}\in T[C,n]\,|\,x_i = t_i\}} \prod_{i=0}^{m} h_l(i,x_i,x_{i+1})\, h_\lambda(i,x_i,x_{i+1})\, h_s(i+1,x_i) \tag{58}$$

$$= \sum_{\{\mathbf{x}\in T[C,n]\,|\,x_i = t_i\}} \prod_{i=0}^{m} h_s(i,x_i)\, h_l(i,x_i,x_{i+1})\, h_\lambda(i,x_i,x_{i+1}), \tag{59}$$

where eq. (59) follows because $h_s(0,x) = h_s(m+1,x) = 1$.

Define Q by the recurrence

$$Q(1,x) = h_l(0,1,x) h_\lambda(0,1,x), \tag{60}$$

$$Q(i,x) = \sum_{y=1}^{x} Q(i-1,y) h_s(i-1,y) h_l(i-1,y,x) h_\lambda(i-1,y,x), \quad 1 < i \leqslant m+1. \tag{61}$$

$Q(i,x)$ is the unnormalized probability corresponding to placing segment i at relative coordinate x but assigning it zero score, together with all preceding segments and loops, summed over all possible placements of the preceding segments. That is, $Q(i,x)$ is $\mu^1_{\langle \mathbf{a},n,C \rangle}$ restricted to segments $i-1$ and below and the substring $\mathbf{a}[x]$ and before. Consequently,

$$\mu^1_{\langle \mathbf{a},n,C,i,t_i \rangle} = Q(i,t_i) h_s(i,t_i) R(i,t_i). \tag{62}$$

4.9.4. Recurrence equations for $\mu^1_{\langle \mathbf{a},n,C,I,T \rangle}$

$$\mu^1_{\langle \mathbf{a},n,C,I,T \rangle} = \sum_{\{\mathbf{x} \in T[C,n] \,|\, j \in I \Rightarrow x_j = t_j\}} \exp\left(-f^1(\mathbf{a}, C, \mathbf{x})\right) P(\mathbf{x}|n, C) \tag{63}$$

$$= \sum_{\{\mathbf{x} \in T[C,n] \,|\, j \in I \Rightarrow x_j = t_j\}} \prod_{i=0}^{m} h_s(i,x_i) h_l(i,x_i,x_{i+1}) h_\lambda(i,x_i,x_{i+1}). \tag{64}$$

Recall that $I = \{i_1, i_2, \ldots, i_k\}$ and $T = \{t_{i_1}, t_{i_2}, \ldots, t_{i_k}\}$. By convention, let $i_0 = 0$ and $t_{i_0} = 1$. Define Q_j by the recurrence

$$Q_j(i_{j-1}+1,x) = h_l(i_{j-1}, t_{i_{j-1}}, x) h_\lambda(i_{j-1}, t_{i_{j-1}}, x) \tag{65}$$

$$Q_j(i,x) = \sum_{y=t_{i_{j-1}}}^{x} Q_j(i-1,y) h_s(i-1,y) h_l(i-1,y,x) h_\lambda(i-1,y,x), \tag{66}$$

$$i_{j-1}+1 < i \leqslant i_j,$$

where $Q_j(i,x)$ is the unnormalized probability corresponding to placing segment i at relative coordinate x but assigning it zero score, together with all preceding segments and loops back to but excluding placing segment i_{j-1} at $t_{i_{j-1}}$, summed over all possible placements of the intervening segments. Consequently, with $k = |I|$,

$$\mu^1_{\langle \mathbf{a},n,C,I,T \rangle} = \left(\prod_{j=1}^{k} Q_j(i_j, t_{i_j}) h_s(i_j, t_{i_j}) \right) R(i_k, t_{i_k}). \tag{67}$$

For use with secondary structure prediction, loop modeling, α-/β-hairpins, etc., observe the special case of

$$\mu_{\langle \mathbf{a},n,C,\{i,i+1\},\{t_i,t_{i+1}\} \rangle} = Q(i,t_i) h_s(i,t_i) h_l(i,t_i,t_{i+1}) h_\lambda(i,t_i,t_{i+1}) h_s(i+1,t_{i+1})$$
$$\times R(i+1, t_{i+1}). \tag{68}$$

4.9.5. Recurrence equation invariants

Useful identities that may be used for diagnostic purposes include

$$\mu^1_{\langle \mathbf{a},n,C\rangle} = \sum_{\mathbf{t} \in T[C,n]} \exp\left(-f^1(\mathbf{a}, C, \mathbf{t})\right) \tag{69}$$

$$= R(0,1) \tag{70}$$

$$= Q(m+1, \tilde{n}) \tag{71}$$

$$= \sum_{t_i=1}^{\tilde{n}} \mu^1_{\langle \mathbf{a},n,C,i,t_i\rangle}. \tag{72}$$

In the interest of correct computer code, an implementation should verify eq. (69) for every small search space, and the rest for every sequence–structure pair considered.

4.10. Recurrence equations for per-residue singleton-only objective functions

Many current proposals further specialize the singleton-only f_s (respectively f_l) to be the sum of the individual sequence residue scores at each element of the segment (respectively loop), as discussed in section 3.7.6. This leads to recurrence relations that are more efficient by a factor of $\tilde{n}$. However, the new recurrences no longer make loop endpoints or lengths explicit, so the uninformative loop prior $P(\mathbf{t}|n, C) = |T[C,n]|^{-1}$ is assumed, a per-loop structural environment $s(\lambda_i)$ is used, and pair interactions between adjacent segments are not allowed. A superscript A indicates these assumptions.

The new recurrences are

$$R^A(m,x) = h^A_l(m,x,\tilde{n})|T[C,n]|^{-1}, \tag{73}$$

$$Q^A(1,x) = h^A_l(0,1,x), \tag{74}$$

$$Q^A_j(i_{j-1}+1,x) = h^A_l(i_{j-1}, t_{i_{j-1}}, x), \tag{75}$$

$$R^A(i,x) = \begin{cases} h^A_l(i,x,x)\, h^A_s(i+1,x)\, R^A(i+1,x) \\ \quad + h^A_a(\mathbf{a}[x+c_i], s(\lambda_i))\, R^A(i,x+1), & 1 \leqslant x \leqslant \tilde{n},\ 0 \leqslant i < m, \\ 0, & \text{otherwise}, \end{cases} \tag{76}$$

$$Q^A(i,x) = \begin{cases} Q^A(i-1,x)\, h^A_s(i-1,x)\, h^A_l(i-1,x,x) \\ \quad + Q^A(i,x-1)h^A_a(\mathbf{a}[x-1], s(\lambda_{i-1})), & 1 \leqslant x \leqslant \tilde{n},\ 1 < i \leqslant m+1, \\ 0, & \text{otherwise}, \end{cases} \tag{77}$$

$$Q^A_j(i,x) = \begin{cases} Q^A_j(i-1,x)h^A_s(i-1,x)h^A_l(i-1,x,x) \\ \quad + Q^A_j(i,x-1)h^A_a(\mathbf{a}[x-1], s(\lambda_{i-1})), & t_{i_{j-1}} \leqslant x \leqslant t_{i_j},\ i_{j-1}+1 < i \leqslant i_j, \\ 0, & \text{otherwise}. \end{cases} \tag{78}$$

Consequently,

$$Z^A_{\langle n,C,\mathbf{t}\rangle} = \prod_{i=0}^{m} H^A_l(i,t_i,t_{i+1}) H^A_s(i,t_i), \tag{79}$$

$$\mu^A_{\langle \mathbf{a},n,C\rangle} = R^A(0,1), \tag{80}$$

$$\mu^A_{\langle \mathbf{a},n,C,i,t_i\rangle} = Q^A(i,t_i)h^A_s(i,t_i) R^A(i,t_i), \tag{81}$$

$$\mu^A_{\langle \mathbf{a},n,C,J,T\rangle} = \left(\prod_{j=1}^{k} Q^A_j(i_j,t_{i_j}) \, h^A_s(i_j,t_{i_j})\right) R^A(i_k,t_{i_k}). \tag{82}$$

5. Analysis – search space tasks

The branch and bound search algorithm, described in detail in section 8.1, works by repeatedly subdividing the search space into smaller subsets, always choosing the most promising subset to split at each step. In order to succeed, branch and bound search here requires the ability to (1) represent the entire search space as a set of possibilities; (2) split any set into subsets; and (3) compute a lower bound on the best score achievable within any subset. Any correct implementation of these three requirements would result in a correct search, but search speed would vary dramatically. The keys to an efficient search are a powerful lower bound and good branch points when splitting sets.

For a given sequence and a specific selected core template, this section develops formulae for lower bound, search space splitting, and various search space parameters.

5.1. Lower bound on scores in threading sets

The branch and bound search exploits a lower bound on the score $f(\mathbf{t})$ attainable by any threading $\mathbf{t}$ in any set $T[\mathbf{b},\mathbf{d}]$. Any correct lower bound would result in correct search behavior, but the stronger the lower bound, the more rapidly the search prunes unwanted sets of threadings and converges to the optimum. Total search time is an engineering trade-off between a polynomial-time lower bound computation and an exponential-time search. As a general rule, stronger lower bounds are more expensive to compute but result in smaller exponent coefficients than do weaker lower bounds. Evaluation of the lower bound occurs in the inner loop of the search algorithm and consumes virtually all of its computation time. Consequently, it is crucial that it be computable efficiently.

For example, one lower bound that is easy to derive and fast to compute can be obtained from eq. (3) by summing lower bounds on each term separately:

$$\min_{\mathbf{t}\in T} f(\mathbf{t}) = \min_{\mathbf{t}\in T} \sum_i \left[g_1(i,t_i) + \sum_{j>i} g_2(i,j,t_i,t_j) \right]$$

$$\geqslant \sum_i \left[\min_{b_i \leqslant x \leqslant d_i} g_1(i,x) + \sum_{j>i} \min_{\substack{b_i \leqslant y \leqslant d_i \\ b_j \leqslant z \leqslant d_j}} g_2(i,j,y,z) \right]. \tag{83}$$

The indicated min operations are computable efficiently using binary trees over sub-intervals of $g_1(i,x)$, and quad-trees or 2D trees [89] over sub-intervals of $g_2(i,j,y,z)$.

This simple formula works well for small cases, and consequently would be useful for threading small super-secondary structure motifs or for testing a prototype branch and bound search implementation. It is sufficiently powerful to provide effective pruning in search space sizes of about 10^9 or 10^{12}.

We have explored several alternative forms of the lower bound [1]. Our current version, denoted $\mathrm{lb}(\mathcal{T})$, is effective in search space sizes up to about 10^{25} or 10^{30}:

$$\min_{t \in \mathcal{T}} f(\mathbf{t}) \geqslant \mathrm{lb}(\mathcal{T})$$

$$= \min_{\mathbf{t} \in \mathcal{T}} \sum_i \left(g_1(i, t_i) + g_2(i-1, i, t_{i-1}, t_i) + \min_{\substack{\mathbf{u} \in \mathcal{T} \\ l_j^{\max} = +\infty}} \sum_{|j-i| > 1} \tfrac{1}{2} g_2(i, j, t_i, u_j) \right).$$

(84)

The enclosing $\min_{\mathbf{t} \in \mathcal{T}}$ ensures that the lower bound will be instantiated on a specific legal threading $\mathbf{t}^{\mathrm{lb}} \in \mathcal{T}$. This will be used in splitting $\mathcal{T}$, below. The equation further ensures that the singleton term, in $g_1(i, t_i)$, remains consistent both with the terms that reflect loop scores, in $g_2(i-1, i, t_{i-1}, t_i)$, and with the other (non-loop) pairwise terms, in $g_2(i, j, t_i, u_j)$. The inner $\min_{\mathbf{u} \in \mathcal{T}}$ allows a different vector $\mathbf{u}$ for each i, but requires $\mathbf{u}$ to be a legal threading. The assumption $l_j^{\max} = +\infty$ supports an efficient implementation. Equation (84) would be a tight lower bound (i.e. actually achieved in $\mathcal{T}$) if we further required that $\mathbf{u} = \mathbf{t}$; but then evaluating the bound would be equivalent to solving the search problem. It is easy to see that if $\mathcal{T}$ is a singleton set, $\{\mathbf{t}\}$, then $\mathrm{lb}(\{\mathbf{t}\}) = f(\mathbf{t})$.

5.1.1. Lower bound invariants
Two useful invariants are (1) $\mathcal{T}_1 \supset \mathcal{T}_2$ implies $\mathrm{lb}(\mathcal{T}_1) \leqslant \mathrm{lb}(\mathcal{T}_2)$, and (2) for any $\mathbf{t}$, $\mathrm{lb}(\{\mathbf{t}\})$ by eq. (84), $f(\mathbf{t})$ by summing g_1 and g_2 by eq. (3), and $f(\mathbf{t})$ by summing vertex, edge, and loop components by eq. (7), all are equal. In the interest of correct computer code, an implementation should verify the first invariant whenever a subset is split, and the second whenever a global optimum threading is found.

5.2. Splitting threading sets

The second key element of branch and bound search as used here is the ability to subdivide sets of threadings successively. A set is split by choosing a single core segment C_i and a single split-point t_i^{split} (see Fig. 5). The interval $[b_i, d_i]$ is divided into three sub-intervals: the points (1) less than the split-point, $[b_i, t_i^{\mathrm{split}} - 1]$; (2) equal to the split-point, $[t_i^{\mathrm{split}}, t_i^{\mathrm{split}}]$; and (3) greater than the split-point, $[t_i^{\mathrm{split}} + 1, d_i]$. This results in three mutually disjoint and exhaustive subsets of the original set. There are many possible ways to choose C_i and t_i^{split}. The choice affects search speed, but not so much as does the choice of lower bound.

Currently, we choose t_i^{split} based on $\mathbf{t}^{\mathrm{lb}} \in \mathcal{T}$. The specific threading t_i^{lb} that instantiates the lower bound in eq. (84) is used to choose $t_i^{\mathrm{split}} = t_i^{\mathrm{lb}}$.

It is less obvious how to select the core segment C_i at which to split. One simple method, easy to implement and appropriate for threading small super-secondary structure motifs, is to split at the segment having the widest interval, i.e. at the i that maximizes

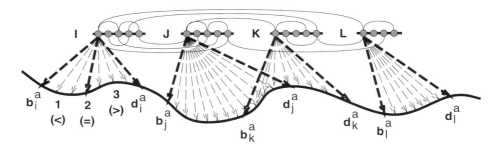

Fig. 5. Defining and splitting sets of threadings. Sets used in the branch-and-bound search are defined by lower and upper limits (dark arrows, labeled b_x^a and d_x^a for segment X) on the sequence amino acid placed into the first core element of each segment. The set consists of all legal threadings such that the first element of each segment X is within the interval $[b_x^a, d_x^a]$. A set is split into subsets by choosing one core segment (here, segment I) and one split point (dark interior arrow). Its interval is split into sub-intervals (1) less than, (2) equal to, and (3) greater than the split point.

the value of $d_i - b_i$. The method we currently use chooses the segment i that has the most negative expected score contribution if its interval were to be split at t_i^{lb}. Specifically, we split at $\langle i, t_i^{lb} \rangle$ where i yields the most negative value of the expression

$$P_1(i, t_i^{lb}) \left[g_1(i, t_i^{lb}) - \overline{g}_i + \sum_{j \neq i} \left(1 - \tfrac{1}{2}\alpha(j) \right) \left(g_2(i, j, t_i^{lb}, t_j^{lb}) - \overline{g}_{i,j} \right) \right]. \qquad (85)$$

Assuming a uniform probability distribution over all legal threadings, $P_1(i, t_i^{lb})$ is the probability that a randomly drawn threading will place C_i at t_i^{lb}; $\overline{g}_i$ is the expected value of $g_1(i, *)$; $\overline{g}_{i,j}$ is the expected value of $g_2(i, j, *, *)$; these terms are defined in section 6. $\alpha(i)$ indicates whether segment i is active (variable) or inactive (fixed) in the set

$$\alpha(i) = \begin{cases} 1, & \text{if } b_i < d_i, \\ 0, & \text{if } b_i = d_i. \end{cases} \qquad (86)$$

In eq. (85), the factor $P_1(i, t_i^{lb})$ biases the choice to prefer combinations of i and t_i^{lb} that are a priori more likely. The terms $g_1(i, t_i^{lb}) - \overline{g}_i$ and $g_2(i, j, t_i^{lb}, t_j^{lb}) - \overline{g}_{i,j}$ bias the choice to prefer scores that are lower than expected. The factor $(1 - \tfrac{1}{2}\alpha(j))$ assigns the entire pairwise term $g_2(i, j, t_i^{lb}, t_j^{lb}) - \overline{g}_{i,j}$ to core segment C_i if C_j is inactive, and shares it evenly between them if both are active.

One-step look-ahead yields a much more effective, but much more expensive, heuristic for splitting sets. In this method, all m possible splitting segments are considered in turn. For each segment, three lower bounds are computed corresponding to its three resulting split sets ($<$, $=$, and $>$). Segments are ranked by the minimum lower bound among their three sets. The segment that maximizes the minimum lower bound is chosen as the splitting segment because it will result in the largest overall increase in lower bound. Because $3m$ lower bound computations must be done this approach is quite slow on a serial computer. In a distributed processing environment each lower bound could be computed simultaneously on a different processor, resulting in enhanced performance for the same elapsed search time.

6. Analysis – search space formulae

Formulae in this section are used when splitting sets of threadings, and also to obtain important statistical information.

6.1. Fast approximate formulae

For the heuristic choice of which core segment to split in eq. (85), speed is important and approximate formulae are acceptable. Fast formulae result from the simplifying assumptions that the entire search space is included ($b_i = 1$ and $d_i = \tilde{n}$), and that loops can be arbitrarily long ($l_i^{\max} = +\infty$). Equations (87)–(89) are exact for this case, and are used to approximate all other cases in eq. (85). Exact algorithms for all other cases are given below in sections 6.2 and 6.3.

   Under these simplifying assumptions the number of legal threadings, or search space size, $S = |T[C, n]|$, is the result of the binomial coefficient function

$$S \approx \binom{\tilde{n} + m - 1}{m}. \tag{87}$$

   Simple formulae also hold for segment placement probabilities. Under a uniform probability distribution on threadings, let $P_1(i, t_i)$ be the probability that segment i occurs at index t_i in a randomly drawn threading, and let $P_2(i, j, t_i, t_j)$ be the probability that segment i occurs at index t_i and simultaneously segment j occurs at index t_j. Let $i < j$ and $t_i \leqslant t_j$. Then,

$$P_1(i, t_i) \approx \binom{t_i + i - 2}{i - 1}\binom{\tilde{n} - t_i + m - i}{m - i} \Big/ \binom{\tilde{n} + m - 1}{m}, \tag{88}$$

$$P_2(i, j, t_i, t_j) \approx \binom{t_i + i - 2}{i - 1}\binom{t_j - t_i + j - i - 1}{j - i - 1}\binom{\tilde{n} - t_j + m - j}{m - j} \Big/ \binom{\tilde{n} + m - 1}{m}. \tag{89}$$

Each factor is always the binomial coefficient corresponding to the number of ways to choose the number of available core segments, out of one less that the number of available sequence indices plus the number of available core segments. The denominator is the approximate search space size from eq. (87). Successive factors in the numerator correspond to the combinatorial number of arrangements between successive pairs of core segments fixed by the arguments. Similar formulae hold for $P_3(i, j, k, t_i, t_j, t_k)$, for $P_4(i, j, k, l, t_i, t_j, t_k, t_l)$, and so forth.

   These relations permit us to estimate the expected singleton and pairwise score components attributable to each segment. The expected singleton contribution for segment i is $\overline{g}_i$, and the expected pairwise contribution for segments i and j is $\overline{g}_{ij}$. Where $i < j$ and $t_i \leqslant t_j$,

$$\overline{g}_i = \sum_x P_1(i,x)g_1(i,x), \tag{90}$$

$$\overline{g}_{ij} = \sum_x \sum_{y \geqslant x} P_2(i,j,x,y)\,g_2(i,j,x,y). \tag{91}$$

6.1.1. Computing P_1 and P_2 efficiently

In practice we compute the logarithm of eqs. (88) and (89), then exponentiate. When loading the system we precompute and store $\log n$ for $n < 1000$ and $\log \binom{n}{m}$ for $m < 50$ and $n < 1000$. Equations (88) and (89) then require only the sum of a few array references plus a transcendental function call. The approximations to $P_1(i,t_i)$, $P_2(i,j,t_i,t_j)$, $\overline{g}_i$, and $\overline{g}_{ij}$, all are constant for a given search space, and are precomputed and stored when each search is begun. The precomputation is fast and the storage required is approximately the size of the g_1 and g_2 arrays. Consequently, eq. (85) requires only sums and products of a few array references.

6.2. Exact search space size, probabilities, and uniform sampling

In practice, external knowledge may constrain core segments to arbitrary sub-intervals or specify maximum loop lengths. This section provides exact formulae for such cases.

6.2.1. Search space size

Let $\mathcal{T}[\mathbf{b},\mathbf{d}] = \{\mathbf{t}|b_i \leqslant t_i \leqslant d_i\}$ be the set of threadings delimited by $\mathbf{b}$ and $\mathbf{d}$, let $S[\mathbf{b},\mathbf{d}]$ be the number of legal threadings it contains, and let $B(i,x)$ be the number of legal threadings of segments i through m when segment i is placed at relative sequence index x or higher. B is given by the recursive formula

$$B(i,x) = \begin{cases} d_m - x + 1, & \text{if } i = m \text{ and } b_m \leqslant x \leqslant d_m; \\ B(i,b_i), & \text{if } 1 \leqslant i \leqslant m \text{ and } x < b_i; \\ B(i,x+1) + B(i+1,x) - B(i+1,x+\tilde{l}_i+1), \\ & \text{if } 1 \leqslant i < m \text{ and } b_i \leqslant x < d_i; \\ 0, & \text{otherwise.} \end{cases} \tag{92}$$

The numbers involved in computing B become combinatorically large; arbitrary precision integer arithmetic is a language primitive in Lisp, and usually available as a subroutine in other languages.

Consequently,

$$S[\mathbf{b},\mathbf{d}] = B(1,b_1) \tag{93}$$

is exact for arbitrary $\mathbf{b}$, $\mathbf{d}$, $\mathbf{l}^{\min}$, and $\mathbf{l}^{\max}$. By applying eq. (92) to $b_i = 1$ and $d_i = \tilde{n}$,

$$S = S[\mathbf{1},\tilde{\mathbf{n}}] \tag{94}$$

gives the exact size of the entire legal search space. This is the exact formula corresponding to the approximate equation (87), and is used for all search space sizes reported here.

6.2.2. Exact segment placement probabilities

Exact formulae for segment placement probabilities are computable as the ratio of the search space sizes corresponding to the constrained and the entire search spaces. The denominator in all cases is the entire search space size given by eq. (94). The numerator corresponding to $P_1(i, t_i)$ arises from the set of threadings that fix C_i at t_i, denoted $T\langle i, t_i \rangle$. Its search space size $S\langle i, t_i \rangle$ may be computed from eq. (92) applied to $b_j = \{\text{if } j < i \text{ then } 1 \text{ else } t_i\}$ and $d_j = \{\text{if } j \leqslant i \text{ then } t_i \text{ else } \tilde{n}\}$. Then

$$P_1(i, t_i) = \frac{S\langle i, t_i \rangle}{S[1, \tilde{n}]} \tag{95}$$

is the exact formula corresponding to the approximate equation (88). Similar formulae hold for $P_2(i, j, t_i, t_j)$, $P_3(i, j, k, t_i, t_j, t_k)$, $P_4(i, j, k, l, t_i, t_j, t_k, t_l)$, and so forth.

6.2.3. Uniform random sampling

Equation (92) also allows us to randomly sample the threadings in any set $T[\mathbf{b}, \mathbf{d}]$, assuming a uniform probability distribution (blind draw) on threadings. Let s be a random integer uniformly drawn between one and $S[\mathbf{b}, \mathbf{d}]$ inclusive; uniform random numbers are a language primitive in Lisp, and usually available as a subroutine in other languages. Convert s to a unique threading as follows:

FOR i FROM 1 TO m DO
 (1) Find x such that $b_i \leqslant x \leqslant d_i$ and $B(i, x+1) < s \leqslant B(i, x)$. (96)
 (2) Set t_i to x.
 (3) Set s to $s - B(i, x+1)$.

It is only necessary to compute $S[\mathbf{b}, \mathbf{d}]$ and B once for each set $T[\mathbf{b}, \mathbf{d}]$.

6.3. Exact analytic search space mean and standard deviation

Let $f(t) = \sum_i g_1(i, t_i) + \sum_i \sum_{i<j} g_2(i, j, t_i, t_j)$ be the threading score function chosen. Then the score distribution mean $\bar{f}$ is

$$\bar{f} = E(f(*)) = \sum_i E(g_1(i, *)) + \sum_i \sum_{i<j} E(g_2(i, j, *, *)), \tag{97}$$

where

$$E(g_1(i, *)) = \sum_x P_1(i, x) g_1(i, x), \tag{98}$$

$$E(g_2(i, j, *, *)) = \sum_x \sum_y P_2(i, j, x, y) g_2(i, j, x, y). \tag{99}$$

The distribution variance is V and the standard deviation is $\sigma = \sqrt{V}$:

$$V = E([\bar{f} - f(*)]^2) = E(\bar{f}^2 - 2\bar{f}f(*) + f^2(*)) = E(f^2(*)) - \bar{f}^2, \tag{100}$$

where

$$E(f^2(*)) = E([\sum_i g_1(i,*) + \sum_i \sum_{i<j} g_2(i,j,*,*)]^2) \tag{101}$$

$$= \sum_i E([g_1(i,*)]^2) + 2\sum_i \sum_{i<j} E(g_1(i,*)g_1(j,*))$$

$$+ \sum_i \sum_{i<j} \sum_k E(g_1(k,*)g_2(i,j,*,*))$$

$$+ \sum_i \sum_{i<j} E([g_2(i,j,*,*)]^2) \tag{102}$$

$$+ 2\sum_i \sum_{i<j} \sum_{j<k} \sum_{k<l} E(g_2(i,j,*,*)g_2(k,l,*,*)),$$

$$E([g_1(i,*)]^2) = \sum_x P_1(i,x)[g_1(i,x)]^2, \tag{103}$$

$$E(g_1(i,*)g_1(j,*)) = \sum_x \sum_y P_2(i,j,x,y)\,g_1(i,x)\,g_1(j,y), \tag{104}$$

$$E(g_1(k,*)g_2(i,j,*,*)) = \sum_x \sum_y \sum_z P_3(i,j,k,x,y,z)\,g_1(k,z)\,g_2(i,j,x,y), \tag{105}$$

$$E([g_2(i,j,*,*)]^2) = \sum_x \sum_y P_2(i,j,x,y)\,[g_2(i,j,x,y)]]^2, \tag{106}$$

$$E(g_2(i,j,*,*)g_2(k,l,*,*))$$

$$= \sum_x \sum_y \sum_z \sum_v P_4(i,j,k,l,x,y,z,v)\,g_2(i,j,x,y)\,g_2(k,l,z,v). \tag{107}$$

The analytic formula for the mean has a computational complexity of $O(m^2\tilde{n}^2)$. The analytic formula for the standard deviation has a computational complexity of $O(m^4\tilde{n}^4)$.

6.4. Computing the exact analytic mean and standard deviation efficiently

In practice, the fourth-power computational complexity of the analytic standard deviation formula is burdensome for most proteins. Consequently, we usually estimate the mean and standard deviation by sampling the search space. It takes only a few seconds to draw and score 10 000 uniformly distributed random threadings using the g_1 and g_2 arrays in conjunction with methods in section 6.2.3. This results in sufficiently accurate estimates for most ordinary purposes. In case an exact value is important, the analytic formulae are available at additional computational cost.

7. Computational complexity

The complexity of finding the optimal sequence-to-structure threading is a function of whether arbitrary length alignment gaps are allowed, and whether or not the score function includes pairwise or higher-order interactions among sequence amino acids. Given these two properties, the threading problem is known to be NP-complete.

Our current knowledge of protein structure and evolution requires the first property, the allowance of alignment gaps of near arbitrary length. In fact, a very wide range of sequences have been observed to have been inserted into various basic folds, particularly at the protein's surface. The extreme case is observed in many multidomain proteins where the linear sequence encoding for one domain's fold has been inserted into that encoding a second independent fold. The necessity of the second property, inclusion of at least long-range pairwise interactions in the scoring schema, is much less clear. It generally is believed to be important, but the issue is far from settled. In the three-dimensional folded form of a native protein, the key characteristics of the local environment in which any amino acid finds itself are determined largely by its contacting neighbors. Interactions between spatially neighboring amino acids are formed by those at arbitrary positions along the sequence.

The general problems of protein folding [90–92], protein threading [62,63], and protein structure comparison [87], all are known to be NP-hard. Thus an exact solution is widely believed to require exponential time (unless P=NP). In the general case, any approach to the information transform shown in Fig. 4 must solve (or approximate) at least one NP-hard problem. Different approaches trade off which NP-hard problems to solve or approximate, and how. We expect that many current approaches will be found to contain computational analogs or approximations to the quantities above. Fast approximate knowledge-intensive solutions often perform well in practice, and a clear understanding of the optimal formulae can help us to evaluate the speed/accuracy trade-offs such approximate solutions must make.

7.1. Computational complexity and NP-completeness

An exhaustive review of computational complexity and NP-completeness is beyond the scope of this chapter, and this section presents only enough material to motivate the discussion in the balance of the section. The interested reader is referred to refs. [93–95] for formal treatment of the subject; and to refs. [90–92] for discussion of its biological relevance.

The analysis of computational complexity is concerned with the question of how the running time of an algorithm grows as the size of its input increases. For a given algorithm, this is made specific by naming some function, f, and asserting that the algorithm's running time grows with inputs of increasing size "no faster" than f grows with arguments of increasing magnitude. Formally, let $\#(A, I)$ denote the running time (number of steps) of algorithm A when started with input I, and let $|I|$ be some reasonable measure of the size of I. Then we write $A = O(f)$, read "A is of the order of f" (or "A is big oh of f"), if there exists any positive constant C such that

$$\lim_{|I| \to \infty} \frac{\#(A, I)}{f(|I|)} \leqslant C$$

Algorithms whose computational complexity is the order of some polynomial ("polynomial-time algorithms") are considered to be formally efficient. All other algorithms have a running time that grows faster than any possible polynomial, and are considered to be inefficient. It is possible, of course, that an exponential-time algorithm with a tiny exponent may terminate rapidly on small- and medium-sized inputs, or that a polynomial-time algorithm applied to a very large input may not. Nonetheless, the distinction is valuable and important in most practical cases.

A "problem" is a class of computational tasks defined in terms of a set of parameters (for example, SEQUENCE is a parameter of the protein threading problem). A problem "instance" results from replacing the parameters by actual values (for example, replacing SEQUENCE by a specific string). An algorithm solves a problem if it terminates correctly on every instance of the problem. A "decision problem" is one to which the answer is either "YES" or "NO". For example, the decision problem addressed in protein threading is, "Does there exist a threading of this sequence into this structure under this score function, such that the threading score is less than K?" This might be the case in which a candidate threading already has been found, and one wishes simply to ask whether or not another threading with a better score exists.

It is customary to identify the computational complexity of a problem with that of the most efficient algorithm that solves it. The class of problems that can be solved by a polynomial-time algorithm is named "P". Problems which belong to this class are formally tractable. The class of decision problems whose solution, *once found or guessed,* may be verified in polynomial time is named "NP". Note that it may not be possible actually to *find* a solution in polynomial time; the condition refers only to the *verification* of a putative solution. Problems in NP are solvable in "non-deterministic polynomial time", meaning that if one could somehow non-deterministically guess and check all possible solutions simultaneously, the solution would be obtained in polynomial time. The practical message of this theoretical condition is that the search for a solution, not the verification step, determines whether a polynomial-time solution is possible or not. Clearly, P is a subset of NP, but it is unknown whether $P = NP$.

There is an important class of problems which belong to NP, and have the property that an algorithm solving any problem in the class can be transformed in polynomial time to solve any other problem in NP. Therefore, a polynomial-time algorithm solving any problem in this class would immediately yield a polynomial-time solution for every problem in NP. These problems are the hardest in NP, and are known as the "NP-complete" problems. It is not known whether or not they have a polynomial-time solution (if so, then $P = NP$ because polynomials are closed under composition). They include many problems deeply central to computer science, and so a great deal of effort by a great many talented people has been expended searching for a polynomial-time solution to any one of them. Because so many talented people have failed, it is widely accepted that no polynomial-time algorithm is likely ever to be found.

In some cases it is possible to prove directly from first principles that a problem at hand is NP-complete, but this is usually quite difficult. Most proofs proceed by constructing a polynomial-time transformation of another problem, already known to be NP-complete, into an instance of the problem at hand. It follows that, if the problem at hand could be solved in polynomial time, so could the other problem, and therefore by extension all of the problems in NP. Consequently, the problem at hand is NP-complete.

For many NP-complete decision problems there is an associated optimization problem, for which the task is to produce an optimal solution. For example, the optimization problem associated with the threading decision problem stated above is, "Find the threading of this sequence into this structure under this score function having the optimal (minimum) score." It is easy to see that the optimization problem cannot be easier than the decision problem. This is because a polynomial-time solution to the optimization problem could be transformed into a polynomial-time solution to the decision problem. Thus, if the optimization problem is solvable in polynomial time, then so is the decision problem. For example, if we could find the optimal threading in polynomial time, then it would be easy to compute its score (also in polynomial time). This would let us answer the decision question of whether there exists a threading with score less than K, by checking to see if the optimal score was less than K, with only one additional step. Search problems bearing this relationship to an NP-complete decision problem are called NP-hard.

Finally, we note that problems can be NP-complete for different reasons. In some cases, a polynomial-time solution fails only because the numbers associated with the problem can become exponentially large in magnitude (for example, perhaps the binary bits specifying an integer are used to encode some other non-numeric information). In most cases these numbers are integers; more complicated numbers are theoretically treated as composites of several distinct integers. If a problem is NP-complete, and remains NP-complete when restricted to problem instances for which the magnitude of the largest integer is bounded by a polynomial in the problem instance size, then the problem is called NP-complete in the strong sense.

7.2. Alignment – informal sketch of proof

The protein threading problem consists of a sequence, a core, and an objective function. The decision problem is whether there exists a threading of the sequence into the core with a score under the objective function of some specified constant K or less. Call this problem "PRO-THREAD". The associated optimization problem is to produce a threading whose score is the global optimum. The bulk of the proof consists of constructing an encoding from a known NP-complete problem into PRO-THREAD. The problem we choose for this is ONE-IN-THREE 3SAT, a variant of SATISFIABILITY. The remainder of this section briefly and informally sketches the encoding. Formal details are in ref. [62].

The canonical (and first) NP-complete problem is SATISFIABILITY. A problem instance consists of a set of Boolean variables, plus a set of Boolean clauses (a clause is a disjunction, or logical OR, of a set of literals; a literal is either one of the variables or the negation of one of the variables). The question is whether any setting (truth-value assignment) of the variables makes all of the clauses true simultaneously. 3SAT is a well-known variant which restricts the clauses to contain exactly three literals. ONE-IN-THREE 3SAT is a further variant of 3SAT which requires that each of the clauses be made true by *exactly one* of the three literals. All these problems are known to be NP-complete [93].

The proof that PRO-THREAD is NP-complete proceeds by showing that we can encode any arbitrary instance of ONE-IN-THREE 3SAT (does there exist a setting of the Boolean variables making all the clauses simultaneously true by exactly one literal?) as an equivalent instance of PRO-THREAD. Threadings with a score of zero encode solutions of the original ONE-IN-THREE 3SAT problem; threadings with positive scores encode failures. The equivalent encoded PRO-THREAD question is: does there exist a threading with a score of zero or less? The answer to this question is "YES" exactly when a solution exists for the original ONE-IN-THREE 3SAT problem.

The essence of the proof is this:
- Amino acids from the sequence can encode whether a Boolean variable is TRUE (by T, a threonine residue) or FALSE (by F, phenalanine); and also which literal makes a Boolean clause true (P, proline, encodes the first; Q, glutamine, the second; and R, arginine, the third, literal). In the encoded problem, the sequence $\mathbf{a}$ to be threaded is

$$\mathbf{a} = PQRPQRPQRPQR \cdots PQRTF \cdots TFTFTFTF$$

where we allot one "PQR" for each clause, and one "TF" for each Boolean variable.
- By making each core segment exactly one element long, it is threaded to exactly one amino acid. Consequently, any given threading assigns every core segment to one of $\{P, Q, R, T, F\}$. (As discussed below, extensions that add "GAP" to this list are also NP-complete.)
- We can use one core segment to encode each Boolean clause, and choose which literal makes it true by threading it to P (= the first literal), Q (= the second), or R (= the third) in the sequence $\mathbf{a}$. Similarly, one core segment encoding each Boolean variable is threaded to T (= TRUE) or F (= FALSE), and thereby chooses truth values.
- Pairs of core elements are taken as neighbors in the core (and recorded as such in the adjacency graph) exactly when the clause encoded by the first element contains a literal naming the variable encoded by the second. The edge environment label assigned is an ordered pair, $d = (i, j)$, that encodes which literal (i = 1, 2, or 3) is involved and whether the variable is negated (j = YES or NO).
- An edge score function can be written that is zero when the edge label d is consistent with the literal choice encoded by amino acid a (as P, Q, or R) and the truth-value encoded by amino acid b (as F or T); and is one otherwise.
- By summing the edge score function over all edges, a threading score function can be written that is zero when a candidate threading encodes a truth-value and literal assignment correctly solving the original problem, and positive otherwise. The question "Does there exist a threading with a score of zero or less?" is now equivalent to the original ONE-IN-THREE 3SAT question.
- Thus, if we could solve the general PRO-THREAD problem in polynomial time, we could solve ONE-IN-THREE 3SAT in polynomial time. PRO-THREAD is NP-complete.

In fact, PRO-THREAD is NP-complete in the strong sense (i.e. is not a number problem), because the only numbers used in the construction are zero and one. The optimization problem, to produce an optimal threading, is NP-hard.

The basic proof can be used to prove that many threading methodology extensions and generalizations are also NP-hard. The general strategy in such cases is first to show that the extended problem remains in NP (because a putative solution can be checked in polynomial time), then to show that the problem has not been made easier (by exhibiting some setting of the extended parameters for which the extended problem can be made to solve PRO-THREAD). Consequently, a polynomial-time solution to the extended problem would imply a polynomial-time solution to the simpler PRO-THREAD. Without producing formal proofs, we sketch this for three cases of interest: allowing a core element to be unoccupied (threaded to a gap), as some dynamic programming methods permit; the inclusion of triplet or higher-order terms; and the presence of constraint equations on environment labels. Suppose we allow unoccupied elements. A method for solving this problem can be made to solve PRO-THREAD by using a score function that assigns any such threading a positive score. Similarly, extensions including triplet or higher-order terms can be made equivalent to PRO-THREAD by employing a score function that assigns all such terms a score of zero. Extensions which admit constraint equations on environment labels can be made to solve PRO-THREAD by adding tautologically true constraint equations to the original PRO-THREAD problem. Generally speaking, any related problem that includes PRO-THREAD as a special case remains NP-hard.

7.3. Selection tasks and Bayes' constants

It is easy to modify the basic proof to show that computing any of the Bayes' selection task global sums is NP-hard if pair interactions are allowed.

7.3.1. Complexity of $\mu_{\langle \mathbf{a},n,C \rangle}$
Combine the edge score functions of a given threading using multiplication instead of addition, and change the edge score function so that threadings that score zero encode failures to the original ONE-IN-THREE 3SAT problem and threadings that score 1 encode solutions. Then $\mu_{\langle \mathbf{a},n,C \rangle}$ is greater than zero exactly when a solution exists to the original ONE-IN-THREE 3SAT problem.

7.3.2. Complexity of $Z_{\langle n,C,\mathbf{t} \rangle}$
Use the same embedding as above. Again, a score of zero corresponds to a failure, and a score of 1 corresponds to a solution, of the original ONE-IN-THREE 3SAT problem. Shorten the sequence so that there is exactly one amino acid per core segment in the encoded problem, hence exactly one threading in the solution search space. Then $Z_{\langle n,C,\mathbf{t} \rangle}$ is greater than zero exactly when a solution exists to the original ONE-IN-THREE 3SAT problem.

7.3.3. Complexity of $\mu_{\langle \mathbf{a},n,C,i,t_i \rangle}$ and $\mu_{\langle \mathbf{a},n,C,I,T \rangle}$
Because $\mu_{\langle \mathbf{a},n,C \rangle} = \sum_{t_i=1}^{\tilde{n}} \mu_{\langle \mathbf{a},n,C,i,t_i \rangle}$, a polynomial-time computation for $\mu_{\langle \mathbf{a},n,C,i,t_i \rangle}$ would imply a polynomial-time computation for $\mu_{\langle \mathbf{a},n,C \rangle}$, which is NP-complete by section 7.3.1. In turn, $\mu_{\langle \mathbf{a},n,C,i,t_i \rangle}$ is a special case of $\mu_{\langle \mathbf{a},n,C,I,T \rangle}$, which therefore cannot be easier.

8. Search algorithm

The search algorithm requires as input the sequence, core template, and score function. In any threading trial, the input sequence, core template, and score function exactly define an abstract mathematical space. Each point in this search space corresponds one-to-one with a distinct alignment between the sequence and the structure. The score function assigns a scalar value (a score or pseudo-energy) to each point. The global minimum score on the resulting pseudo-energy landscape is the lowest score achieved by any point in the space. The global optimum alignment(s) is exactly the point(s) that achieves the global minimum score. These are well-defined objects of independent mathematical interest. They are fixed, in an exact mathematical sense, once the input sequence, core template, and score function are known.

Although thereby determined, the landscape features generally are unknown. The sole task performed by the search algorithm is to report the value of the global minimum score, and to identify the global optimum threading(s) that instantiate(s) it. In contrast to approximate search methods, the branch and bound search algorithm here either finds the mathematically exact answer or it first exhausts time or space resources. The version discussed here never returns an approximate or inexact result, although fast approximate versions are possible. Of course, even for the same sequence and core template, different input score functions will produce different landscapes and different global minima. Different scoring landscapes affect the time required by such an algorithm (and hence, whether or not it converges within a specific time limit), but they cannot change the fact that the algorithm here always either finds the mathematically exact global minimum or fails to converge. The particular values of the global minimum and the best alignment, therefore, are a function only of the input; while our ability to identify them is a function of the search algorithm. Consequently, for an exact search algorithm, any agreement – or lack of it – between optimal and native alignments is a property only of the input, and not of the search algorithm.

8.1. Branch and bound algorithm

Branch and bound search [96,97] is a computational method of finding the mathematically exact global optimum in large complex search spaces. In the best case it exploits constraints from the problem to prune the search space, so that most potential solutions are never actually examined. In the worst case it performs no better than exhaustive search. Here it is used to find the global optimum threading when both variable-length gaps and pairwise interactions are allowed [1]. Given a fixed core template, sequence, and score function, the branch and bound search algorithm here is guaranteed to find the optimal threading first, and thereafter to enumerate successive candidate threadings in score order. It provides a mathematically exact implementation for the gapped block alignment threading methodology.

The search begins with a single set containing all legal threadings. At each step, the algorithm chooses the set with the currently lowest lower bound and splits it into several subsets. The entire search space always is represented explicitly as the union of the sets created so far. After some finite number of steps, the chosen set will contain only one threading. Its score equals its lower bound. Every other set had an equal or greater lower bound, and so every other threading must have an equal or greater score. Consequently, this is the desired global optimum threading.

A set is pruned whenever its lower bound is above the global minimum score, because the global optimum threading will be discovered before that set is ever considered again. The global minimum score is unknown until the search terminates, and so pruning is implicit. If the search space may be pruned rapidly, then the search may be relatively short. Cases of multiple threadings with the same optimal score occur very rarely, and are detected automatically by continuing the search.

8.2. Algorithm pseudo-code

INITIALIZATION
(1) Compute a lower bound for the set of all threadings.
(2) Initialize a sorted list to contain one entry, the set of all threadings with its lower bound.
ITERATION
(1) Remove from the list the set having the lowest lower bound.
(2) If the set contains only one threading, stop and announce success. This is a global optimum threading. The procedure later may be continued from this point to enumerate successive candidate threadings in score order.
(3) Otherwise, split the set into smaller subsets.
(4) Compute a lower bound for each new subset.
(5) Merge the new subsets into the list, sorted by lower bound.

The sorted list is implemented as a priority queue, or heap [98], for rapid access to the currently lowest lower bound.

8.3. Lower bound implementation

Efficient calculation of a strong lower bound is the essence of the branch and bound algorithm. The first part of this section describes an efficient implementation strategy. The second part describes a practical caching scheme that avoids much of the computation.

8.3.1. Implementation
This subsection describes an implementation of the lower bound $\mathrm{lb}(\mathcal{T})$ on the possible scores achieved by threadings within a set $\mathcal{T}$. As in eq. (86), say that a search space axis i (i.e. the placement of core segment C_i in the sequence) is "active" in $\mathcal{T}$ if $b_i < d_i$ (i.e. the placement of core segment C_i in the sequence may vary within $\mathcal{T}$), and "inactive" if $b_i = d_i$ (C_i is fixed in $\mathcal{T}$). Note that this does not refer to pairwise or singleton contributions; both active and inactive segments may have contributions from both pairwise and singleton sources.

Separate the lower bound $\mathrm{lb}(\mathcal{T})$ into an inactive part $q(\mathcal{T})$ and an active part $r(\mathcal{T})$. These satisfy $\mathrm{lb}(\mathcal{T}) = q(\mathcal{T}) + r(\mathcal{T})$. The inactive part $q(\mathcal{T})$ sums the contributions that can be determined by knowing the exact placement of the inactive axes. These are the singleton contributions from each inactive axis, plus the pairwise contributions from each pair of inactive axes. For each subset created during the search, $q(\mathcal{T})$ is stored with the m-vectors $\mathbf{b}$ and $\mathbf{d}$ and updated at each split. The active part $r(\mathcal{T})$ estimates a lower bound on the contribution from the active axes plus their pairwise interactions with the inactive axes. It is recomputed each time the lower bound computation is done.

Use $\alpha(i)$ to indicate whether axis i is active (eq. 86), and $\beta(i, j)$ to indicate whether either of axes i or j are active. Let

$$\beta(i, j) = \begin{cases} 1, & \text{if either axis } i \text{ or } j \text{ is active;} \\ 0, & \text{otherwise.} \end{cases} \tag{108}$$

β is related to α by

$$\beta(i, j) = \alpha(i) + \alpha(j) - \alpha(i)\alpha(j) \tag{109}$$

$$= \alpha(i)\left(1 - \tfrac{1}{2}\alpha(j)\right) + \alpha(j)\left(1 - \tfrac{1}{2}\alpha(i)\right). \tag{110}$$

Then define

$$q(\mathcal{T}) = \sum_i (1 - \alpha(i)) \left[g_1(i, b_i) + \sum_{j > i} (1 - \beta(i, j)) g_2(i, j, b_i, b_j) \right], \tag{111}$$

$$r(\mathcal{T}) = \min_{\mathbf{t} \in \mathcal{T}} \sum_i \Big[\alpha(i) g_1(i, t_i) + \beta(i - 1, i) g_2(i - 1, i, t_{i-1}, t_i)$$

$$+ \alpha(i) \min_{\substack{\mathbf{u} \in \mathcal{T} \\ t_j^{\max} = +\infty}} \sum_{|j - i| > 1} \left(1 - \tfrac{1}{2}\alpha(j)\right) g_2(i, j, t_i, u_j) \Big]. \tag{112}$$

Note that in the inner $\min_{\mathbf{u} \in \mathcal{T}}$, the ordering constraints imply that

$$j < i \Rightarrow u_j \leqslant t_i \quad \text{and} \quad j > i \Rightarrow u_j \geqslant t_i,$$

as otherwise $g_2(i, j, t_i, u_j) = +\infty$. By convention, $g_2(j, i, t_j, t_i) = g_2(i, j, t_i, t_j)$.

Recall that $\mathrm{lb}(\mathcal{T}) = q(\mathcal{T}) + r(\mathcal{T})$. The terms in eqs. (111) and (112) have the same meanings as in eq. (84). The inactive part $q(\mathcal{T})$ is easy to update after each split simply by accounting for newly inactive axes. The remainder of this section describes a recursive formulation of $r(\mathcal{T})$ which leads to an efficient implementation.

Define K as

$$K(i, t_i) = \alpha(i) \Big[g_1(i, t_i) + K^*(i, m, t_i, d_m) \Big]$$

$$+ \min_{\substack{x \geqslant \max(b_{i-1}, t_i - \tilde{l}_{i-1}) \\ x \leqslant \min(d_{i-1}, t_i)}} (K(i - 1, x) + \beta(i - 1, i) g_2(i - 1, i, x, t_i)), \tag{113}$$

where $g_1(i, t_i)$ accounts for singleton terms, $g_2(i - 1, i, x, t_i)$ forces pairwise terms containing loop scores to be consistent between $i - 1$ and i, and $K^*(i, m, t_i, d_m)$ bounds the contribution from non-loop pairwise terms at $\langle i, t_i \rangle$. K^* is defined as

$$K^*(i, k, t_i, x)$$

$$= \begin{cases} \min\left(\left[K^*(i, k - 1, t_i, x) + \left(1 - \tfrac{1}{2}\alpha(k)\right) g_2(i, k, t_i, x)\right], K^*(i, k, t_i, x - 1)\right), \\ \qquad \text{if } k < i - 1 \text{ or } k > i + 1; \\ K^*(i, k - 1, t_i, x), \text{ if } i - 1 \leqslant k \leqslant i + 1; \\ +\infty, \qquad \text{if } x < b_k, \ x > d_k, \ k < i \text{ and } x > t_i, \text{ or } k > i \text{ and } x < t_i; \\ 0, \qquad \text{otherwise.} \end{cases} \tag{114}$$

Equation (114) treats i and t_i as parameters, and uses the assumption that $l_j^{\max} = +\infty$. From eq. (114) it follows that

$$K^*(i, m, t_i, d_m) = \min_{\substack{\mathbf{u} \in \mathcal{T} \\ l_j^{\max} = +\infty}} \sum_{|j-i|>1} \left(1 - \tfrac{1}{2}\alpha(j)\right) g_2(i, j, t_i, u_j) \tag{115}$$

and consequently

$$r(\mathcal{T}) = \min_x K(m, x) \tag{116}$$

as desired.

One important aspect of this lower bound computation is that the lower bound actually is instantiated on a specific threading $\mathbf{t}^{lb}$ in the outermost $\min_{\mathbf{t} \in \mathcal{T}}$ of eq. (112). By keeping track of the indices x at which the minimum was actually achieved in eq. (113), it is possible to follow the backtrace from the x minimizing eq. (116) in order to produce $\mathbf{t}^{lb}$. This plays an important role in choosing the next split point.

A reasonably efficient implementation results from holding K and K^* in arrays and iteratively computing the array values using dynamic programming techniques. The formal computational complexity of the lower bound computation is $O(m^2 \tilde{n}^2)$, but this can be reduced as described next. An open problem is to devise a clever tree-structured embedding that avoids brute-force iteration, much as binary trees avoid brute-force iteration when finding the minimum value on an interval [89]. A second open problem is to strengthen the current lower bound. A third is to generalize it to higher-order core segment interactions.

8.3.2. Computing the lower bound efficiently

Most of the time is expended while computing K^* for use in computing K. However, most values of K are so bad that we actually do not need the strong bound given by K^*. In most cases, we can substitute

$$J^*(i, t_i) = \sum_{|j-i|>1} \left(1 - \tfrac{1}{2}\alpha(j)\right) \min_{1 \leq x \leq \tilde{n}} g_2(i, j, t_i, x). \tag{117}$$

The fact that $J^*(i, t_i) \leq K^*(i, m, t_i, d_m)$ guarantees that the result is a valid lower bound. Computing $J^*(i, t_i)$ is very fast because $\min_x g_2(i, j, t_i, x)$ can be precomputed and stored for each (i, j, t_i), and the computation then reduces to sums of a few array references.

In fact, it is sufficient if we ensure that $\mathbf{t}^{lb}$ and the value of its associated lower bound are computed using K^*; all other cases may use J^*. To do this, we record all indices $\langle i, t_i \rangle$ that have ever appeared in $\mathbf{t}^{lb}$ during any lower bound computation. Equation (113) is computed using K^* (eq. 114) for each such $\langle i, t_i \rangle$, and using J^* (eq. 117) otherwise. Specifically, let

$$\gamma(i, t_i) = \begin{cases} 1, & \text{if } \langle i, t_i \rangle \text{ ever appeared in any } \mathbf{t}^{lb}; \\ 0, & \text{otherwise.} \end{cases} \tag{118}$$

$$K^{\text{fast}}(i, t_i) = \alpha(i) \left[g_1(i, t_i) + \gamma(i, t_i) K^*(i, m, t_i, d_m) + (1 - \gamma(i, t_i)) J^*(i, t_i) \right]$$
$$+ \min_{\substack{x \geq \max(b_{i-1}, t_i - \tilde{l}_{i-1}) \\ x \leq \min(d_{i-1}, t_i)}} \left(K^{\text{fast}}(i-1, x) + \beta(i-1, i) g_2(i-1, i, x, t_i) \right). \tag{119}$$

272

Equation (119) is used in place of eq. (113) in order to avoid most invocations of eq. (114).

It remains to ensure that the current computation did not reach a new $\langle i, t_i \rangle$ appearing in the current $\mathbf{t}^{lb}$ for the first time, by checking $\gamma(i, t_i^{lb})$ for each $\langle i, t_i^{lb} \rangle$. If $\gamma(i, t_i^{lb}) = 0$ for any i, then that $\gamma(i, t_i^{lb})$ must be set to 1 and the lower bound computation repeated. In practice, only a few such $\langle i, t_i \rangle$ ever appear. Because most values of K are sufficiently bad, the difference between K^* and J^* does not matter in most cases. Cases where it does matter typically are identified early on, and subsequently very little repeat computation is done.

An efficient implementation might scale and round the input in order to use fast integer arithmetic; keep arrays as nested pointers in order to avoid multi-dimensional array references; lay out large arrays in small pieces in order to minimize disk paging; precompute or cache values where possible; and so on. A parallel MIMD implementation could distribute subsets among arbitrarily many processors. A parallel SIMD implementation could embed the array computations in a connected grid of processors.

9. Computational experiments

This section presents the branch and bound search algorithm's computational behavior and current limits. It shows that the search can succeed in many practical cases, and illustrates the relationship between problem size and computational resources required. Detailed computational analyses are based on the score function of Bryant and Lawrence [28], because it has the highest convergence rate found (99.8%) and thus gives a picture of performance spanning thirty orders of magnitude in search space size ($<10^1$ to $>10^{31}$). Five score functions [28–30,56,59] are used to illustrate general trends. Every example described has been run under all five score functions employed, and yields the same qualitative behavior (often with substantial variation in detail).

Two of the five score functions shown below [28,59] directly provide loop (or loop reference state) score terms as part of their score function. The other three [29,30,56] here require an auxiliary loop score function. This was set to zero for the timing analysis, which therefore depends only on previously published values or theories. For biological examples [1] we set it proportional to a simple log-odds ratio, $\log[P(a|\text{loop})/P(a)]$, summed over all amino acids a in the loop. Here $P(a)$ is the prior probability of a and $P(a|\text{loop})$ is the probability of observing a in a loop region.

9.1. Core template library

We developed a library of core templates taken from 58 non-homologous, monomeric, single-domain, soluble, globular proteins representing diverse structure types (described in Table 2). We believe this to be one of the simplest interesting test cases: statistical artifacts arising from much smaller test sets are avoided, and the proteins require no arbitrary decisions about hydrophobic face packing on domain or multimer boundaries. In order to avoid any subjective bias in core definition, core segments were exactly the main-chain plus beta-carbon atoms (inferred for glycine) of alpha-helices and beta-strands taken from the Brookhaven Protein Data Bank feature tables [99], or if not present were computed from atomic coordinates using DSSP [100] (smoothed as in Stultz et al. [60]). All side-chains were replaced by alanine, in order to assign core template environments independent of the original amino acid identities. Loops were discarded. The resulting core templates were equivalent to a backbone trace plus beta-carbons of the core secondary structure, annotated as required by the score function. An even more abstract template would use ideal coordinates. We sought to reduce residual traces of the structure's original primary sequence (sequence memory) and loop lengths (gap memory), as otherwise threading alignment accuracy on distant structural homologs may suffer (see discussions by refs. [41,42,101]).

We exhaustively threaded every library sequence through every library core template. This created 3364 sequence–template pairs, each consisting of a single fixed sequence and core template. Template loops assigned length zero or one by the crystallographer were treated as fixed-length because they usually reflect constrained "kinks" in the secondary structure. In all other cases we considered all physically realizable loop lengths that maintained core segment topological order. Any loop length that could be proven to break the main-chain or violate excluded atomic volumes was discarded as illegal. Consequently, 833 sequence–template pairs were discarded a priori because the sequence was too short to occupy the template under any legal loop assignment. With the remaining 2531 admissable pairs we searched for the global optimum threading under all five score functions considered. This resulted in a total of 12 655 legal trials, where each trial corresponded to a search for the global optimum threading given a fixed sequence, core template, and score function. Trials were run on a desktop workstation DEC Alpha 3000-M8000, using public-domain CMU Common Lisp [102], and were terminated at the computational limit of two hours. For each trial, we computed the size of the search space of legal threadings and recorded the elapsed time required to find its global optimum threading.

9.2. Problem size and computation time

In a total of 12 109 trials (96%) the search converged within two hours; in 488 trials (4%) time was exhausted first; and in 58 trials (0.5%) space was exhausted first. Figure 6 shows the time required to find the global optimum in every convergent trial under all five score functions, as a function of search space size. On a DEC Alpha 3000-M8000 desktop workstation running Lisp, we have identified the global optimum threading in NP-hard search spaces as large as 9.6×10^{31} at rates ranging as high as 6.8×10^{28} equivalent threadings per second, most of which were pruned before they were ever explicitly examined.

Table 2 shows detailed timing results for self-threading each sequence into its own core template using the Bryant and Lawrence (1993) score function [28], abbreviated "BL93". Protein size is stated in terms of sequence length and number of core segments; search space growth is exponential in number of core segments, but in practice proteins are roughly one-half secondary structure and so the two measures are roughly proportional. Total elapsed time is resolved into initialization and search components, showing that

274

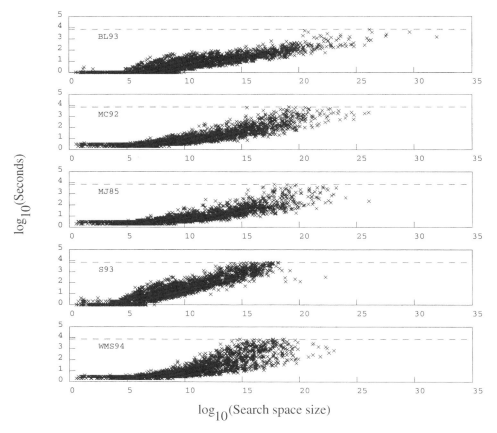

Fig. 6. Slow exponential growth. The time required to find the global minimum is shown on log–log axes as a function of search space size. All sequences and all core templates in our library were threaded through each other under every score function considered. All physically realizable loops were considered. Occasionally the crystallographer assigns a loop length of zero or one; this usually reflects a constrained "kink" in secondary structure, not a true loop, and was left unchanged. The graph for each score function shows every trial that converged under that score function. PDB codes are shown in table 2. Score functions: "BL93" = Bryant & Lawrence (1993) [28]; "MC92" = Maiorov & Crippen (1992) [29]; "MJ85" = Miyazawa & Jernigan (1985) [56]; "S93" = Sippl (1990, 1993) [30]; "WMS94" = White et al. (1994) [59]. Timing resolution is one second. The dashed line corresponds to our computational limit of two hours. All physically realizable loop lengths were admitted, but gaps provably breaking the chain or violating excluded atomic volumes were prohibited. Trials were performed using CMU Common Lisp [102] running on a DEC Alpha 3000-M8000 desktop workstation.

the fast search does not require a prohibitively long initialization. The data in Table 2 appear together with all non-self-threading trials in Fig. 6 in the plot labeled "BL93", transformed by $x = \log_{10}$("Search space size") and $y = \log_{10}$("Total seconds").

Table 3 shows the fraction of trials that converged in each case, the total and per-trial time required, and the log–log regression slopes and intercepts, across all five score functions used. It compares native and non-native threadings for each graph in Fig. 6,

Table 2
Timing details for self-threading (Bryant and Lawrence (1993), on DEC Alpha), part 1

PDB code	Protein length	Number of core segments	Search space size	Number of search iterations	Total (search-only) seconds	Equivalent threadings Per iteration	Per second
256b	106	5	6.19e+3	6	1 (1)	1.03e+3	6.19e+3
1end	137	3	4.79e+4	6	1 (1)	7.98e+3	4.79e+4
1rcb	129	4	5.89e+4	7	1 (1)	8.41e+3	5.89e+4
2mhr	118	4	9.14e+4	7	1 (1)	1.31e+4	9.14e+4
351c	82	4	1.12e+5	5	1 (1)	2.24e+4	1.12e+5
1bgc	174	4	1.63e+5	6	1 (1)	2.72e+4	1.63e+5
1ubq	76	5	1.70e+5	6	1 (1)	2.83e+4	1.70e+5
1mbd	153	8	1.77e+5	10	1 (1)	1.77e+4	1.77e+5
1lis	136	5	5.02e+5	7	1 (1)	7.17e+4	5.02e+5
1aep	161	5	5.76e+5	13	1 (1)	4.43e+4	5.76e+5
1hoe	74	6	7.36e+5	8	1 (1)	9.20e+4	7.36e+5
2hpr	87	6	1.34e+6	8	1 (1)	1.68e+5	1.34e+6
5cyt	103	5	1.37e+6	8	1 (1)	1.71e+5	1.37e+6
1bp2	123	5	1.53e+6	8	1 (1)	1.92e+5	1.53e+6
1aba	87	7	1.95e+6	13	1 (1)	1.50e+5	1.95e+6
1cew	108	6	2.32e+6	8	1 (1)	2.91e+5	2.32e+6
5cpv	108	5	2.60e+6	6	1 (1)	4.33e+5	2.60e+6
2mcm	112	10	1.31e+7	15	1 (1)	8.75e+5	1.31e+7
5fd1	106	5	2.25e+7	12	1 (1)	1.88e+6	2.25e+7
1plc	99	6	3.63e+7	10	1 (1)	3.63e+6	3.63e+7
1alc	123	6	1.70e+8	10	2 (1)	1.70e+7	8.51e+7
1yat	113	7	2.03e+8	8	1 (1)	2.54e+7	2.03e+8
7rsa	124	10	2.54e+8	12	1 (1)	2.12e+7	2.54e+8
3fxn	138	9	7.09e+8	12	2 (1)	5.91e+7	3.54e+8
9rnt	104	8	7.53e+8	21	2 (1)	3.58e+7	3.76e+8
2sns	149	8	2.19e+9	14	4 (1)	1.56e+8	5.47e+8
1ifc	132	12	2.31e+9	87	2 (1)	2.66e+7	1.16e+9
2lzm	164	12	3.16e+9	37	2 (1)	8.54e+7	1.58e+9
3chy	128	10	4.08e+9	45	1 (1)	9.06e+7	4.08e+9
1pkp	150	9	5.32e+9	20	3 (1)	2.66e+8	1.77e+9
1aak	152	8	2.34e+10	10	3 (1)	2.34e+9	7.82e+9
8dfr	189	10	1.45e+11	25	7 (1)	5.78e+9	2.06e+10
1cde	212	13	1.51e+11	38	5 (1)	3.99e+9	3.03e+10
2cpl	165	10	1.82e+11	17	5 (1)	1.07e+10	3.65e+10
3adk	194	13	1.89e+12	66	3 (1)	2.86e+10	6.30e+11
1rec	201	10	3.54e+12	30	4 (1)	1.18e+11	8.85e+11
2cyp	294	10	3.55e+12	181	20 (4)	1.96e+10	1.78e+11
1f3g	161	16	5.17e+12	45	6 (1)	1.15e+11	8.61e+11
4fgf	146	12	1.06e+13	48	4 (1)	2.22e+11	2.66e+12

continued on next page

Table 2, *continued*

PDB code	Protein length	Number of core segments	Search space size	Number of search iterations	Total (search-only) seconds	Equivalent threadings Per iteration	Equivalent threadings Per second
1baa	243	9	1.53e+13	64	10 (2)	2.39e+11	1.53e+12
2act	220	11	1.12e+14	34	7 (1)	3.30e+12	1.60e+13
1dhr	241	14	4.56e+14	51	5 (1)	8.94e+12	9.12e+13
1mat	264	11	5.25e+14	100	15 (2)	5.25e+12	3.50e+13
1tie	172	12	1.19e+15	394	20 (9)	3.03e+12	5.96e+13
3est	240	13	1.92e+15	1946	47 (36)	9.85e+11	4.08e+13
2ca2	259	10	4.51e+15	100	20 (2)	4.51e+13	2.25e+14
1byh	214	14	1.07e+16	95	12 (4)	1.12e+14	8.90e+14
1apa	266	14	3.56e+17	141	18 (6)	2.52e+15	1.98e+16
4tgl	269	14	5.86e+18	361	22 (7)	1.62e+16	2.66e+17
5tmn	316	14	6.51e+18	164	28 (7)	3.97e+16	2.32e+17
1lec	242	15	7.01e+18	320	26 (12)	2.19e+16	2.70e+17
1nar	290	17	2.33e+19	3984	208 (183)	5.85e+15	1.12e+17
1s01	275	15	4.36e+19	541	32 (13)	8.05e+16	1.36e+18
5cpa	307	16	1.22e+20	1089	72 (50)	1.12e+17	1.69e+18
9api	384	17	1.95e+22	290	57 (25)	6.71e+19	3.41e+20
2had	310	19	2.57e+22	4027	201 (179)	6.39e+18	1.28e+20
2cpp	414	20	6.37e+24	3068	205 (164)	2.08e+21	3.11e+22
6taa	478	23	9.63e+31	4917	1409 (1267)	1.96e+28	6.83e+28

Table 3
Convergence rates, total hours, slopes, and intercepts

Potentials	Native or non-native	Searches converged %	Searches converged #	Total hours	Average time per search (s)	Regression Slope	Regression Intercept
BL93	Native	100%	(58/58)	0.7	43.3	0.12	−0.74
	Non-native	99.8%	(2467/2473)	42.3	61.5	0.13	−0.47
MC92	Native	98%	(57/58)	3.2	199.1	0.13	−0.50
	Non-native	98%	(2426/2473)	167.5	243.8	0.14	−0.44
MJ85	Native	98%	(57/58)	3.3	204.5	0.12	−0.42
	Non-native	99%	(2446/2473)	101.6	148.0	0.13	−0.35
S93	Native	90%	(52/58)	13.7	853.1	0.16	−0.43
	Non-native	89%	(2189/2473)	749.7	1091.4	0.24	−0.81
WMS94	Native	88%	(51/58)	18.4	1143.7	0.18	−0.74
	Non-native	93%	(2306/2473)	455.6	663.2	0.18	−0.62
Pooled	Pooled	96%	(12109/12655)	1556.0	442.6	0.15	−0.40

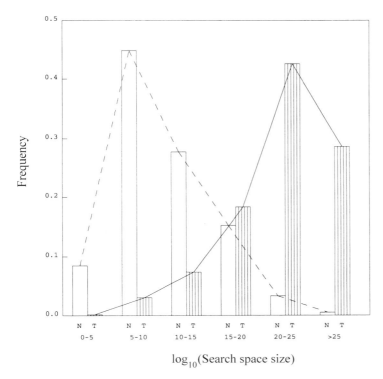

$\log_{10}$(Search space size)

Fig. 7. Histograms of number of trials ("N", white bars, dotted line) and total time expended ("T", striped bars, solid line), grouped by search space size and expressed as fractions of the total. Trials and search space sizes reflect the 2531 legal sequence-template pairs that result from threading every sequence through every core template in our library (PDB codes in table 2). Time expended reflects trials using the Bryant & Lawrence (1993) score function [28], as shown in figure 6, BL93. Each non-convergent trial expended two hours. The histograms group trials according to $\log_{10}$(Search space size). For example, "0–5" indicates the trials such that the search space size is between 10^0 and 10^5, "N" indicates the fraction of total trials having search space sizes in that range, and "T" indicates the fraction of the total time that was expended on them.

and gives the pooled results of all trials. Table 3 summarizes Table 2 in the row labeled "BL93 Native".

Figure 7 shows histograms of number of trials and total time expended finding optimal threadings under Bryant and Lawrence [28], grouped by search space size. In 81% of all trials, the search space contained fewer than 10^{15} legal threadings. However, the searches in those same trials expended only 11% of the total time. Conversely, only 4% of all trials involved a search space that contained more than 10^{20} threadings, but their searches expended 71% of the total time. Figure 7 corresponds to Fig. 6 in the graph labeled "BL93".

10. Discussion

It is clear from Table 2 that the search algorithm is successful at drastically reducing the portion of search space that actually need be examined. This allows the algorithm to find the optimal threading in vastly larger search spaces than heretofore possible. Figure 6 shows that this behavior is characteristic of a wide variety of score functions. Larger search spaces do require more time, as expected, but in most cases examined an exact search could be accomplished within reasonable limits.

Because the algorithm search behavior depends on the input score function, sequence, and core template, it is difficult to give an analytic statement of its average-case time complexity. Figure 6 shows that the log–log relationships remain approximately linear over nearly twenty-five orders of magnitude. For the range of data considered, the regression slopes in Table 3 were between 0.12 and 0.24 and the intercepts were between -1.16 and -0.39. Changes in raw speed due to different computer languages or hardware should affect the intercept but leave the slope unchanged, producing a constant vertical offset in Fig. 6. Because $\log y = a \log x + b$ implies $y = e^b x^a$, the search time in seconds was approximately proportional to between the fourth and eighth root of the search space size and the proportionality constant was between 0.3 and 0.7. Differences in the underlying search space landscapes probably give rise to the considerable scatter about the central tendency and the variation in timing behavior between score functions. We expect that further speed increases can be achieved by parallelizing the algorithm. It has efficient implementation strategies for both single instruction multiple data (SIMD) and multiple instruction multiple data (MIMD) parallel computers. Tighter lower bound formulae (cf. eq. 84) would decrease the slope in Fig. 6, which would lead to greater leverage in larger search spaces.

Due to the exponential nature of the search, most of the time is expended during the few trials that search the very largest search spaces. However, most trials involve search spaces that are substantially smaller, hence more quickly searched. As Fig. 7 shows, most results can be obtained for relatively little computational effort. All five score functions converged quickly on almost all trials that involved a small to medium sized search space, e.g. of size 10^{20} or less. An important implication is that core templates at the level of super-secondary structure or domain motifs should thread very quickly. This result should greatly encourage efforts to recognize protein structure building blocks and assemble them hierarchically.

10.1. Scoring schemes

Our recent experiences and those of others [1,37] on current threading structural environment descriptors and scoring potential functions strongly suggest a need for better protein structural environment definitions and objective functions. There are at least two potential problems in all of the current scoring schemes. The first is a standard statistical problem associated with limited data or low counts. Due in part to the limited size of the current database, some amino acid pairs are sparsely populated in particular structural environments. For example, if there are four or more structural environments associated with each modeled fold position, then there are nearly four hundred times sixteen different independent pairwise score terms to estimate. However, some amino acids are quite rare, and currently there are only a hundred or so completely independent single fold protein structures determined [87,103–106]. Thus, for many neighboring pairs in particular environments there are very few or zero observations. In such cases the threading score functions are sensitive to fluctuations in the observed number of occurrences. For example, methods that use the logarithms of probabilities or derived odds ratios are sensitive to small frequencies because the logarithm becomes numerically unstable as its argument approaches zero. This rare observation problem compounds the testing or validation of any threading score function because score functions may tend to "memorize" the particular proteins in the data set used to derive score function parameters. This is, of course, due to the fact that a rare event can provide sequence-to-structure memory in the score function, thus invalidating the test. Cross-validated tests (jack-knife, hold-one out, etc.) are critical (see the discussion by Ouzounis et al. [41]).

The second potential problem (and likely the more limiting) is in the definition of neighbor and/or pairwise contact environments. The question appears to be, when are two amino acid positions meaningful neighbors? If they are physically close? If they have a particular geometric relationship to one another? If their side-chain atoms interact energetically? The simplest side-chain-independent definition of neighboring structural positions is one that depends only on the physical distance between the alpha- or beta-carbon pairs. Thus in a number of scoring schemes, the pairwise preferences of amino acids are obtained by counting the occurrence frequencies at which each type of amino acid's beta-carbon is within a particular distance of another type of amino acid's beta-carbon. This simple definition is not obviously related to the physical pairwise interactions expected to affect protein folding. Note, two amino acids on opposite sides of a beta-sheet might satisfy the distance conditions, but fail to make any physical contact. More to the point is the fact that two close amino acids, even on the same side of a beta-sheet or alpha-helix, may make little or no physical energetic contact due to the positioning of their side-chain rotamers. The rotamer of an amino acid defines the direction in which the rest of its side-chain atoms point. Most amino acids have beta-carbons with four atoms bonded in a tetrahedral configuration, with freedom to rotate about the alpha/beta-axis when in solution. Two very close amino acids whose side-chains point away from each other normally will have no significant contact. Thus any simple spatial definition of neighborness used in calculating a threading scoring scheme directly from the occupation frequencies of nearby positions will be a mix of non-interacting "chance" neighbors and truly interacting neighbors. That would give a potential noise to signal ratio of at least two to one.

11. Conclusions

One of the most probable limiting factors of the threading predictability appears to be current score functions. There are obviously strong correlations between what particular type of amino acid is found in a given structural position and who its neighbors are. However, one needs to distinguish between the fact that a buried amino acid is most likely to be strongly hydrophobic and the fact that most of its neighbors also being buried will be hydrophobic – that is, amino acids sharing similar local environments will have correlated properties, but not necessarily because they interact. Nearly all of the current threading scoring functions contain both local and non-local terms, the former being the local environment as a function of the position only within the overall structure, and the latter being a function of what other amino acids are brought into the neighborhood by their alignment with the rest of the structure. Currently, in most cases, under current score functions a small fraction of the non-native threadings in the search space continue to score better than the native. Because the search space may be combinatorically large, however, this small fraction may include very many individual threadings.

Why try to continue to work on a method that seems to have so many problems? For one thing, it may work, even with the current limitations. In any particular test case the pairwise noise may happen to "average out" and the local environmental preferences alone may provide good predictions; in fact, this often happens in practice. In addition, there are many extensions to the threading approach that should prove very useful. By combining functional diagnostic patterns of amino acids with threading models, one should be able to extract information as to the most likely positions of those key residues. By extending the threading concept to the threading or alignment of amino acid sequences with partially determined X-ray electron densities [2] or NMR data, one should be able to speed up structure determinations. By using threading as a design tool, one should be able to engineer sequences for particular folds.

Acknowledgments

Ilya Muchnik helped develop the probabilistic framework within which this work is situated. We thank Melissa Cline, Lisa Tucker-Kellogg, Loredana Lo Conte, Sophia Zarakhovich and Srikar Rao for their work on core modelling and score functions; Gene Myers and Jim Knight for discussions of the mathematical formalism; Tomás Lozano-Perez and Patrick Winston for discussions of computational protein folding; Janice Glasgow, David Haussler and Alan Lapedes for applications and extensions; and Steve Bryant, Gordon Crippen, Chip Lawrence, Vladimir Maiorov, and Manfred Sippl for discussions of their score functions. Comments from Nick Steffen improved the presentation. Special thanks to all crystallographers who deposited their coordinates in the international scientific databases.

This chapter describes research performed at the Department of Information and Computer Science of the University of California, Irvine; the Artificial Intelligence Laboratory of the Massachusetts Institute of Technology; and the BioMolecular Engineering Research Center of Boston University. Support for the first author is provided in part by a CAREER grant from the National Science Foundation. Support for the Artificial Intelligence Laboratory's research is provided in part by the Advanced Research Projects Agency of the Department of Defense under Office of Naval Research contract N00014-91-J-4038. Support for the BioMolecular Engineering Research Center is provided in part by the National Library of Medicine under grant number P41 LM05205-14, and by the National Science Foundation under grant number BIR-9121546. The contents of this chapter are solely the responsibility of the authors and do not necessarily represent the official views of the granting agencies.

References

[1] Lathrop, R.H. and Smith, T.F. (1996) J. Mol. Biol. 255, 641–665.

[2] Baxter, K., Steeg, E., Lathrop, R.H., Glasgow, J. and Fortier, S. (1996) From electron density and sequence to structure: Integrating protein image analysis and threading for structure determination. In: D.J. States, P. Agarwal, T. Gaasterland, L. Hunter and L. Smith (Eds.), Proc. Int. Conf. on Intelligent Systems and Molecular Biology. AAAI Press, Menlo Park, CA, pp. 25–33.

[3] Smith, T.F., Lo Conte, L., Bienkowska, J., Rogers, B., Gaitatzes, C. and Lathrop, R.H. (1997) The threading approach to the inverse folding problem. In: S. Istrail, R. Karp, T. Lengauer, P. Pevzner, R. Shamir and M. Waterman (Eds.), Proc. Int. Conf. on Computational Molecular Biology. ACM Press, New York, pp. 287–292.

[4] Smith, T.F., Lo Conte, L., Bienkowska, J., Gaitatzes, C., Rogers Jr, R.G. and Lathrop, R.H. (1997) J. Comput. Biol. 4, 217–225.

[5] Lathrop, R.H., Rogers Jr, R.G., Smith, T.F. and White, J.V. A Bayes-optimal probability theory that unifies protein sequence-structure recognition and alignment. To be published.

[6] Smith, T.F., Lathrop, R.H. and Cohen, F.E. (1996) The identification of protein functional patterns. In: J. Collado-Vides, B. Magasanik and Smith, T.F. (Eds.), Integrative Approaches to Molecular Biology. MIT Press, Cambridge, MA, pp. 29–61.

[7] Kolata, G. (1986) Science 233, 1037–1039.

[8] Herzfeld, C. (chair) et al. (1990) Grand challenges: High performance computing and communications. Technical Report by the Committee on Physical, Mathematical, and Engineering Sciences of the U.S. Office of Science and Technology Policy.

[9] Weiner, S.J., Kollman, P.A., Case, D.A., Singh, U.C., Ghio, C., Alagona, G., Profeta, S. and Weiner, P. (1984) J. Am. Chem. Soc. 106, 765–784.

[10] Brooks, C.L., Karplus, M. and Pettitt, B.M. (1990) Proteins: A Theoretical Perspective of Dynamics, Structure, and Thermodynamics. Wiley, New York.

[11] Creighton, T.E. (1983) Biopolymers 22, 49.

[12] Novotný, J., Rashin, A.A. and Bruccoleri, R.E. (1988) Proteins: Struct. Func. Genet. 4, 19–30.

[13] Moult, J., Pedersen, J.T., Judson, R. and Fidelis, K. (1995) Proteins: Struct. Func. Genet. 23, ii–iv.

[14] Srinivasan, R. and Rose, G.D. (1995) Proteins: Struct. Func. Genet. 22, 81–99.

[15] Skolnick, J., Kolinski, A. and Ortiz, A.R. (1997) J. Mol. Biol. 265, 217–241.

[16] Chothia, C. (1992) Nature 357, 543–544.

[17] Orengo, C.A., Jones, D.T. and Thornton, J.M. (1994) Nature 372, 631–634.

[18] Greer, J. (1990) Proteins: Struct. Func. Genet. 7, 317–333.

[19] Zheng, Q., Rosenfeld, R., Vajda, S. and DeLisi, C. (1993) Protein Sci. 2, 1242–1248.

[20] Desmet, J., De Maeyer, M., Hazes, B. and Lasters, I. (1992) Nature 356, 539–542.

[21] Mandal, C. and Linthicum, D.S. (1993) J. Computer-Aided Mol. Design 7, 199–224.

[22] Sankof, D. and Kruskal, J.B. (Eds.) (1983) Time Warps, String Edits and Macromolecules. Addison-Wesley, Reading, MA.

[23] Pennisi, E. (1997) Science 277, 1432–1434.

[24] Holm, L. and Sander, C. (1993) J. Mol. Biol. 233, 123–138.

[25] Holm, L. and Sander, C. (1994) Nucleic Acids Res. 22, 3600–3609.

[26] Richardson, J.S. (1981) Adv. Protein Chem. 34, 167–339.

[27] Lüthy, R., Bowie, J.U. and Eisenberg, D. (1992) Nature 356, 83–85.

[28] Bryant, S.H. and Lawrence, C.E. (1993) Proteins: Struct. Func. Genet. 16, 92–112.

[29] Maiorov, V.N. and Crippen, G.M. (1992) J. Mol. Biol. 227, 876–888.

[30] Sippl, M.J. (1993) J. Computer-Aided Mol. Design 7, 473–501.

[31] Bowie, J. and Eisenberg, D. (1993) Curr. Opin. Struct. Biol. 3, 437–444.

[32] Bryant, S.H. and Altschul, S.F. (1995) Curr. Opin. Struct. Biol. 5, 236–244.

[33] Fetrow, J.S. and Bryant, S.H. (1993) Bio/Technology 11, 479–484.

[34] Jernigan, R.L. and Bahar, I. (1996) Curr. Opin. Struct. Biol. 6, 195–209.

[35] Jones, D.T. and Thornton, J.M. (1993) J. Computer-Aided Mol. Design 7, 439–456.

[36] Jones, D.T. and Thornton, J.M. (1996) Curr. Opin. Struct. Biol. 6, 210–216.

[37] Lemer, C.M.-R., Rooman, M.J. and Wodak, S.J. (1995) Proteins: Struct. Func. Genet. 23, 337–355.

[38] Sippl, M.J. (1995) Curr. Opin. Struct. Biol. 5, 229–235.

[39] Wodak, S.J. and Rooman, M.J. (1993) Curr. Opin. Struct. Biol. 3, 247–259.

282

[40] Crippen, G.M. (1996) Proteins 26, 167–71.

[41] Ouzounis, C., Sander, C., Scharf, M. and Schneider, R. (1993) J. Mol. Biol. 232, 805–825.

[42] Russell, R.B. and Barton, G.J. (1994) J. Mol. Biol. 244, 332–350.

[43] Thomas, P.D. and Dill, K.A. (1996) J. Mol. Biol. 257, 457–469.

[44] William of Ockham (ca. 1319) Commentary on the Sentences of Peter Lombard (the Reportario). Cited by G. Leff, G. (1975) William of Ockham. Manchester University Press, Manchester, UK, p. 35n.

[45] Sippl, M.J., Hendlich, M. and Lackner, P. (1992) Protein Sci. 1, 625–640.

[46] Kolinski, A., Skolnick, J. and Godzik, A. (1996) An algorithm for prediction of structural elements in small proteins. In: L. Hunter and T. Klein (Eds.), Proc. Pacific Symp. on Biocomputing '96. World Scientific, Singapore, pp. 446–460.

[47] Simons, K.T., Kooperberg, C., Huang, E. and Baker, D. (1997) J. Mol. Biol. 268, 209–225.

[48] Smith, T.F. (1995) Science 268, 958–959.

[49] Abagyan, R., Frishman, D. and Argos, P. (1994) Proteins: Struct. Func. Genet. 19, 132–140.

[50] Bauer, A. and Beyer, A. (1994) Proteins: Struct. Func. Genet. 18, 254–261.

[51] Bowie, F.U., Lüthy, R. and Eisenberg, D. (1991) Science 253, 164–170.

[52] Godzik, A., Kolinski, A. and Skolnick, J. (1992) J. Mol. Biol. 227, 227–238.

[53] Hendlich, M., Lackner, P., Weitckus, S., Flöckner, H., Froschauer, R., Gottsbacher, K., Casari, G. and Sippl, M.J. (1990) J. Mol. Biol. 216, 167–180.

[54] Huang, E.S., Subbiah, S. and Levitt, M. (1995) J. Mol. Biol. 252, 709–720.

[55] Kocher, J.-P.A., Rooman, M.J. and Wodak, S.J. (1994) J. Mol. Biol. 235, 1598–1613.

[56] Miyazawa, S. and Jernigan, R.L. (1985) Macromolecules 18, 534–552.

[57] Wang, Y., Lai, L., Han, Y., Xu, X. and Tang, Y. (1995) Proteins: Struct. Func. Genet. 21, 127–129.

[58] Wilmanns, M. and Eisenberg, D. (1993) Proc. Natl. Acad. Sci. USA 90, 1379–1383.

[59] White, J.V., Muchnik, I. and Smith, T.F. (1994) Math. Biosci. 124, 149–179.

[60] Stultz, C.M., Nambudripad, R., Lathrop, R.H. and White, J.V. Predicting protein structure with probabilistic models. In: N. Allewell and C. Woodward (Eds.), Protein Folding and Stability. JAI Press, Greenwich. In press.

[61] Sippl, M.J. (1990) J. Mol. Biol. 213, 859–883.

[62] Lathrop, R.H. (1994) Protein Eng. 7, 1059–1068.

[63] Akutsu, T. and Miyano, S. (1997) On the approximation of protein threading. In: S. Istrail, R. Karp, T. Lengauer, P. Pevzner, R. Shamir and M. Waterman (Eds.), Proc. Int. Conf. on Computational Molecular Biology. ACM Press, New York, pp. 3–8.

[64] Orengo, C.A. and Taylor, W.R. (1990) J. Theor. Biol. 147, 517–551.

[65] Taylor, W.R. and Orengo, C.A. (1989) J. Mol. Biol. 208, 1–22.

[66] Lawrence, C.E., Altschul, S.F., Boguski, M.S., Liu, J.S., Neuwald, A.F. and Wootton, J.C. (1993) Science 262, 208–214.

[67] Xu, Y. and Uberbacher, E. (1996) CABIOS 12, 511–517.

[68] Jones, D.T., Taylor, W.R. and Thornton, J.M. (1992) Nature 358, 86–89.

[69] Flöckner, H., Braxenthaler, M., Lackner, P., Jaritz, M., Ortner, M. and Sippl, M.J. (1995) Proteins: Struct. Func. Genet. 23, 376–386.

[70] Maiorov, V.N. and Crippen, G.M. (1994) Proteins: Struct. Func. Genet. 20, 167–173.

[71] Benner, S.A., Cohen, M.A. and Gonnet, G.H. (1993) J. Mol. Biol. 229, 1065–1082.

[72] Finkelstein, A.V. and Reva, B. (1991) Nature 351, 497–499.

[73] Madej, T., Gibrat, J.-F. and Bryant, S.H. (1995) Proteins: Struct. Func. Genet. 23, 356–369.

[74] Bagley, S.C., Wei, L., Cheng, C. and Altman, R.B. (1995) Characterizing oriented protein structural sites using biochemical properties. In: C. Rawlings, D. Clark, R. Altman, L. Hunter, T. Lengauer and S. Wodak (Eds.), Proc. 3rd Int. Conf. on Intelligent Systems for Molecular Biology. AAAI Press, Menlo Park, CA, pp. 12–20.

[75] Grossman, T., Farber, R. and Lapedes, A. (1995) Neural net representations of empirical protein potentials. In: C. Rawlings, D. Clark, R. Altman, L. Hunter, T. Lengauer and S. Wodak (Eds.), Proc. 3rd Int. Conf. on Intelligent Systems for Molecular Biology. AAAI Press, Menlo Park, CA, pp. 154–161.

[76] Tropsha, A., Singh, R.K., Vaisman, I.I. and Zheng, W. (1996) Statistical geometry analysis of proteins:

Implications for inverted structure prediction. In: L. Hunter and T. Klein (Eds.), Proc. Pacific Symp. on Biocomputing '96. World Scientific, Singapore, pp. 614–623.

[77] Munson, P.J. and Singh, R.K. (1997) Multi-body interactions within the graph of protein structure. In: T. Gaasterland, P. Karp, K. Karplus, C. Ouzounis, C. Sander and A. Valencia (Eds.), Proc. 5th Int. Conf. on Intelligent Systems for Molecular Biology. AAAI Press, Menlo Park, CA, pp. 198–201.

[78] Bayes, T. (1764) Philos. Trans. R. Soc. London 53, 370–418. Reprinted in: E.S. Pearson and M.G. Kendall (Eds.) (1970) Studies in the History of Statistics and Probability. Charles Griffin, London, pp. 131–153.

[79] Box, G.E. and Tiao, G.C. (1973) Bayesian Inference in Statistical Analysis. Addison-Wesley, Reading, MA.

[80] Hartigan, J.A. (1983) Bayes Theory. Springer, New York.

[81] Arnold, G.E., Dunker, A.K., Johns, S.J. and Douthart, R.J. (1992) Proteins: Struct. Func. Genet. 12, 382–399.

[82] Dunbrack Jr, R.L. and Cohen, F.E. (1997) Protein Sci. 6, 1661–1681.

[83] Thompson, M.J. and Goldstein, R.A. (1996) Proteins: Struct. Func. Genet. 25, 38–47.

[84] Hunter, L. and States, D.J. (1992) IEEE Expert 7, 67–75.

[85] White, J.V., Stultz, C.M. and Smith, T.F. (1994) Math. Biosci. 191, 35–75.

[86] Murzin, A.G., Brener, S.E., Hubbard, T. and Chothia, C. (1995) J. Mol. Biol. 247, 536–540.

[87] Holm, L. and Sander, C. (1996) Science 273, 595–602.

[88] Steele Jr, G.L. (1990) Common Lisp: The Language. Digital Press, Bedford, MA.

[89] Sedgewick, R. (1990) Algorithms in C. Addison-Wesley, Reading, MA.

[90] Fraenkel, A.S. (1993) Bull. Math. Biol. 55, 1199–1210.

[91] Ngo, J.T. and Marks, J. (1992) Protein Eng. 5, 313–321.

[92] Unger, R. and Moult, J. (1993) Bull. Math. Biol. 55, 1183–1198.

[93] Garey, M.R. and Johnson, D.S. (1979) Computers and Intractability: A Guide to the Theory of NP-Completeness. W.H. Freeman & Co, New York.

[94] Hopcroft, J.E. and Ullman, J.D. (1979) Introduction to Automata Theory, Languages, and Computation. Addison-Wesley, Reading, MA.

[95] Lewis, H.R. and Papadimitriou, C.H. (1979) Elements of the Theory of Computation. Prentice-Hall, Englewood Cliffs, NJ.

[96] Winston, P.H. (1993) Artificial Intelligence, 3rd edition. Addison-Wesley, Reading, MA.

[97] Kumar, V. (1992) Search, branch-and-bound. In: S.C. Shapiro (Ed.), Encyclopedia of Artificial Intelligence, Vol. 2. Wiley, New York, pp. 1468–1472.

[98] Aho, A.V., Hopcroft, J.E. and Ullman, J.D. (1982) Data Structures and Algorithms. Addison-Wesley, Reading, MA.

[99] Bernstein, F.C., Koetzle, T.F., G.J.B. Williams, Meyer, E.F., Brice, M.D., Rodgers, J.R., Kennard, O., Shimanouchi, T. and Tasumi, M. (1977) J. Mol. Biol. 112, 535–542.

[100] Kabsch, W. and Sander, C. (1983) Biopolymers 22, 2577–2637.

[101] Rost, B. and Sander, C. (1994) Proteins: Struct. Func. Genet. 20, 216–226.

[102] MacLachlan, R.A. (1992) CMU Common Lisp user's manual, Technical report. School of Computer Science, Carnegie Mellon University, Pittsburgh, PA
CMU Common Lisp source code and executables are freely available via anonymous FTP from
`lisp-rt1.slisp.cs.cmu.edu (128.2.217.9)` and
`lisp-rt2.slisp.cs.cmu.edu (128.2.217.10)`.

[103] Orengo, C.A., Michie, A.D., Jones, S., Jones, D.T., Swindells, M.B. and Thornton, J.M. (1997) Structure 5, 1093–1108.

[104] Brenner, S.E., Chothia, C. and Hubbard, T. (1997) Curr. Opin. Struct. Biol. 7, 369–376.

[105] Holm, L. and Sander, C. (1997) Nucleic Acids Res. 25, 231–234.

[106] Michie, A.D., Orengo, C.A. and Thornton, J.M. (1996) J. Mol. Biol. 262, 168–185.

S.L. Salzberg, D.B. Searls, S. Kasif (Eds.), *Computational Methods in Molecular Biology*
© 1998 Elsevier Science B.V.   All rights reserved

CHAPTER 13

THREADER: protein sequence threading by double dynamic programming

David Jones

Department of Biological Sciences, University of Warwick (UK)

1. Introduction

The prediction of protein tertiary structure from sequence may be expressed symbolically by expressing the folding process as a mathematical function

$$C = F(S),$$

where

$$S = [s_1, s_2, \ldots, s_n], \qquad C = [\theta_1, \theta_2, \ldots, \theta_{3n-2}], \qquad s \in \{\text{Ala}, \text{Arg}, \ldots, \text{Val}\}.$$

In this case the main chain conformation of the protein chain S is represented as a vector of main chain torsion angles C, with the chain itself being defined as a vector of elements corresponding to members of the set of 20 standard amino acids. The folding process is therefore defined as a function which takes an amino acid sequence and computes from it a sequence of main chain torsion angles. The choice of representation of the folded chain conformation in torsion space is arbitrary, and the problem can just as readily be expressed in terms of relative orthogonal 3D coordinates, or with some indeterminacy in chirality, interatomic distances.

The protein folding problem can thus be considered a search for the folding function F. It is probable, however, that no simple representation of the folding function exists, and that even if the function exists in any form whatsoever, the only device capable of performing the required function evaluation is the protein chain itself. Conceptually, the simplest way to arrange for a protein sequence to code for its own native 3D structure is to arrange for the native structure to be the global minimum of the protein chain's free energy. The folding process is therefore transformed into an energy function minimization process, where the energy function could take as input the protein sequence vector S, and the vector of torsion angles C. Given a particular sequence S, the folding process is therefore transformed into a *search* through the set of all corresponding vectors of torsion angles C for the minimum of an energy function E, where E is defined thus:

$$E(S, C_{\text{native}}) < E(S, C_{\text{non-native}}).$$

The exact form of this energy function is as yet unknown, but it is reasonable to assume that it would incorporate terms pertaining to the types of interactions observed in protein structures, such as hydrogen bonding and van der Waals effects. The conceptual simplicity of this model for protein folding stimulated much research into *ab initio* tertiary structure prediction. A successful *ab initio* approach necessitates the solution of two problems. The first problem to solve is to find a potential function for which the above inequality at least generally holds. The second problem is to construct an algorithm capable of finding the global minimum of this function. To date, these problems remain essentially unsolved, though some progress has been made, particularly with the construction of efficient minimization algorithms.

It is unlikely that proteins really locate the global minimum of a free energy function in order to fold into their native conformation. The case against proteins searching conformational space for the global minimum of free energy was argued by Levinthal [1]. The *Levinthal paradox*, as it is now known, can be demonstrated fairly easily. If we consider a protein chain of N residues, we can estimate the size of its conformational space as roughly 10^N states. This assumes that the main chain conformation of a protein may be adequately represented by a suitable choice from just 10 main chain torsion angle triplets for each residue. In fact, Rooman et al. [2] have shown that just 7 states are sufficient. This of course neglects the additional conformational space provided by the side chain torsion angles, but is a reasonable rough estimate, albeit an underestimate. The paradox comes from estimating the time required for a protein chain to search its conformational space for the global energy minimum. Taking a typical protein chain of length 100 residues, it is clear that no physically achievable search rate would enable this chain to complete its folding process. Even if the atoms in the chain were able to move at the speed of light, it would take the chain around 10^{82} seconds to search the entire conformational space, which compares rather unfavorably to the estimated age of the Universe (10^{17} seconds).

Clearly proteins do not fold by searching their entire conformational space. There are many ways of explaining away Levinthal's paradox. A highly plausible mechanism for protein folding is that of encoding a *folding pathway* in the protein sequence. Despite the fact that chains of significant length cannot find their global energy minimum, short chain segments (5–7 residues) could quite easily locate their global energy minimum within the average lifetime of a protein, and it is therefore plausible that the location of the native fold is driven by the folding of such short fragments [3]. Levinthal's paradox is only a paradox if the free energy function forms a highly convoluted energy surface, with no obvious downhill paths leading to the global minimum. The folding of a short fragment can be envisaged as the traversal of a small downhill segment of the free energy surface, and if these paths eventually converge on the global energy minimum, then the protein is provided with a simple means of rapidly locating its native fold.

One subtle point to make about the relationship between the minimization of a protein's free energy and protein folding is that the native conformation need not correspond to the global minimum of free energy. One possibility is that the folding pathways initially locate a local minimum, but a local minimum which provides stability for the average lifetime of the protein. In this case, the protein in question would always be observed with a free energy slightly higher than the global minimum *in vivo*, but would eventually locate its global minimum if isolated and left long enough *in vitro* – though the location of the global minimum could take many years. Thus, a biologically active protein could in fact be in a *metastable* state, rather than a stable one.

2. A limited number of folds

Many fragments of evidence point towards there being a limited number of *naturally occurring* protein folds. If we consider a chain of length 50 residues we might naively calculate the number of possible main chain conformations as 7^{50} ($\approx 10^{42}$). Clearly most of these conformations will not be stable folds, and many will not be even physically possible. In order to form a compact globular structure a protein chain necessarily has to form regular secondary structures [4,5], and it is this constraint, along with the constraints imposed from a requirement to effectively pack the secondary structures formed that limit the number of stable conformational states for a protein chain. In addition to the constraints imposed from physical effects on protein stability, there are also evolutionary constraints on the number of occurring folds. Where do new proteins come from? The answer according to Doolittle [6] is of course from other proteins. In other words the folding patterns we observe today are the result of the evolution of a set of ancestral protein folds.

If the number of possible folds is limited, then this fact should be apparent in the presently known protein structures. Do folds recur in apparently unrelated proteins? The answer appears to be a definite "yes" [4]. Reports of these "fold analogies" are becoming more and more common in the literature, though whether this is due to a real saturation effect where the probability of the fold of a newly solved structure matching an existing one increases due to the increase in the number of known folds, or whether this is simply due to an increased awareness of the possibility (and the increased use of structural comparison programs) is a matter of debate.

A limited number of folds and the recurrence of folds in protein which share no significant sequence similarity offer a "short-cut" to protein tertiary structure prediction. As already described, it is impractical to attempt tertiary structure prediction by searching a protein's entire conformational space for the minimum energy structure, but if we know that there could be as few as 1000 possible protein folds [7,8], then the intelligent way to search a protein's conformational space would be to simply consider only those regions which correspond to this predefined set. This is analogous to the difference between an exam requiring the writing of an essay and an exam requiring multiple-choice questions to be answered. Clearly a person with no knowledge of the subject at hand has a much greater chance of achieving success with the multiple-choice paper than with the essay paper.

Suppose we had derived a practical potential function for which the native conformational energy was lower than that of any other conformation, and that we had identified M possible chain folds, then we would have the basis of a useful tertiary structure prediction scheme. In order to predict the conformation of a given protein chain S, the chain would be folded into each of the M known chain conformations ($C_1, ..., C_M$), and the energy of each conformation calculated. The predicted chain conformation would be the conformation with the lowest value of the potential function. The term generally applied to schemes of this type is *fold recognition*, where instead of trying to predict the fold of a protein chain *ab initio*, we attempt to recognize the correct chain fold from a list of alternatives.

Figure 1 shows an outline of the fold recognition approach to protein structure prediction, and identifies three clear aspects of the problem that need consideration: a

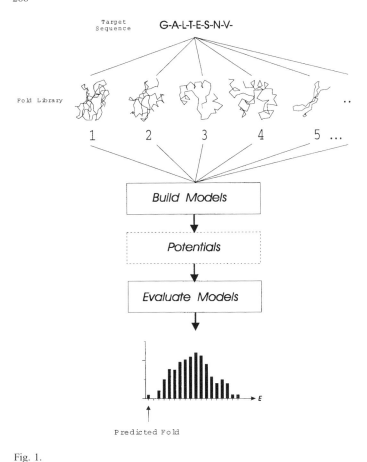

Fig. 1.

fold library, a method for modelling the object sequence on each fold, and a means for assessing the goodness-of-fit between the sequence and the structure.

3. The fold library

A suitable representative library of folds needs to be found. These folds can be observed in crystal structures, NMR structures, or even theoretical model structures. If the library is limited to observed structures then the method will evidently be capable of recognizing only previously observed folds. As has been already discussed, the frequency of occurrence of similar folds between proteins sharing no significant sequence similarity would seem to indicate that creating a library entirely out of known folds is perfectly reasonable. A possible future development for fold recognition might, however, involve the generation of putative model folds, based on folding rules derived from the known structures. Several groups have already attempted the generation of putative folds with some success [9,10], however, the structures created were only very approximate models of real proteins.

4. The modelling process

The central process in structure-based fold recognition is the fitting of a given sequence onto a structural template. One way of visualizing this process is to imagine the side chains of the object protein being fitted onto the backbone structure of the template protein. This process is of course almost identical to the process of comparative modelling. The standard modelling process consists of three basic steps. Firstly, at least one suitable homologous template structure needs to be found. Secondly, an optimal alignment needs to be generated between the sequence of the template structure (the source sequence) and the sequence of unknown structure (the object sequence). Thirdly, the framework structure of the template is "mapped" onto the object sequence. After several stages of energy minimization, the model is ready for critical evaluation.

Each step in the modelling process has its associated problems, though the first two steps are the most critical overall. Evidently, if no homologous structure can be found, the process cannot even be started, and even when a homologous structure is available (perhaps selected on the basis of functional similarity), the degree of homology may be so low as to render the alignment of the sequences impossible by normal means ("by eye" or by automatic alignment). More recently, pattern matching methods have been developed which offer far greater sensitivity than that offered by simple pairwise sequence alignment [12–15]. These methods in one way or another generate a consensus pattern based on the multiple alignment of several homologous sequences. For example, a globin template [14] may be constructed by aligning the many available globin sequences against a known globin structure, identifying the conserved amino acid properties at each position in the template. Though these methods are capable of inferring reasonably distant homologies, allowing, for example, the modelling of HIV protease based on the aspartyl proteinases [16], they are limited by their dependence on the availability of several homologous sequences, and on the ability of multiple alignment algorithms to successfully align them.

The previously described methods for detecting homology work by increasing the sensitivity of standard sequence comparison algorithms. The general assumption is that some residual sequence similarity exists between the template sequence and the sequence under investigation, which is often not the case. Clearly, therefore, the ideal modelling method would not make this assumption, and work with cases where there is no detectable sequence similarity between the object sequence and the source protein.

A method capable of aligning a sequence with a structural template without reference to the sequence of the template protein is clearly the goal here, but for reasons that will be discussed later, this is a computationally hard problem.

5. Evaluating the models

The lack of ability of standard atomic force-fields in the detection of misfolded proteins was first demonstrated by Novotny et al. [18]. Their test problem was simple, and yet serves as a good illustration. In this study, the sequences of myohemerythrin and an immunoglobulin domain of identical length were swapped. Both the two native structures, and the two "misfolded" proteins were then energy minimized using the CHARMm [19] force-field. The results were somewhat surprising in that it was impossible to distinguish between the native and misfolded structures on the basis of the calculated energy sums. Novotny et al. realized that the reason for this failure was the neglect of solvation effects in the force-field. In a later study [20], the force-field was modified to approximate some of the effects of solvent and in this case the misfolded structures could be identified reasonably well. The work of Novotny et al. encouraged several studies into effective methods for evaluating the correctness of protein models, which will now be briefly reviewed.

Eisenberg and McLachlan [21] distinguished correct models from misfolded models by using an elegantly simple solvation energy model alone. By calculating a solvation free energy for each amino acid type and calculating the degree of solvent accessibility for each residue in a given model structure, the correctly folded models were clearly distinguished from the misfolded.

Baumann et al. [22] also used a solvation term to recognize misfolded protein chains, along with a large number of other general statistical properties of sequences forming stable protein folds. Holm and Sander [23] have proposed another solvation model, which appears to be very able at detecting misfolded proteins, even those proteins which have shifts of their sequence on their correct native structure. Interestingly enough a sequence–structure mismatch can quite easily occur not just in theoretically derived models, but even in crystallographically derived models. For example one of the xylose-isomerase structures in the current Brookhaven database has in part a clearly mistraced chain. Such errors can be detected by use of a suitable solvation-based model evaluation procedure.

A very widely known method for testing the overall quality of a protein model is that proposed by Lüthy et al. [24], who used a rather more precise definition of residue environment to assess models. This method will be discussed more fully later.

6. Statistically derived pairwise potentials

Several groups have used statistically derived pairwise potentials to identify incorrectly folded proteins. Using a simplified side chain definition, Gregoret and Cohen [5] derived a contact preference matrix and attempted to identify correct myoglobin models from a set of automatically generated models with incorrect topology, yet quite reasonable core packing.

Hendlich et al. [25] used potentials of mean force, first described by Sippl [26], not only to correctly identify the misfolded protein models of Novotny and Karplus [18], but also to identify the native fold of a protein amongst a large number of decoy conformations generated from a database of structures. In this latter case, the sequence of interest was fitted to all contiguous structural fragments taken from a library of highly resolved structures, and the pairwise energy terms summed in each case. For example, consider a protein sequence of 100 residues being fitted to a structure of length 200 residues. The structure would offer 101 possible conformations for this sequence, starting with the sequence being fitted to the first 100 residues of the structure, and finishing with the sequence being fitted to the last 100. Taking care to eliminate the test protein from the calculation of potentials, Hendlich et al. [25] correctly identified 41 out of 65 chain folds. Using a factor analysis method, Casari and Sippl [27] found that the principal component of their potentials of mean force behaved like a hydrophobic potential of simple form. This principal component potential alone is found to be almost as successful as the full set of potentials in identifying correct folds.

In a similar study to that performed by Hendlich et al. [25], Crippen [28] used simple discrete contact potentials to identify a protein's native fold from all contiguous structural fragments of equal length extracted from a library of highly resolved structures. The success rate (45 out of 56) was marginally higher than that of Hendlich et al. [25] due to the fact that the contact parameters in this case were optimized against a "training set" of correct and incorrect model structures. Maiorov and Crippen [29] improved upon these results using a continuous contact potential, with the new contact function correctly identifying virtually all chain folds defined as being "compact".

Both the work of Hendlich et al. and Crippen demonstrates a very restricted example of fold recognition, whereby sequences are matched against suitably sized contiguous fragments in a template structure. A much harder recognition problem arises when more complex ways of fitting a sequence to a structure are considered i.e. by allowing for relative insertions and deletions between the object sequence and the template structure. Suitable treatment of insertions and deletions is essential to a generalized method for protein fold recognition.

6.1. Ponder and Richards (1987)

The first true example of a fold recognition attempt was the template approach of Ponder and Richards [30] where they concerned themselves with the inverse folding problem. Ponder and Richards tried to enumerate sequences that could be compatible with a given backbone structure. The evaluation potential in this case was a simple van der Waals potential, and so models were effectively scored on the degree of overlap between side chain atoms. A further requirement was for the core to be well-packed, which was achieved by considering the conservation of side chain volume. In order to fit the side chains of a given sequence onto the backbone an exhaustive search was made through a "rotamer library" of side chain conformations. If after searching rotamer space the side chains could not be fitted successfully into the protein core, then the sequence was deemed incompatible with the given fold. As a sensitive fold recognition method, however, this method was not successful. Without allowing for backbone shifts, the packing requirement of a given protein backbone was found to be far too specific. Only sequences very similar to the native sequence could be fitted successfully to the fixed backbone.

6.2. Bowie et al. (1990)

A more successful attempt at fold recognition was made by Bowie et al. [31]. The first stage of this method involves the prediction of residue accessibility from multiple sequence alignments. In essence, alignment positions with high average hydrophobicity and high conservation are predicted to be buried and relatively polar variable positions predicted to be exposed to solvent. The degree of predicted exposure at each position of the aligned sequence family is then encoded as a string. This string is then matched against a library of similarly encoded strings, based, however, not on predicted accessibilities but on *real* accessibilities calculated from structural data. Several successful recognition examples were demonstrated using this method. Of particular note was the matching of an aligned set of Ef Tu sequences with the structure of flavodoxin. The similarity between Ef Tu and flavodoxin is not readily apparent even from structure [32] and so this result is really quite impressive.

6.3. Bowie et al. (1991)

Bowie, Lüthy and Eisenberg [33] attempted to match sequences to folds by describing the fold not just in terms of solvent accessibility, but in terms of the *environment* of each residue location in the structure. In this case, the environment is described in terms of local secondary structure (3 states: α, β and coil), solvent accessibility (3 states: buried, partially buried and exposed), and the degree of burial by polar rather than apolar atoms. The environment of a particular residue defined in this way tends to be more highly conserved than the identity of the residue itself, and so the method is able to detect more distant sequence–structure relationships than purely sequence based methods. The authors describe this method as a 1D–3D profile method, in that a 3D structure is encoded as a 1D string of amino acids, which can then be aligned using traditional dynamic programming algorithms (e.g. ref. [11]). Bowie et al. have applied the 1D–3D profile method to the inverse folding problem and have shown that the method can indeed detect fairly remote matches, but in the cases shown the hits have still retained some sequence similarity with the search protein, even though in the case of actin and the 70 kD heat-shock protein the sequence similarity is very weak [34]. Environment-based methods appear to be incapable of detecting structural similarities between the most divergent proteins, and between proteins sharing a common fold through probable convergent evolution – environment only appears to be conserved up to a point. Consider a buried polar residue in one structure that is found to be located in a polar environment. Buried polar residues tend to be functionally important residues, and so it is not surprising then that a protein with a similar structure but with an entirely different function would choose to place a hydrophobic residue at this position in an apolar environment. A further problem with environment-based methods is that they are sensitive to the multimeric state of a protein. Residues buried in a subunit interface of a multimeric protein will not be buried at an equivalent position in a monomeric protein of similar fold. In fact, the above authors went on to use their method to successfully evaluate protein models [24], and demonstrated that the method was capable of detecting a previously identified chain tracing error in a structure solved in their own laboratory.

6.4. Finkelstein and Reva (1991)

Finkelstein and Reva [35] used a simplified lattice representation of protein structure for their work on fold recognition, where the problem they considered was that of matching a sequence to one of the 60 possible 8-stranded β-sandwich topologies. Each strand has 3 associated variables: length, position in the sequence and spatial position in the lattice Z direction. The force-field used by Finkelstein and Reva includes both short-range and long-range components, both based on physical terms rather than statistically derived terms. The short-range component is simply based on the beta-coil transition constants for single amino acids, similar in many respects to the standard Chou–Fasman propensities [36]. The long-range interaction component has a very simple functional form. For a pair of contacting residues, it is defined simply as the sum of their solvent transfer energies as calculated by Fauchere and Pliska [37].

The configurational energy of the 8 strands in this simple force-field is minimized by a simple iterative algorithm. At the heart of the method is a probability matrix (a 3-dimensional matrix in this case) for each of the strands, where each matrix cell represents one triplet of the strand variables i.e. length, sequence position and spatial position. The values in each cell represent the probability of observing the strand with the values associated with the cell. The novel aspect of this optimization strategy is that the strands themselves do not physically move in the force-field, only the probabilities change. At the start of the first iteration the strand positional probabilities are assigned some arbitrary value, either all equal, or set close to their expected values (the first strand is unlikely to be positioned near the end of the sequence for example). A new set of probabilities is then calculated using the current mean field and the inverse Boltzmann equation (see later for more about the inverse Boltzmann equation). As more iterations are executed it is to be hoped that most of the probabilities will collapse to zero, and that eventually a stable "self-consistent" state will be reached. Finkelstein and Reva found that the most probable configurations corresponded to the correct alignment of the 8-stranded model with the given sequence, and that when the process was repeated for each of the 60 topologies, in some cases the most probable configuration of the native topology had the highest probability of all.

The simplicity of the lattice representation used here and the uncomplicated force-field are probably critical to the success of this method. A more detailed interresidue potential would prevent the system from reaching a self-consistent state, and would be left either in a single local minimum or more likely oscillating between a number of local minima. In addition, whilst it is quite practical to represent β-sheets on a lattice, it is not clear how α-helices could be reasonably represented, though in later work, the authors have used highly simplified real 3D protein structures as pseudo-lattices.

7. *Optimal sequence threading*

The method described in this chapter has something in common both with the method of Bowie, Lüthy and Eisenberg, and that of Finkelstein and Reva. Despite the obvious computational advantages of using residue environments, it is clear that the fold of a protein chain is governed by fairly specific protein–protein and protein–solvent atomic interactions. A given protein fold is therefore better modelled in terms of a "network" of pairwise interatomic energy terms, with the structural role of any given residue described in terms of its interactions. Classifying such a set of interactions into one environmental class such as "buried alpha-helical" will inevitably result in the loss of useful information, reducing the *specificity* of sequence–structure matches evaluated in this way. The main difficulty in the use of environments alone for recognizing protein folds is that helices look like other helices, and strands like other strands. A sequence that folds into one helix of particular structure, will probably easily fold into any other helix of similar

```
      APRKF---------------FVGGNWKMNGKRKSLGELIHTLDGAKLSADTEVVCGAPS
TIM   *9992---------------0000103032*8*400*10*61262*957*261000002
      .....---------------PPPPP..B...HHHHHHHHHHHHH....SS.PPPPP..T

      ..HHHHH....S.......SSPPPPP..----SHHHHHHHHHHHHTTT..S--PPPPP.S.
LDH   *7*****96*********3*61000000----443020006200*77104--10000299
      ATLKDKLIGHLATSQEPRSYNKITVVGV----GAVGMACAISILMKDLAD--EVALVDVM

      IYLDFARQKLDAK---------IGVAAQNCYKVPKGAFTGEIS------------PAMI
TIM   0000304*71688---------010000101547*14401110------------0300
      THHHHHHHHS.TT---------PPPPPP...SSSSBS.SS...------------HHHH

      HHHHHHHHHHHHHHTGGG...S.PPPPSSGGGGTT.SPPPP.......TT..HHHHHHHHH
LDH   ***0*4327*26*15**2*09*1220*92540440700002141*8**845925100800
      EDKLKGEMMDLQHGSLFLHTAKIVSGKDYSVSAGSKLVVITAGARQQEGESRLNLVQRNV

      KDIGAA-----------WVILGH--SERRHVFGESDELIGQKVAHALAEGLGVIACIGEK
TIM   *71205-----------201000--0203*655246*300700330195500000000109
      HHHT..-----------PPPP..--HHHHHHH...HHHHHHHHHHHHHHTT..PPPPPPP.

      HHHHHHHHHHHHHH.TT.PPPP..SS----------HHHHHHHHHHHHT..GGGPPE.TT-
LDH   5608*104502*507*000000083----------000002003*42615974000100-
      NIFKFIIPNIVKHSPDCIILVVSNP----------VDVLTYVAWKLSGLPMHRIIGSGC-

      LDEREAGITEKVVFQETKAIADNVKDWSKVVLAYEP---------VWAIGTGKTAT----
TIM   3*85*83528*104*20*102*41*62860000000---------1227955**24----
      HHHHHHTTHHHHHHHHHHHHHHHH....TTPPPPPPP---------GGGSSSSS...----

      --------HHHHHHHHHHHHHHHHTS.TTTPP..B.BSSSTT..B.GGG.AATTAAHHHHS
LDH   --------12004507*300**776*3750606000251*600113220338*86337*9
      --------NLDSARFRYLMGERLGVHSCSCHGWVIGEHGDSVPSVWSGMNVASIKLHPLD

      -----------PQQAQEVHEKLRGWLKTHVSDAVAVQS-----------RIIYGGSVTG
TIM   -----------3*6029008*03340*9*44*710770-----------1001018045
      -----------HHHHHHHHHHHHHHHHHHH.HHHHHHS------------PPPP.S...T

      S..SSSSSTHHHHHHHHHHHHHHHHHHSS..HHHHHHHHHHHHHHHHTT..AAAAAAAA.T
LDH   6615***7456039401841**48**85930*310*1005003002*7778610000107
      GTNKDKQDWKKLHKDVVDSAYEVIKLKGYTSWAIGLSVADLAETIMKNLCRVHPVSTMVK

      GNCKELASQHDVDGFLVGGASLKP-----------EFVDIINAKH-------------
TIM   440*70152*400001015207*7-----------50290151**-------------
      THHHHHHTSTT..PPPPSGGGGST-----------HHHHHHT...-------------

      TSSS..SS---.AAAAAAAAATTAAEAA......HHHHHHHHHHHHHHHHH...S...
LDH   *5350*54---00000002026*024*35*3*288706*906*007509*12*3****
      DFYGIKDN---VFLSLPCVLNDHGISNIVKMKLKPNEEQQLQKSATTLWDIQKDLKFF
```

Fig. 2. Manually derived alignment of triose phosphate isomerase (TIM) with lactate dehydrogenase based on residue environments. Line 1 (TIM)/3 (LDH): amino acid sequence, Line 2: residue accessibility (0 = 0–9%, 9 = 90–99%, * > 99%), Line 3 (TIM)/1 (LDH): secondary structure (H = α-helix, A = antiparallel strand, P = parallel strand, G = 3/10 helix, otherwise coil).

length. A very good example of two topologies which cannot be distinguished after encoding into environmental classes is an $(\alpha\beta)_8$ barrel (a "TIM barrel") and a parallel $\alpha\beta$ sandwich (a Rossmann fold). In this case both topologies comprise alternating α and β structure, where the strands are mostly inaccessible to solvent. Providing that the $\alpha\beta$ sandwich is of sufficient size, or if flanking domain regions provide additional secondary structural elements (the Rossmann domain itself typically has only 6 strands), then the 1D descriptors of the two structures are almost identical. This is illustrated in Fig. 2, where the secondary structure and accessibility of TIM (triose phosphate isomerase) has been manually aligned with those of lactate dehydrogenase.

The factor that limits the scope of the search for a stable threading is packing. Whilst the sequence of any isolated helix could substitute for any other, the sequences for a packed pair of helices are much more highly constrained. For a complete protein structure, solvation effects also come into play. In general, then, for a globular protein, the threading of its sequence onto its structure is constrained by local interactions (in the example given, the required formation of a helix), long-range pairwise interactions (helix–helix packing for example) and solvation effects, which are primarily governed by the periodic accessibilities of exposed helices and strands.

In view of this, we should like to match a sequence to a structure by considering the plethora of detailed pairwise interactions, rather than averaging them into a crude environmental class. However, incorporation of such non-local interactions into standard alignment methods such as the algorithm of Needleman and Wunsch [11], has hitherto proved computationally impractical. Possible solutions to this computational problem will be discussed later.

8. Formulating a model evaluation function

The general approach described here employs a set of information theoretic potentials of mean force [25,26]. These potentials associate event probabilities with statistical free energy. If a certain event is observed with probability p (say the occurrence of a leucine residue α-carbon and an alanine α-carbon at a separation of 5 Å) we can associate an "energy" with this event by the application of the inverse Boltzmann formula:

$$E = -kT \ln(p) .$$

The constant $-kT$ may be ignored, in which case the units are no longer those of free energy but of *information* (in units of nats). For simplicity, we have also ignored the additional term Z, known as the Boltzmann sum, a clear explanation of why this is acceptable is given by Sippl [26]. The important point about both free energy and information entropy formulations of probability is that the resulting values are additive. Consider two independent events with probabilities p and q, respectively. The probability of both events occurring together is simply pq, but multiplication is difficult to implement in pattern matching algorithms. Transforming the combined probability pq by taking logs provides the following useful result:

$$\ln(pq) = \ln(p) + \ln(q) .$$

Therefore the important part of the calculation of potentials of mean force, and the related techniques of information theory is simply converting probabilities to log-likelihoods.

The real computational key to this approach to transforming probabilities into energy-like parameters is really so that we can transform the problem of multiplying probabilities into the problem of adding related terms. In general it is relatively easy to handle additive terms algorithmically, but very hard to handle values which need to be multiplied. If it makes it easier to follow, readers who are happier working with units of information than units of energy can simply erase the $-kT$ terms from the equations which follow, and even change the base of the logarithms to base-2 so that the scores can be expressed in units of "bits".

Typically we are interested in relative rather than absolute probabilities. Taking the above example, it is of little interest to know how probable it is that a leucine α-carbon and an alanine α-carbon are found to be separated by $5\,\text{Å}$. Of much greater interest is the question of how probable this leucine–alanine separation is in comparison with other residue pairs. If the probability of *any* residue pair having an α-carbon separation of s is $f(s)$ and the frequency of occurrence for residue pair ab is $f_{ab}(s)$ then we can write down the potential of mean force as follows:

$$\Delta E_{ab}(s) = -kT \ln \left[\frac{f_{ab}(s)}{f(s)} \right].$$

Sippl divides this potential into a set of potentials relating to different topological levels $1, \ldots, k$, which is simply the residue pair sequence separation. For the tripeptide sequence MFP, $k = 1$ for residue pairs MF and FP, with $k = 2$ for residue pair MP. In reality, probability density functions $f_k(s)$ and $f_k^{ab}(s)$ are unknown and must be replaced by the relative frequencies observed in the available structural database denoted $g_k(s)$ and $g_k^{ab}(s)$, respectively, where s is typically divided into 20 intervals for sampling. As there are 400 residue pairs (sequence asymmetry is assumed) and only some 15 000–20 000 residues in the set of non-homologous protein structures, the observed frequency distributions $g_k^{ab}(s)$ are only weak approximations of the true probability densities and must therefore be corrected to allow for the very small sample size. By considering the observation process as the collection of information quanta, Sippl suggests the following transformation:

$$f_k^{\ ab}(s) \approx \frac{1}{1+m\sigma} g_k(s) + \frac{m\sigma}{1+m\sigma} g_k^{\ ab}(s),$$

where m is the number of pairs ab observed at topological level k and σ is the weight given to each observation. As $m \to \infty$ this transformation has the required property that the right and left-hand sides of the equation become equal as $g_k^{ab}(s) \to f_k^{ab}(s)$. Given the number of residues in the database and the small number of histogram sampling intervals it is assumed that $f_k(s) \approx g_k(s)$. From the previous two equations the following formula may be derived:

$$\Delta E_k^{\ ab} = kT \ln(1+m\sigma) - kT \ln \left[1 + m_{ab}\sigma \frac{g_k^{\ ab}(s)}{g_k(s)} \right].$$

The potentials used in this work are calculated exactly as described by Hendlich et al. [25] where pairwise interatomic potentials are derived from a set of non-homologous proteins. The following interatomic potentials are calculated between the main chain N, O, and side chain Cβ: Cβ → Cβ, Cβ → N, Cβ → O, N → Cβ, N → O, O → Cβ, and O → N. In all, 7 pairwise interactions are considered between each pair of residues i, j. By excluding interactions between atoms beyond the Cβ atom in each residue, the potentials are rendered independent of specific side chain conformation. Dummy Cβ atoms were constructed for glycine residues and other residues with missing Cβ atoms.

A possible criticism of the mean force potentials proposed by Sippl is that there exists in the force-field a dependence on protein size. The problem lies in the fact that interactions even as distant as 80 Å are taken into account in the calculation of the potentials, and so consequently, the bulk of data for these large distances is derived from large proteins. This was recognized by Hendlich et al. [25], where it was suggested that the ideal case would be for the potentials to be calculated from proteins of roughly equal size to the protein of interest. Unfortunately, this simple solution is generally impractical. The data set used to generate the mean force potentials is already sparse, even before subdivision into size ranges.

In order to render the mean force potentials less dependent on protein chain length, these long-distance interactions must be replaced by a size independent parameter. The first requirement in replacing these interaction parameters is to determine a suitable dividing line which separates short-distance from long-distance interactions. The next step is then to determine the nature of the information encoded by these interactions. Finally, a suitable size-independent parameter can be sought to replace this information.

Consider two protein atoms separated by a distance d. Clearly if d is large there will be no significant physical interaction between these atoms. Conversely, if d is small then we might expect there to be some influence, whether it be a hydrophobic effect, an electrostatic effect or even a covalent interaction. If such an influence exists, then we might also expect there to be some residue preferences for the residues containing these atoms, and consequently we would expect some kind of correlation between the two residue identities. This provides a possible way to determine a cut-off distance for meaningful residue–residue interactions. If the identities of two residues can be considered to be independent variables, then these residues (or more correctly, the residue side chains) will probably not be involved in a significant physical interaction.

To determine the degree of dependency between residues separated by a particular distance, some measure of statistical association is required. The method selected here is based on *statistical entropy*, a common concept in statistical physics and information theory. The entropy of a system with I states, where each state occurs with probability p_i, is defined as

$$H(x) = -\sum_{i=1}^{I} p_i \ln p_i.$$

Consider two experiments x and y with I and J possible outcomes respectively, each of which occurs with a probability $p_{i\cdot}$ $(i = 1, \ldots, I)$, and $p_{\cdot j}$ $(j = 1, \ldots, J)$. The entropy H of these systems is defined as

$$H(x) = -\sum_{i=1}^{I} p_{i\cdot} \ln p_{i\cdot}, \qquad H(y) = -\sum_{j=1}^{J} p_{\cdot j} \ln p_{\cdot j}.$$

Entropy in this case is essentially defined as the degree of freedom of choice, or more strictly in this case, the degree of equiprobability. If the outcome probabilities in each experiment are equal then the statistical entropy is maximized, as this represents the maximum freedom of choice. If the probability of one outcome is unity (the others of course being zero) then zero entropy is achieved, corresponding to a lack of choice whatsoever.

If we link both experiments, then we can represent the overall outcomes in the form of a *contingency table*. An example of such a linked pair of experiments is the throwing of a pair of dice, in which case the contingency table would have 6 rows and 6 columns, representing the 6 possible outcomes for each die.

The entropy of the combined experiment is:

$$H(x, y) = -\sum_{i,j} p_{ij} \ln p_{ij}.$$

To determine the statistical association between experiments x and y, the entropy of y *given* x and x *given* y may be derived. If a knowledge of the outcome of experiment x allows a wholly accurate prediction of the outcome of experiment y, then the entropy of y *given* x must be zero. Conversely, if a knowledge of experiment x is found to be of no benefit whatsoever in the prediction of y, then the conditional entropy in this case is maximized.

The entropy of y *given* x is as follows:

$$H(y|x) = \sum_{i} p_i \sum \frac{p_{ij}}{p_{i\cdot}} \ln \left(\frac{p_{ij}}{p_{i\cdot}} \right) = \sum_{i,j} p_{ij} \ln \left(\frac{p_{ij}}{p_{i\cdot}} \right)$$

and the entropy of x *given* y:

$$H(x|y) = \sum_{i} p_i \sum \frac{p_{ij}}{p_{\cdot j}} \ln \left(\frac{p_{ij}}{p_{\cdot j}} \right) = \sum_{i,j} p_{ij} \ln \left(\frac{p_{ij}}{p_{\cdot j}} \right).$$

Finally a suitable symmetric measure of interdependence (known as the *uncertainty*) between x and y is defined thus:

$$U(x,y) \equiv 2 \frac{H(y) + H(x) - H(x,y)}{H(x) + H(y)}.$$

An uncertainty between x and y of zero indicates that the two experimental variables are totally independent ($H(x,y) = H(x) + H(y)$), whereas an uncertainty of one ($H(x) = H(y) = H(x,y)$) indicates that the two variables are totally dependent. One would hope that in the case of the two dice experiment previously described, that $U(x,y)$ would be found to be close to zero for a large number of trials, though gluing the dice together would be a sure way of forcing $U(x,y)$ to unity.

Using the uncertainty measure, it is now possible to evaluate residue correlations in protein structures. In this case, the two experimental variables are the identities of two residues separated by a given distance in a particular structure. Using a set of

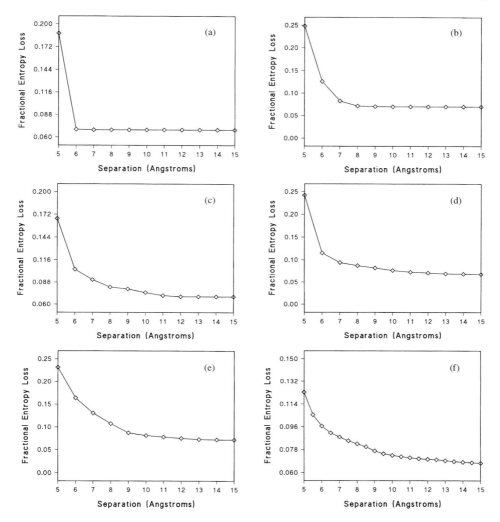

Fig. 3. Uncertainty coefficient (fractional loss of statistical entropy) for residue identities over sequence separations: (a) 1, (b) 2, (c) 3, (d) 4, (e) 5, (f) > 10. The maximum observed distances for each sequence separation are as follows: (a) 6.36 Å, (b) 9.72 Å, (c) 13.08 Å, (d) 16.45 Å, (e) 19.43 Å, (f) > 32.97 Å. Points beyond these distances have no meaning.

102 chains as listed in Jones et al. [38], six 20×20 contingency tables were set up for each distance range. The first 5 tables were constructed by counting residue pairs with sequence separations of 1 to 5 (short-*range* interactions), the other being constructed by counting all pairs with sequence separations > 10 (long-range). Values in each table were converted to relative frequencies by normalization.

The plots in Fig. 3 clearly show the ranges over which short-range and long-range effects can be detected statistically. As might be expected, the strongest sequence specific

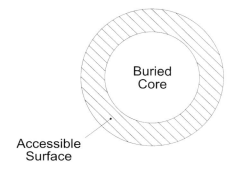

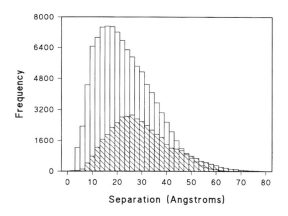

Fig. 4. Distance distributions for accessible (shaded) and inaccessible residue pairs in monomeric protein structures. Buried residues are taken to be those with relative accessibilities $< 5\%$; accessible residues with relative accessibilities $> 50\%$.

effects are observed across short sequential separations, where steric and covalent effects predominate. Most importantly, both the short- and long-range interactions become undetectable when averaged over distances greater than around 12 Å (though it must be realized that for the very short separations, 1 and 2, it is impossible for the separation to exceed 10 Å). It must be stressed that this does not necessarily imply that the physical forces themselves do not act beyond this distance, only that the effects do not manifest themselves in the selection of amino acid residues.

Bearing in mind the observable extent of detectable sequence specific effects, the calculation of mean force potentials was modified from the method described by Sippl [26]. Rather than taking into account all interactions up to around 80 Å, only atom pairs separated by 10 Å or less were used. However, much useful information remains in the long-distance distributions. Considering a protein molecule as a globule comprising an inner hydrophobic core it is readily apparent that the bulk of the longer pairwise distances will originate from residue pairs distributed on the surface of the globule, which is illustrated in Fig. 4.

As the excluded long-distance potentials clearly only encode information pertaining to the hydrophobic effect, the most logical replacement for these interactions must be a potential based on the solvent accessibility of the amino acid residues in a structure.

9. Calculation of potentials

For the short-range potentials, minimum and maximum pairwise distances were determined for each type of atomic pairing at each topological level from a small set of very highly resolved crystal structures. These distance ranges were subdivided into 20 intervals. For the medium and long-range potentials, interactions were sampled over the range 0–10 Å with a fixed sampling interval of 2 Å. A small selection of the pairwise interaction potentials is shown in Fig. 5.

As discussed in the previous section, in addition to the pairwise potentials (and in place of the long-range, long-distance interactions), a solvation potential was also incorporated.

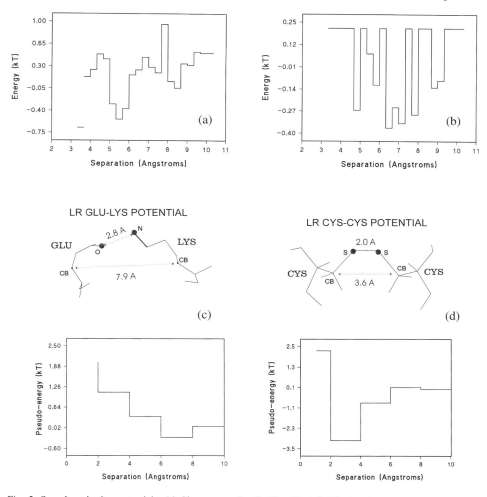

Fig. 5. Sample pairwise potentials: (a) Short-range ($k = 3$) Ala–Ala Cβ–Cβ, (b) Short-range ($k = 3$) Phe–Tyr Cβ–Cβ, (c) Long-range ($k > 30$) Glu–Lys Cβ–Cβ, shown below a diagram of the typical geometry for a Glu–Lys salt-bridge, (d) Long-range ($k > 30$) Cys–Cys Cβ–Cβ, shown below a diagram of the typical geometry for a disulphide bond.

302

This potential simply measures the frequency with which each amino acid species is found with a certain degree of solvation, approximated by the residue solvent accessible surface area. The solvation potential for amino acid residue a is defined as follows:

$$\Delta E^a_{\text{solv.}}(r) = -kT \ln \left[\frac{f^a(r)}{f(r)} \right],$$

where r is the % residue accessibility (relative to residue accessibility in GGXGG extended pentapeptide). Residue accessibilities were calculated using the DSSP program of Kabsch and Sander [39]. The solvation potentials were generated with a histogram sampling interval of 5%. To ensure that subunit or domain interactions did not affect the results, only monomeric proteins were used in the calculation. These solvation potentials clearly show the hydropathic nature of the amino acids and prove to be a more sensitive measure of the likelihood of finding a particular amino acid with a given relative solvent accessibility than the long-distance interaction potentials they are designed to replace.

10. Searching for the optimal threading

Given an efficient means for the evaluation of a hypothetical sequence threading relationship, the problem of finding the optimal threading must be considered. For a protein sequence of length L and a template structure of which M residues are in regular secondary structures, the total number of possible threadings is given by

$$\binom{L}{M} \equiv \frac{L!}{(L - M)! \; M!}.$$

The scale of the search problem for locating the optimal threading of a sequence on a structure amongst all possible threadings may be appreciated by considering bovine pancreatic trypsin inhibitor (Brookhaven code 5PTI) as an example. Of the 58 residues of 5PTI, 30 are found to be in regular secondary structure. Using the above expression the total number of threadings for 5PTI is calculated as 2.9×10^{16}. Given that 5PTI is a very small protein, it is clear that the search for optimal threading is a non-trivial problem.

One way to reduce the scale of the problem is to restrict insertions and deletions (indels) to the loop regions. By excluding indels from secondary structural elements, the problem reduces to a search for the optimal set of loop lengths for a protein. Under these conditions threading a sequence on a structure may be visualized as the sliding of beads on an abacus wire, where the beads are the secondary structures, and the exposed wire the remaining loop regions. The restricted threading of a small four helix bundle protein (myohemerythrin) is depicted in Fig. 6. Restricting indels to loop regions in this way reduces the search space in this case from 1.3×10^{33} to just 44 100.

ADLEDDMQTLNDNLKVIEKABBZKANDAALVKMRAAALNAQKATPPKLEDNSQPMKDFRHGFDILVEGIDDALKLANEGKVKEAQAAAEQLKTTRNAYHQKYR

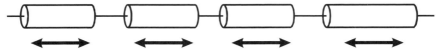

Fig. 6. Diagram illustrating the exhaustive threading procedure. Secondary structures are effectively moved along the sequence like abacus beads. On average, loop lengths vary between 2 and 12.

Unfortunately, even with this extreme restriction on the threading patterns, the search space again becomes unreasonably large for proteins of even average size. The exact number of threadings with secondary structure indels disallowed is a complex function of both sequence length and the number and lengths of constituent secondary structural elements. As a rule of thumb, however, it is found that for N secondary structural elements, with loops of average length (say between 2 and 12 residues), the number of threadings is $O(10^N)$. For typical proteins comprising 10–20 secondary structures, it is clearly not possible to locate the optimal threading by an exhaustive search in a reasonable period of time.

11. Methods for combinatorial optimization

Various means have been investigated for locating the optimal threading of a sequence on a structure. The methods are briefly detailed below.

11.1. Exhaustive search

As demonstrated earlier, for small proteins of <6 secondary structural elements, and disallowing secondary structure indels, it is practical to simply search through all possible threadings in order to locate the threading of lowest energy. Unfortunately, the evaluation function is tailored towards average sized globular proteins with hydrophobic cores, whereas the small proteins tend to be less globular and typically lack a hydrophobic core.

11.2. Monte Carlo methods

Monte Carlo methods have been often exploited for conformation calculations on proteins. Two *directed* search procedures have been used in my laboratory: *simulated annealing*, and *genetic algorithms*. Simulated annealing has been exploited in the alignment of protein structures [40], and in the optimization of side chain packing in protein structures [41]. Simulated annealing is a simple random search process. In this instance, random threadings are generated and evaluated using the evaluation function described earlier. Where a proposed threading has a lower energy than the current threading, the proposed threading is accepted. In the case where a proposed threading has a higher energy than the current, it is accepted with probability p, where

$$p = \exp\left(-\frac{\Delta E}{kT}\right),$$

and where ΔE is the difference between the current and the proposed threading energy and T is the current annealing "temperature". After a predefined number of accepted changes, the temperature is slightly reduced. This whole procedure is repeated until no further reduction in threading energy is achieved, at which point the system is said to be frozen. The schedule of cooling is critical to the success of simulated annealing.

Genetic algorithms [42] are similar in concept to simulated annealing, though their model of operation is different. Whereas simulated annealing is loosely based on the principles of statistical mechanics, genetic algorithms are based on the principles of natural selection. The variables to be optimized are encoded as a string of binary digits, and a *population* of random strings is created. This population is then subjected to the genetic operators of selection, mutation and crossover. The probability of a string surviving from one generation to the next relates to its fitness. In this case, low energy threadings are deemed to be fitter than those with higher energies. Each string may be randomly changed in two ways. The mutation operator simply selects and changes a random bit in the string. An alternative means for generating new strings is the crossover operator. Here a randomly selected portion of one string is exchanged with a similar portion from another member of the string population. The crossover operator gives genetic search the ability to combine moderately good solutions so that "super-individuals" may be created.

In use, these methods prove to be capable of locating the optimal threading, but with no guarantee that they will do so in any given run of the threading program. Ideally the results from many runs should be pooled and the best result extracted, which is of course time consuming. A further problem is that the control parameters (the cooling schedule in the case of simulated annealing and the selection, mutation and crossover probabilities in the case of genetic search) need adjustment to match each threading problem individually. Parameters found suitable for threading a protein with 10 secondary structures will generally not be suitable for threading a protein with 20 secondary structures for example. The methods are typically plagued by "unreliability", yet are found to be highly robust. Given a sufficiently slow cooling rate in the case of simulated annealing, or a sufficiently large population of strings in the case of genetic algorithms, and in both cases a sufficient number of runs, very low energy threadings will be found providing they exist at all in the given search space.

11.3. Dynamic programming

It should be apparent that there exists a clear similarity between optimizing the threading of a sequence on a structural template and finding the optimal alignment of two sequences. In such terms, threading is simply the alignment of an amino acid sequence against a sequence of positions in space. At first sight it might well appear that the same dynamic programming methods used in sequence alignment (e.g. ref. [11]) could easily be applied to the threading problem. Unfortunately, this is not the case. In a typical alignment algorithm a score matrix is constructed according to the following recurrence formula:

$$S_{ij} = D_{ij} + \max \begin{cases} S_{i+1,j+1}; \\ \max_{k=i+2 \to N_A} S_{k,j+1} - g; \\ \max_{l=j+2 \to N_B} S_{i+1,l} - g; \end{cases}$$

where S_{ij} is an element of the score matrix, D_{ij} is a measure of similarity between residues i and j in sequences of length N_A and N_B, respectively, and g is a gap penalty which may be either a constant or a function of, for example, gap length. By tracing the highest scoring path through the finished matrix the mathematically optimum alignment between the two sequences may be found for the given scoring scheme. In the special case where D_{ij} is a function only of elements i and j, dynamic programming alignment algorithms have execution times proportional to the product of the sequence lengths. However, if D_{ij} is defined in terms of non-local sequence positions in addition to i and j, dynamic programming no longer offers any advantage; the alignment effectively requires a full combinatorial search of all possible pairings. In the case of the evaluation function defined here, in order to determine the energy for a particular residue, all pairwise interactions between the residue in question and every other residue in the protein need to be calculated. In other words, in order to evaluate the threading potentials in order to fix the location of a single residue, the location of every other residue needs to have been fixed beforehand.

By excluding the medium and long-range ($k > 10$) pairwise terms from the evaluation function and by considering only interactions between residues in the same secondary structural element, dynamic programming can be applied to the problem. For example, consider a case where a template comprising a single 10 residue helical segment is being matched against a 100 residue sequence. Discounting the possibility of indels in the helix itself, there are 91 possible alignments between the helical template and the sequence. As indels may not occur in the helix, for any given position in the sequence ($i = 1, ..., 91$), all possible interhelical pairwise interactions are defined. However, this simplification allows only local conformational effects to be considered. Packing between secondary structures may be evaluated only by means of the solvation potentials and not by any pairwise terms. Clearly it would be ideal to devise an efficient dynamic programming method capable of taking non-local pairwise terms into account.

11.4. Double dynamic programming

The requirement here to match pairwise interactions relates to the requirement of structural comparison methods. The *potential environment* of a residue i is defined here as being the sum of all pairwise potential terms involving i and all other residues $j \neq i$. This is a similar definition to that of a residue's *structural environment*, as described by Taylor and Orengo [43]. In the simplest case, a residue's structural environment is defined as being the set of all inter-Cα distances between residue i and all other residues $j \neq i$. Taylor and Orengo propose a novel dynamic programming algorithm (known as double dynamic programming) for the comparison of residue structural environments, and it is a derivative of this method that I have used for the effective comparison of residue potential environments.

Let T_m ($m = 1, ..., M$) be the elements of a structural template, and S_n ($n = 1, ..., N$) be the residues in the sequence to be optimally fitted to the template. We wish to determine a score $Q(T_m, S_n)$ for the location of residue n at template position m. In order to achieve this, the optimal interaction between residue n and all residues $q \neq n$ conditional on the matching of T_m and S_n is calculated by the application of the standard dynamic programming algorithm. We define two matrices: a low-level matrix L (more precisely a *set* of low-level matrices), and a high-level matrix H into which the best paths through each of the low-level matrices are accumulated. The calculation of a single low-level matrix L is illustrated in Fig. 7.

For each m, n, the total potential of mean force, $Z(m, n, p, q)$, may be calculated for each p, q where p is again an element in the structural template, and q a residue in the object sequence:

$$Z(m, n, p, q) = \Delta E^{S_q}_{\text{solv.}}(A_p) + \begin{cases} \Delta E^{S_n S_q}_{(q-n)}(d_{mp}), & q > n, \ p > m; \\ \Delta E^{S_q S_n}_{(n-q)}(d_{pm}), & q < n, \ p < m; \\ 0, & q = n, \ p = m \quad \text{or} \quad q = n, p \neq m; \\ U, & q < n, \ p > m \quad \text{or} \quad q > n, \ p < m; \end{cases}$$

here U is a large positive constant penalty which forces the final path to incorporate pair m, n, A_p is the accessibility of template position p, d_{mp} and d_{pm} are elements of the template interatomic distance matrix and the pairwise, $\Delta E^a_k b(r)$, and solvation, $\Delta E^a_{\text{solv}}(s)$, terms are as defined previously. Pairwise terms are summed over all required atom pairs ($C\beta \rightarrow C\beta$, $C\beta \rightarrow N$ for example), using appropriate values from the distance matrix, though typically, for computational efficiency, the low-level matrices are calculated using the $C\beta \rightarrow C\beta$ potential alone.

The low-level matrix L is then calculated using the standard NW algorithm:

$$L_{pq} = Z(m, n, p, q) + \min \begin{cases} L_{p+1, q+1}; \\ \min_{r = p+2 \rightarrow N_A} L_{r, q+1} + g(S_p); \\ \min_{s = q+2 \rightarrow N_B} L_{p+1, s} + g(S_p); \end{cases}$$

where S_p is the secondary structural class (helix, strand, coil) of template position p. $g(S_p)$ is a simple secondary structure dependent gap penalty function:

$$g(S_p) = \begin{cases} G_s, & S_p = \text{helix, strand}, \\ G_c, & S_p = \text{coil}, \end{cases}$$

where G_s and G_c are both positive constants, and $G_s \gg G_c$. As with the application of the algorithm to structure comparison, it is found to be advantageous to accumulate the low-level matrix scores along each suggested low-level path, and so the paths from each low-level matching, conditional on each proposed match between T and S, for which the path scores exceed a preset cut-off, are accumulated into H thus:

$$H'_{pq} = H_{pq} + \min \begin{cases} L_{p+1, q+1}; \\ \min_{r = p+2 \rightarrow N_A} L_{r, q+1}; \\ \min_{s = q+2 \rightarrow N_B} L_{p+1, s}; \end{cases}$$

for all p, q along the optimum traceback path in L.

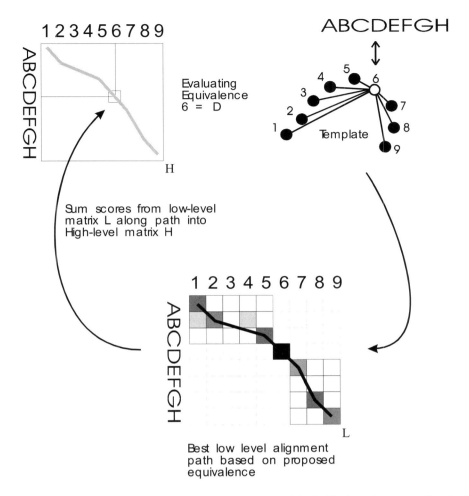

Fig. 7. Calculation of a single low-level matrix path as part of the double dynamic programming algorithm.

The overall operation may be thought of as the matching of a distance matrix calculated from the template coordinates with a *probability* matrix (in practice, an information matrix) calculated from the object sequence.

The final alignment (matrix F) is generated by finding the best path through the final high-level matrix thus:

$$F_{pq} = H_{pq} + \min \begin{cases} F_{p+1,q+1}; \\ \min_{r=p+2\to N_A} F_{r,q+1} + g(S_p); \\ \min_{s=q+2\to N_B} F_{p+1,s} + g(S_p). \end{cases}$$

In the above expressions, each instance of the NW algorithm has been formulated in terms of *minimizing* a cost function, as the scores in this application are energies. In practice, however, these expressions could be converted trivially to a form where a score is maximized, simply by changing the sign of the calculated energy values, or by just leaving the interaction propensities in units of information. Where mention is made of a *high-scoring* path (which is rather more familiar terminology in sequence comparison) in the following pages, then this should be taken as referring to a path with *low* energy.

As described, the double dynamic programming algorithm is too slow to be useful for fold recognition. The efficient algorithm by Gotoh [44] for calculating the NW score matrix is O(MN) where MN is the product of the two sequence lengths. Double dynamic programming involves the use of this algorithm for all MN possible equivalent pairs of residues, giving an overall algorithmic complexity of O(M^2N^2). On a typical present day workstation, a single instance of the Gotoh algorithm for two sequences of length 100 can be performed in around 0.1 CPU s. Multiplying this time by 100^2 provides an estimate of 1000 CPU s to complete a single double dynamic programming comparison, which is clearly too slow to be applied to the fold recognition problem, where many hundreds of instances of the double dynamic programming algorithm would be required. Furthermore, many comparisons would involve sequences or structures longer than 100 residues in length. The absurdity of the problem becomes apparent when it is realized that to compare a single sequence 500 residues in length with a structure of similar size, roughly 12 CPU h would be required. Even for a fold library with a few hundred folds, a single search would take several months to complete.

Clearly, if the double dynamic programming algorithm is to be of use, short-cuts must be taken. The most straightforward short-cut to take is to apply a window to both levels of the algorithm. This is very helpful, but still insufficient to make the algorithm convenient for general use. Orengo and Taylor proposed the use of a prefiltering (*residue selection*) pass to exclude unlikely equivalences before going on to calculate a low-level path. In the case of structural comparison, Orengo and Taylor initially select pairs of residues from the two structures primarily on the basis of similarities in their relative solvent accessibility and main chain torsion angles. Residue pairs found to be in different local conformations, and with differing degrees of burial are clearly unlikely to be equivalenced in the final structural alignment, and consequently should be excluded as early as possible in the double dynamic programming process. Unfortunately for sequence–structure alignment, it is not possible to select residue pairs on real measured quantities such as accessibility or torsion angles. However, these quantities could in principle be *predicted* for the sequence under consideration, and these predicted values then compared with the real values observed in the template structure. In practice, these values are not actually predicted directly, but the method proposed here certainly makes use of the same principles that might be employed in their prediction.

12. Residue selection for sequence–structure alignments

The residue selection stage of optimal sequence threading involves the summation of local interaction potential terms $\Delta E_k^a b$ over all residue pairs in overlapping windows of length L, where L is a small constant odd-number over the range, say, 5–31. Similarly the solvation terms are summed for each of the L residues. The window is clipped appropriately if it spans either the N- or C-terminus of either the sequence or the structure, for example the window length for the first and last residue in either cases would be $\frac{1}{2}(L+1)$. To equalize the contribution of the pairwise terms and the solvation terms, the average energy is calculated and summed for both, giving a total energy for the sequence–structure fragment of:

$$S_{mn} = E(\text{fragment}) = \frac{2\sum_{i=1}^{L-1}\sum_{j=i+1}^{L} E_{\text{pair}}^{ij}}{L(L-1)} + \frac{\sum_{k=1}^{L} E_{\text{solv}}^{k}}{L}.$$

Energies are calculated for every sequence fragment threaded onto every structure fragment, and the results stored in a selection matrix S. Using the residue selection step, the number of initial pairs (m,n) for which low-level paths need to be calculated is reduced up to 100-fold.

13. Evaluating the method

In the original paper describing the idea of optimal sequence threading [38] the method was applied to 10 examples with good results, in particular it proved able to detect the similarity between the globins and the phycocyanins, and was the first sequence analysis method to achieve this. Also of particular note were the results for some $(\alpha\beta)_8$ (TIM) barrel enzymes and also the b-trefoil folds: trypsin inhibitor DE-3 and interleukin 1β. The degree of sequence similarity between different $(\alpha\beta)_8$ barrel enzyme families and between trypsin inhibitor DE-3 and interleukin 1β is extremely low (5–10%), and as a consequence of this, sequence comparison methods had not proven able to detect these folds.

Although these results are good, it is fair to argue that in all cases the correct answers were already known and so it is not clear how well they would perform in real situations where the answers are not known at the time the predictions are made. The results of a very ambitious world wide experiment were published in a special issue of the journal "PROTEINS", where an attempt was made to find out how successful different prediction methods were when rigorously blind-tested (i.e., applied to problems where the answer was not known at the time). In 1994, John Moult and colleagues approached X-ray crystallographers and NMR spectroscopists around the world and asked them to deposit the sequences for any structures they were close to solving in a database. Before these structures were made public, various teams around the world were then challenged with the task of predicting each structure. The results of this experiment were announced at a meeting held at Asilomar in California, and this ambitious experiment has now become widely known as the Asilomar Experiment (or more commonly the Asilomar Competition).

The results for the comparative modelling and *ab initio* sections offered few surprises, in that the *ab initio* methods were reasonably successful in predicting secondary structure but not tertiary, and homology modelling worked well when the proteins concerned had very high sequence similarity. The results for the fold recognition section [45], however, showed great promise. Overall, roughly half of the structures in this part of the competition were found to have previously observed folds. Almost all of these structures were correctly predicted by at least one of the teams. The threading method described in this chapter proved to be the most successful method [46], with 5 out of 9 folds correctly identified, and with a looser definition of structural similarity, 8 out of 11 correct.

14. Software availability

A program, called THREADER, which implements the threading approach to protein fold recognition described here has been made widely available to the academic community free of charge, and can be downloaded over the Internet from the URL
`http://globin.bio.warwick.ac.uk/~jones/threader.html`

References

[1] Levinthal, C. (1968) Chim. Phys. 65, 44–45.
[2] Rooman, M.J., Kocher, J.P.A. and Wodak, S.J. (1991) J. Mol. Biol. 221, 961–979.
[3] Moult, J. and Unger, R. (1991) Biochemistry 30, 3816–3824.
[4] Chan, H.S. and Dill, K.A. (1990) Proc. Natl. Acad. Sci. USA 87, 6388–6392.
[5] Gregoret, L.M. and Cohen, F.E. (1990) J. Mol. Biol. 211, 959–974.
[6] Doolittle, R.F. (1992) Prot. Sci. 1, 191–200.
[7] Orengo, C.A., Jones, D.T. and Thornton, J.M. (1994) Nature 372, 631–634.
[8] Chothia, C. (1992) Nature 357, 543–544.
[9] Cohen, F.E., Sternberg, M.J.E. and Taylor, W.R. (1982) J. Mol. Biol. 156, 821–862.
[10] Taylor, W.R. (1991) Prot. Eng. 4, 853–870.
[11] Needleman, S.B. and Wunsch, C.D. (1970) J. Mol. Biol. 48, 443–453.
[12] Taylor, W.R. (1986a) J. Mol. Biol. 188, 233–258.
[13] Gribskov, M., Lüthy, R. and Eisenberg, D. (1990) Methods Enzymol. 188, 146–159.
[14] Bashford, D., Chothia, C. and Lesk, A.M. (1987) J. Mol. Biol. 196, 199–216.
[15] Barton, G.J. (1990) Methods Enzymol. 188, 403–428.
[16] Pearl, L.H. and Taylor, W.R. (1987) Nature 328, 351–354.
[17] Bernstein, F.C., Koetzle, T.F., Williams, G.J.B., Meyer, E.F., Brice, M.D., Rodgers, J.R., Kennard, O., Shimanouchi, T. and Tasumi, M. (1977) J. Mol. Biol. 112, 535–542.
[18] Novotny, J., Bruccoleri, R.E. and Karplus, M. (1984) J. Mol. Biol. 177, 787–818.
[19] Brooks, B., Bruccoleri, R.E., Olafson, B.D., States, D.J., Swaminathan, S. and Karplus, M. (1983) J. Comput. Chem. 4, 187–217.
[20] Novotny, J., Rashin, A.A. and Bruccoleri, R.E. (1988) Proteins 4, 19–30.
[21] Eisenberg, D. and McLachlan, A.D. (1986) Nature 319, 199–203.
[22] Baumann, G., Frommel, C. and Sander, C. (1989) Prot. Eng. 2, 329–334.
[23] Holm, L. and Sander, C. (1992) J. Mol. Biol. 225, 93–105.
[24] Lüthy, R., Bowie, J.U. and Eisenberg, D. (1992) Nature 356, 83–85.
[25] Hendlich, M., Lackner, P., Weitckus, S., Floeckner, H., Froschauer, R., Gottsbacher, K., Casari, G. and Sippl, M.J. (1990) J. Mol. Biol. 216, 167–180.
[26] Sippl, M.J. (1990) J. Mol. Biol. 213, 859–883.
[27] Casari, G., Sippl, M.J. (1992) J. Mol. Biol. 224, 725-732.
[28] Crippen, G.M. (1991) Biochemistry 30, 4232–4237.

[29] Maiorov, V.N. and Crippen, G.M. (1992) J. Mol. Biol. 227, 876–888.
[30] Ponder, J.W. and Richards, F.M. (1987) J. Mol. Biol. 193, 775–791.
[31] Bowie, J.U., Clarke, N.D., Pabo, C.O. and Sauer, R.T. (1990) Proteins 7, 257–264.
[32] Orengo, C.A., Brown, N.P. and Taylor, W.R. (1992) Proteins 14, 139–167.
[33] Bowie, J.U., Lüthy, R. and Eisenberg, D. (1991) Science 253, 164–170.
[34] Bork, P., Sander, C. and Valencia, A. (1992) Proc. Natl. Acad. Sci. USA 89, 7290–7294.
[35] Finkelstein, A.V. and Reva, B.A. (1991) Nature 351, 497–499.
[36] Chou, P.Y. and Fasman, G.D. (1974) Biochemistry 13, 212–245.
[37] Fauchere, J.L. and Pliska, V.E. (1983) Eur. J. Med. Chem. 18, 369–375.
[38] Jones, D.T., Taylor, W.R. and Thornton, J.M. (1992) Nature 358, 86–89.
[39] Kabsch, W. and Sander. C. (1983) Biopolymers 22, 2577–2637.
[40] Šali, A. and Blundell, T.L. (1990) J. Mol. Biol. 212, 403–428.
[41] Lee, C. and Subbiah, S. (1991) J. Mol. Biol. 217, 373–388.
[42] Goldberg, D.E. (1989) Genetic Algorithms in Search, Optimization, and Machine Learning. Addison Wesley, Reading, MA.
[43] Taylor, W.R. and Orengo, C.A. (1989) J. Mol. Biol. 208, 1–22.
[44] Gotoh, O. (1982) J. Mol. Biol. 162, 705–708.
[45] Lemer, C.M.R., Rooman, M.J. and Wodak, S.J. (1995) Proteins 23, 337–355.
[46] Jones, D.T., Miller, R.T. and Thornton, J.M. (1995) Proteins 23, 387–397.

S.L. Salzberg, D.B. Searls, S. Kasif (Eds.), *Computational Methods in Molecular Biology*
© 1998 Elsevier Science B.V.   All rights reserved

CHAPTER 14

From computer vision to protein structure and association

Haim J. Wolfson[a] and Ruth Nussinov[b]

*[a]Computer Science Department, School of Mathematical Sciences, Tel Aviv University,
Tel Aviv 69978, Israel. Telefax: +972-3-640-6476, e-mail: wolfson@math.tau.ac.il.
[b]Sackler Inst. of Molecular Medicine, Faculty of Medicine, Tel Aviv University,
Tel Aviv 69978, Israel. Fax: +972-3-641-4245, e-mail: ruthnu@post.tau.ac.il.
and Laboratory of Experimental and Computational Biology, SAIC, NCI-FCRDC, Bldg. 469, rm. 151,
Frederick, MD 21702, USA. Tel.: (301) 846 - 5579, Fax: (301) 846 - 5598, e-mail: ruthn@fcrfv1.ncifcrf.gov.*

1. Introduction

In this chapter we discuss the application of novel pattern matching methods to the biomolecular recognition and docking problem.

The problem of receptor–ligand recognition and binding is encountered in a very large number of biological processes. For example, it is a prerequisite in cell–cell recognition and in enzyme catalysis and inhibition. The type of molecules involved is very diverse as well: proteins, nucleic acids (DNA and RNA), carbohydrates and lipids. Whereas the receptor molecules are usually large (often proteins), the ligands can be fairly large (e.g. proteins) or small, such as drugs or cofactors. Efficient docking algorithms of drug molecules to target receptor proteins is one of the major ingredients in a computer assisted drug design scheme.

The docking task usually requires shape complementarity of the participating molecules. This complementarity reflects the effect of the van der Waals interactions, which are very sharp at short distances. Thus, there are two simultaneous geometric criteria for a good docking solution. First, the docked molecules should not penetrate each other, and, second, one expects a fairly large interface of tightly matched molecular surfaces. Our task is to detect appropriate conformations of both molecules, so that there is a three-dimensional rotation and translation of one of them which brings it into close fit with the other molecule according to the above mentioned criteria.

Enumerating over all the degrees of freedom of the participating molecules is a formidable task, unless the molecules are very small. Consequently, the first generation of automatic docking methods assumed that the molecules behave as rigid objects, thus, freezing all the degrees of freedom except for the three rotational and three translational degrees of freedom which are necessary to move one molecule into close fit with the other. This is the, so-called, "key-in-lock" principle. For proteins, it has been argued [1,2] that the rigid approximation is justified by the similarity of the crystallographic structures between the bound and the unbound proteins[1] while many other cases exist where more

---

[1] The structure of a single protein in solution is referred here as "unbound", while the structure of the same protein in the docked complex, is called "bound".

substantial conformational changes have been observed. In the last decade, a number of docking methods have been developed which tackle the rigid docking problem with considerable success, especially, in the "bound" experiments. Kuntz et al. [3,4] developed the DOCK program, Connolly [5] docked molecular surfaces by fitting quadruples of topographical extrema, Jiang and Kim [6] suggested the "soft docking" approach, Katchalski-Katzir et al. [7] exploited the Fourier transform technique for fast enumeration of all possible 3D translations of one molecule with regard to the other. Our group suggested two main methods for the rigid docking task. The first is the Computer-Vision-based Geometric Hashing method [8,9], which can efficiently dock ligands to moderate size receptors, without any *a priori* knowledge of the binding site location. The second method [10,11] was especially geared towards the docking of large protein–protein complexes and is based on matching pairs of sparsely sampled extremal points of the molecular surface. These points are similar to the ones used by Connolly [5]. We shall present these two methods in detail in the following sections.

In the last few years, a considerable effort was made to develop techniques for the docking of conformationally flexible ligands to rigid receptors. One of the first attempts was to introduce limited flexibility into the DOCK technique [12,13]. Lengauer's group developed the FLEXX method [14], which is especially geared towards docking of flexible drug molecules into previously known binding sites. Jones et al. [15] tackle the flexible docking task by genetic algorithms. Our group [16,17] developed an efficient algorithm for docking of either flexible ligands or flexible receptors possessing a small number of internal rotational bonds. Lengauer and Rarey [18] have compiled a very good survey of the recent docking methods.

In this chapter we focus on the rigid docking techniques that we have developed. We shall discuss only the geometric aspect of shape complementarity. Our computational approach is based on the *Geometric Hashing* paradigm, originally developed for object recognition tasks in Computer Vision [19,20]. Although coming from an entirely different application field the underlying geometrical and computational principles bear a similar flavor. Let us sketch briefly the model-based object recognition problem in Computer Vision. There one is presented with a database of shapes (models) and the task is to detect if any of these shapes appear in an observed scene, even if they are partially occluded by other objects. In addition, we do not know *a priori* in what locations of the scene the desired shape might appear. In case a shape does appear in the scene, one has to compute the transformation that has to be applied to its model in the database, so that it matches the shape's appearance in the scene. Thus, one has to match shape characteristics of the models to the shape characteristics of the scene. Let us now consider the docking task in an analogous framework. Assume that we are given a database of ligands, and a receptor. The task is to decide whether there is a transformation of a ligand, which results in sufficient shape complementarity with the receptor without penetrating it. In the general case, we do not have *a priori* knowledge on the location of the binding site of the receptor. Already these two formulations alone make the analog quite clear. The ligand database is analogous to the model database. The receptor is analogous to the scene, and the partial occlusion and lack of knowledge of the desired model location of the scene are analogous to the facts that not all of the ligand's molecular surface is complementary to the receptor surface, and that we do not know in advance the location of the binding site[2]. One should bear in mind, though, that while in object recognition one has to match shape characteristics seeking for their similarity, in docking we seek shape complementarity. In this sense the analog between object recognition in Computer Vision and structural protein comparison is even more direct [21].

We shall present two methods for the rigid docking task. In section 2 we formulate the precise mathematical task that will be tackled. Section 3 discusses the molecular surface representation and the shape primitives, that we will use in our docking algorithms. In section 4 we reformulate the molecular surface matching problem as a point + normal set matching problem. Section 5 represents the sparse feature large protein–protein docking algorithm, while section 6 presents the Geometric-Hashing-algorithm-based docking algorithm. Finally, in section 7 we discuss further developments and future tasks.

2. Problem formulation

In this section we describe the mathematical model of the docking problem for which our algorithms are developed. We restrict our discussion to the docking of rigid molecules. One of them, usually the larger one, is called the receptor and the other is called the ligand. Each molecule is regarded as a collection of spherical atoms. Thus, our input is a set of atomic centers together with their van der Waals radii.

The biomolecular recognition or docking problem is defined as follows.

Problem 1: *Given the coordinates of the atomic centers of the receptor and the ligand molecules together with their atomic radii, find those rotations and translations of the ligand, which result in a large contact area between the molecular surfaces of the ligand and the receptor, without causing intermolecular penetration.*

There are two major aspects to the docking problem – geometric and chemical. Geometrically the two molecules should interlock like two pieces of a 3D jigsaw puzzle. From the chemical standpoint, the candidate docking configuration should be energetically favorable, so that the predicted complex would be stable. This presentation deals only with the geometric aspect of the problem which can be directly stated as a pattern recognition problem. Since detailed energy evaluation calculations are time consuming, our goal is to develop geometric matching algorithms which are fast, result in few tens or hundreds of candidate solutions (with the correct one among them), thus enabling a further chemical verification step to filter these remaining candidates, which could further be submitted to human inspection.

3. Molecular surface representation and interest feature extraction

Since molecules dock at their surfaces, we have to define the geometric model of the molecular surface. We have also to define how do we compute complementarity of

---

[2] In case we do know this site, it can be readily incorporated in the algorithms to reduce the amount of computation.

molecular surfaces, which might be of very intricate form. Our solution is to represent the surfaces by sets of discrete *interest points* with their associated normals, and then seek for transformations which best superimpose complementary sets of these interest features. In this section we discuss several methods to compute such *interest points*.

As was mentioned in the previous section, a molecule is modeled by a set of balls, centered at the atomic coordinates of the molecule. The radii of these balls are their van der Waals radii. Richards [22] defined the molecular surface as the surface being created by virtually rolling a probe ball over the van der Waals spheres of the molecule's atoms. Richards' definition has been implemented by Connolly in his MS-DOT program [23], which computes the analytic representation of the molecular surface. The probe mimics the size of a water molecule and is usually chosen to be of 1.4 Å radius. This procedure fills narrow crevices and defines the surface which is actually accessible to atoms of the size of the probe ball. A probe ball can touch the van der Waals surface only in one, two or three points resulting in only three types of molecular surface faces – convex, toroidal or concave (see ref. [24]). A convex face is a connected patch of the atom's van der Waals sphere touched by the probe ball (one contact point). A toroidal face is traced by the probe ball when it rolls along the groove between two atoms (two contact points). When the probe ball is suspended by three atoms, the part of this ball surface mounting the space among the three atoms becomes a concave face (three contact points).

Connolly's program itself returns a discrete set of points with associated normals which samples the molecular surface at the density requested by the user. One could use this set of points as the *interest features* of the docking algorithm. However, a coarse sampling may give an unreliable representation of the molecular surface, while a dense sampling may result in thousands of points to represent the surface of a relatively small molecule, making the computational burden not feasible.

In our research we have used two alternative *interest feature* definitions, which we shortly describe below.

3.1. Caps, pits, and belts

In ref. [25] we extract *interest points* and their associated normals only at key locations of the surface, covering the crests of the exposed atoms and the seams and dents between them. Specifically, we compute one such point at the center of each molecular face and label them as "cap" if the face is convex, "pit" if the face is concave, and "belt" if the face is toroidal. Normals associated with these points are computed as well.

By examining the interfaces of known receptor–ligand complexes we have found hundreds of pairs of our *interest points* to be at excellent complementarity, in both the point location and the normal orientation. The abundance and quality of these pairs provides the desired shape representation for our Geometric-Hashing-based matching algorithm (see ref. [9,26] for details).

The points obtained are considerably more sparse than the original Connolly's MS representation, yet they reliably represent the overall shape of the molecular surface. However, this representation is still somewhat too dense. In particular, the distances between pairs of points should be larger than the error thresholds allowed in the matching.

These error thresholds should be liberal ones, due to the inexact fit of docked molecules. Thus, it is necessary to prune the obtained surface points. We have adopted several pruning mechanisms, such as: (1) selecting points by label; (2) thinning by fusing close points; (3) eliminating points representing faces with too small or too large areas. The output of the pruning operation results in a less accurate surface representation, yet dozens of *interest point* pairs still remain in a typical binding site supplying enough information for the matching phase.

This molecular surface representation appeared both in rigid docking [9], flexible docking [17], and molecular surface superposition [27] algorithms.

3.2. Knobs and holes

The cap, pit and belt interest features provide a reliable molecular shape representation, without requiring a dense sampling of the Connolly surface, yet for large molecules their number is still very big. If we know that the binding interface is expected to be relatively large by itself it would be enough to sample the surface at sparse interest points, which still preserve some salient features of the molecular shape. The properties of such points should be also invariant to the rotation and translation of the molecules.

In a docking algorithm developed by Connolly [5] he suggested to use the convex-ity/concavity extrema of the molecular shape as interest points, which he nicknamed as knobs and holes.

We compute these points as follows. First, the molecular surfaces of the receptor and ligand molecules are approximated by densely sampled dots, using the previously mentioned MS algorithm [23]. Then, at each MS dot, a sphere of a fixed radius is constructed (we use 6 Å which is an approximation of the radius of an amino acid). A shape function at such a point is defined as the volume of the intersection of the molecule with the sphere. This shape function measures the local convexity/concavity of the surface. When the shape function value is small, the surface is convex (knob); when the value of the shape function is high, the surface is concave (hole). Using a non-extrema suppression technique we choose a relatively small set of critical knobs and holes, which are local minima and maxima of this shape function.

For each critical point we also compute the normal vector. At the end of the process, for each molecule, we have lists of knobs and holes along with their associated normal vectors. Note that the definition of these points is invariant to rotation and translation, since they are dependent only on the intrinsic shape of the molecule.

4. Interest point correspondence

In section 3 we have shown that the molecular surface can be represented by sets of interest points coupled with their associated normals. One of the suggested representations is reliable, representing all the major geometric characteristics of the surface, while the other representation was a sparse one, representing only the most salient geometric features of the surface. An important property of the selected points is that their existence and intrinsic location on the molecular surface is not affected by translations and rotations that the molecule may undergo. The need for these interest-point-based representations stems from our desire to reduce the molecular surface matching task to a point matching task. The reason for this reduction is our ability to develop relatively efficient algorithms for the point matching task which provide the solutions for the original molecular surface matching.

Let us formulate this new task for the ligand–receptor docking. Assume that both molecules are represented by sets of characteristic (interest) points with their associated normals.

Problem 2: *Given the coordinates of the molecular surface interest points + normals of the receptor and the ligand, find those rotations and translations of the ligand, which result in a large set of the ligand interest points superimposed on the receptor interest points with their normals in opposite directions. Only transformations which do not cause penetration of the ligand into the receptor are acceptable.*

One should note several important facts. First, since the molecules interlock only at the binding site, not all the interest points of either the ligand or the receptor will have a matched counterpart, even in an ideal situation. Second, since we do not assume a priori knowledge of the binding site, we cannot know in advance which of the interest points might have a matching counterpart. Third, due to the inherent "inexact match" of molecular surfaces, as well as inaccuracies in the computation of interest points and normals, superimposition does not mean an exact match. It means that in a vicinity of a transformed ligand interest point (sometimes nicknamed "error region") there is a receptor interest point. Since the computation of normal directions is even more noisy than the computation of interest point locations, quite a liberal error threshold is assigned for opposing normal directions. Fourth, it is important to understand that one cannot restrict himself to the examination of only one transformation giving the maximal superimposed set of ligand–receptor points. Due to all the fuzziness inherent in our computations, the maximal superimposed set might not be the correct one, while a somewhat smaller set, defined by a different transformation may hit the target. This can be even examined on real complexes, where quite often the "correct" solution, appearing in the PDB does not score at the top of the list, although, it almost always scores quite high. Thus one has to examine all the transformations which suggest a "large enough" superimposition.

It is quite common to divide point set matching tasks into two subtasks – the point correspondence problem, and the corresponding point superposition problem. The correspondence problem is the tough one. Once a correspondence hypothesis (match) among the points is established, there are well known methods to find the rigid transformation which superimposes these points with minimal least squares distance among the point sets (see, e.g. ref. [28]).

Let us examine the point set correspondence problem more closely. Assume that we have n interest points on the receptor surface and m interest points on the ligand surface and we have to establish a correspondence among them. Without loss of generality we may assume that $m \leqslant n$. If there are no mutual constraints among the points, the first ligand point can match n receptor points, the second can match $n-1$ receptor points and so far. Thus, there are $n(n-1)(n-2)\cdots(n-m+2)(n-m+1)$ candidate solutions. This is an enormous enumeration space, e.g. for $n = 200$ and $m = 100$ the number of hypotheses is of the order of 10^{217}. Luckily enough, our problem is much more constrained, since we are dealing with rigid body motion, so when some of the points are rotated and translated, the whole ensemble of points performs the same transformation. Let us examine, what is the minimal number of point correspondences required to fully determine the transformation of the ligand vis a vis the receptor.

Kuntz et al. [3] and Connolly [5] have used four point set correspondences to define an unambiguous rotation and translation of the ligand. However, it is well know that to determine unambiguously a rotation and translation that superimposes one rigid body onto another one needs a minimum of three non-collinear ordered matching-point pairs. The only constraint is that the pairwise distances among the matching points should be identical. Namely, for each pair of two non-degenerate *congruent* triangles $\triangle ABC$ on the receptor and $\triangle abc$ on the ligand, one can define a unique rotation and translation. First, translate the ligand so the point a is superimposed on A. Then, rotate it until b is superimposed on B. Finally, rotate it again around the ab-axis until c touches C. This construction is unique. Now, the number of candidate solutions is at most $n(n-1)(n-2) \times m(m-1)(m-2)$, which is the number of possible triangle correspondences. In the previous example of $n = 200$ and $m = 100$, the number of hypotheses is at most 7.6×10^{12}. The number of congruent matching triangles is expected to be much smaller.

Still, one would like to reduce the size of the minimal matching set for two major reasons. First, it will reduce the size of the solution space, and second, it will enable recovery of solutions where we do not have a triplet of matching interest points at the interface. It is easy to see that if we exploit only point information, a pair of matching points in the ligand ab and the receptor AB are not enough to recover the transformation, since we are left with the freedom of rotation around the ab-axis (see explanation above). However, in the molecular surface representations that we have seen in section 3 we extract not only points, but also their associated normals. Thus, if the length of the segment ab on the ligand surface is equal to the length of AB on the receptor surface, we can superimpose the segments, and the only remaining degree of rotational freedom is fixed by the requirement that the normal at a should be an opposite direction to the normal at A. The same requirement for the normals at the points b, B supplies an additional constraint. Now, there are at most $n(n-1) \times m(m-1)$ candidate solutions, which is about 3.4×10^8 in our previous example.

5. Large protein–protein docking

5.1. The algorithm

The technique we describe below was especially designed to handle large protein–protein complexes. In retrospective it performed very well for most of the standard protein–ligand cases, where the ligand is a small molecule, as well. Especially encouraging was its relatively good performance on "unbound" examples, which are the real-life tests of docking techniques.

Following are the major steps of this method.

(i)   Computation of the molecular surface of the ligand and the receptor using Connolly's MS algorithm.

(ii) Representation of these molecular surfaces by sparse "interest point + normal" sets of knobs and holes.

(iii) Matching of pairs of ligand interest points + normals to compatible pairs on the receptor.

(iv) Clustering close transformations.

(v)   Rejection of candidate transformation clusters, which cause significant penetration of the ligand into the receptor.

(vi) Scoring of the remaining transformations by contact surface area favoring (optionally) contacts of hydrophobic residues and solutions which result in spatially connected binding patches on the molecular surfaces.

For details the reader is referred to refs. [10,11]. Here we briefly sketch some of the above mentioned steps.

5.1.1. Knob and hole extraction

Knobs and holes are extracted by the technique described in section 3. The criterion for a knob is that the value of the shape function is below $\frac{1}{3}V$, where V is the volume of the intersecting ball, while a hole's shape function should be above $\frac{2}{3}V$. Non-extrema suppression on 12 nearest neighbors is applied.

5.1.2. Interest point matching

For each pair of critical points + normals from the ligand compute the transformation with each compatible pair from the receptor as described in section 4. Note that compatibility requires that the distances between the points be almost equal (up to an error threshold), that a knob should be paired only with a hole (and vice versa), and that the appropriate normals should be in opposite directions. Since protein–protein interactions are localized, and generally do not span distant regions of the proteins without covering close-range ones, we restrict ourselves to critical point pairs which are closer than 20 Å to each other. To enable fast pruning of incompatible critical points, we assign to each pair of points on the ligand/receptor a signature, which has a symbolic component (knob/hole, hydrophobicity) and a geometric component, which includes the distance between the pair of points, the angles that are formed by the line segment connecting the points with each of the normals, and the torsion angle between the two planes formed by this line segment and the respective normals. Only pairs with compatible signatures are considered as a match. Compatibility is defined by appropriate error thresholds (see refs. [10,11] for the details).

5.1.3. Clustering of transformations

Since the same (or nearly identical) transformation can be achieved using different matching critical-point/normal pairs, the candidate transformations are clustered to reduce their number to the effective one. This reduces the workload of the next verification and scoring phase.

5.1.4. Verification of penetration and scoring

For each resulting candidate transformation the ligand is transformed to fit the receptor. If the ligand atom centers penetrate inside the receptor, the transformation is discarded. Otherwise, the transformation is scored favorably for ligand surface dots, which are in close vicinity of the receptor molecular surface and penalized for surface dots penetrating into the receptor. Optional scoring techniques take into account only the dots belonging to large connected interface patches and give an additional weight to dots belonging to hydrophobic residues.

5.2. Experimental results

We have done extensive experimentation with our algorithm both for bound and for the much more difficult unbound cases. In ref. [11] 26 examples of complex prediction are given. The data was taken from the PDB and includes mainly proteinases, immunoglobulins, protein–inhibitor and chain–chain docking examples. It is important to note that all these cases have been solved by one set of program parameters. A sample of the above mentioned examples is given in Table 1, where the PDB name of the complex, receptor and ligand are given as well as the resolution of the crystal structures.

These complexes contain relatively large ligands as can be seen from the information in Table 2, where the number of amino acids, the number of interior and exterior (molecular surface) atoms are given as well as the number of the computed knobs and holes.

Since the tool developed here is a predictive one, in order to evaluate its efficacy we present in Table 3 the best rms distance from the "correct" (PDB complex) solution among the highest ranking 10, 50 and 100 solution clusters, respectively. Columns 3, 4, 5 of the table show the lowest rms for the solutions without the application of a criterion favoring a binding site with a small number of connected components (connectivity filter), while columns 6, 7, 8 present the results which have been subjected to this criterion. One can see that already among the 10 top candidate solutions in these complexes 4 out of 7 have an rms less than 2.0 Å, while among the 100 top candidate solutions all but one have an rms less or equal 2.0 Å. The runtimes of the procedure on a 66 MHz Intel486 PC clone are given in Table 4.

Table 1
Bound examples

	PDB	Receptor name	Ligand name	Res. in Å
1	1cho	alpha-chymotrypsin (E)	turkey 2 ovomucoid third domain (I)	1.8
2	2hfl	IG*G1 Fab fragment (LH)	lysozyme (Y)	2.5
3	2mhb	hemoglobin α-chain (A)	β-chain (B)	2.0
4	2sni	subtilisin novo (E)	chymotrypsin inhibitor (I)	2.1
5	3hfm	IG*G1 Fab fragment (LH)	lysozyme (Y)	3.0
6	4sgb	serine proteinase (E)	potato inhibitor pci-1 (I)	2.1
7	4tpi	trypsinogen (Z)	pancreatic trypsin inhibitor (I)	2.2

Table 2
Size of bound examples

PDB	Receptor[a]					Ligand[a]				
	aa	int at	ext at	holes	knobs	aa	int at	ext at	holes	knobs
1 1cho	146	655	392	57	144	53	469	232	31	108
2 2hfl	424	1606	1621	220	270	129	517	483	60	93
3 2mhb	141	581	487	84	118	146	618	515	87	119
4 2sni	275	832	1105	112	155	63	306	206	30	65
5 3hfm	429	1552	1741	214	262	129	516	484	59	110
6 4sgb	227	603	706	60	97	51	247	132	17	56
7 4tpi	230	836	792	108	141	58	279	176	23	60

[a]Abbreviations: aa, number of amino acids; int at, number of interior molecular surface atoms; ext at, number of exterior molecular surface atoms.

Table 3
Best rms among top 10, 50 and 100 clusters of the bound examples

PDB	Fast scoring			With connectivity		
	Top 10	Top 50	Top 100	Top 10	Top 50	Top 100
1 1cho	0.53	0.53	0.53	0.53	0.53	0.53
2 2hfl	48.71	14.43	2.07	14.43	2.07	2.07
3 2mhb	0.69	0.69	0.69	0.69	0.69	0.69
4 2sni	1.16	1.16	1.16	1.16	1.16	1.16
5 3hfm	36.44	18.09	0.94	36.44	0.94	0.94
6 4sgb	13.28	4.50	4.50	3.59	3.59	3.59
7 4tpi	1.40	1.40	1.40	1.40	1.40	1.40

Table 4
CPU time on a 66 MHz Intel486 clone

PDB	Match	CPU time in minutes	
		Scoring with connectivity	Scoring without connectivity
1 1cho	3.3	40.6	12.6
2 2hfl	20.8	283.6	114.2
3 2mhb	14.3	227.1	97.2
4 2sni	4.4	82.3	26.8
5 3hfm	21.3	332.2	137.4
6 4sgb	1.8	26.5	7.7
7 4tpi	4.1	54.2	17.3

Table 5
Unbound examples

	PDB	Receptor name	Res. in Å	Ligand name	Res. in Å
1	1hfm-1lym(A)	IG*G1 fv fragment	model	lysozyme (A)	2.5
2	1tgn-4pti	trypsinogen	1.6	trypsin inhibitor	1.5
3	1tld-4pti	beta-trypsin	1.5	trypsin inhibitor	1.5
4	2hfl-6lyz	IG*G1 Fab fragment	2.5	lysozyme	2.0
5	2pka-4pti	kallikrein a	2.0	trypsin inhibitor	1.5
6	2ptn-4pti	trypsin	1.5	trypsin inhibitor	1.5
7	2sbt-2ci2	subtilisin novo	2.8	chymotrypsin inhibitor	2.0
8	5cha(A)-2ovo	alpha-chymotrypsin (A)	1.7	ovomucoid third domain	1.5

Table 6
Size of the unbound examples

	PDB	receptor[a]					ligand[a]				
		aa	int at	ext at	holes	knobs	aa	int at	ext at	holes	knobs
1	1hfm-1lym(A)	113	884	830	176	161	129	459	542	109	198
2	1tgn-4pti	218	817	804	139	240	58	182	271	59	22
3	1tld-4pti	220	851	778	145	202	58	182	271	59	22
4	2hfl-6lyz	213	1593	1627	270	223	129	479	522	109	72
5	2pka-4pti	223	915	884	131	153	58	182	271	59	22
6	2ptn-4pti	220	857	772	140	249	58	182	271	59	22
7	2sbt-2ci2	275	1061	873	160	129	65	205	316	64	27
8	5cha(A)-2ovo	237	894	841	135	102	56	136	282	69	22

[a] Abbreviations: aa, number of amino acids; int at, number of interior molecular surface atoms; ext at, number of exterior molecular surface atoms.

Our success with the reconstruction of docked configurations where the molecules have been crystallized together, has encouraged us to turn to the far more difficult but realistic case, that of the "unbound" proteins. Here the receptor and the ligand have been determined separately. When the molecules dock, their surfaces adjust to each other, so using our (and other rigid docking) techniques one should expect molecular penetrations. This requires relaxation of the rejection criteria in the penetration test, resulting in a potentially large increase in the number of docked configurations. To test our method, we have docked 19 examples of unbound proteins, in which the "correct" complex is also known. This gave us a possibility to compare our results with the real-life solution. Table 5 lists the names of eight docked proteins as well as their crystal resolution. The examples were chosen as large proteins (see Table 6, that have been crystallized both in complex and as separate molecules (unbound proteins).

The lowest rms solutions among the top ranking 10, 50 and 100 solution clusters are given in Table 7. When looking for the top 100 clusters notice that 3 solutions have an rms

324

Table 7
Best rms among top 10, 50 and 100 clusters of the bound examples

PDB	Fast scoring			With connectivity		
	Top 10	Top 50	Top 100	Top 10	Top 50	Top 100
1 1hfm–1lym(A)	27.32	15.29	11.75	34.84	13.67	11.75
2 1tgn–4pti	8.24	8.24	2.75	8.24	8.24	1.85
3 1tld–4pti	10.03	6.26	6.26	13.20	6.26	6.26
4 2hfl–6lyz	28.02	24.83	23.33	28.65	9.53	1.21
5 2pka–4pti	8.11	5.65	5.65	9.90	5.65	5.53
6 2ptn–4pti	10.01	6.14	5.85	6.14	6.14	3.53
7 2sbt–2ci2	10.38	10.11	10.11	16.40	10.11	5.64
8 5cha(A)–2ovo	6.45	1.49	1.49	8.81	1.49	1.49

less than 2 Å, and 4 additional solutions are with an rms below 6.5 Å, however, among the 50 top ranking solutions only one is below 2 Å. This is an unparalleled performance for the "unbound" case which was by now almost untackled by a purely geometric approach. Introduction of potential energy calculations on our top ranking examples should lead to a significant improvement in the fit. Since energy calculations are very time consuming, the reduction of the transformation space to less than 100 candidates gives a solid starting point for such an approach.

6. The Geometric-Hashing-based docking method

6.1. The algorithm

In this section we outline our Geometric Hashing surface matching algorithm which is based on ideas originated in Computer Vision [19]. Here we use the cap–pit–belt shape representation which is described in section 3.1. Given this surface representation our task is the same as in the previous section, namely, to find a rotation and translation of the ligand, so that a large portion of its surface "cap" interest points are in the vicinity of corresponding receptor surface "pit" interest points[3] having normals which are almost in the opposite direction. The resulting transformation should not cause inter-penetration of the ligand and receptor atoms. Since the number of cap/pit interest points is significantly larger than the knob/hole points, we applied a computationally more efficient matching algorithm for this case.

Below, we briefly overview the underlying principles of the Geometric Hashing technique with special emphasis on the salient features of the docking application. For additional details on the Geometric Hashing and its applications to structural comparison of molecules the reader is referred to refs. [21,29–31].

Assume that we are given a ligand with m caps and a receptor with n pits. In the method described in section 5 there are $n(n-1) \times m(m-1)$ transformations which are candidate

---

[3] Of course, the roles of "pits" and "caps" can be reversed.

solutions. Some of these candidates are immediately rejected since the signatures of the compared point pairs are incompatible. However, for the remaining candidates a time consuming verification stage, which includes penetration check and surface contact scoring is initiated. Our goal is twofold. First, to reduce the number of candidate solutions which pass to the verification stage and, second, to develop a more efficient matching procedure to detect those candidates. To achieve this we use an associative memory type approach. We learn the geometric attributes of the ligand interest points and store the points in a memory (hash-table) which is indexed by these attributes. Then, given the receptor interest point set, we compute their geometric attributes and retrieve from the memory those ligand's points having similar properties. The idea behind this approach in Computer Vision was triggered by the human visual performance. Assume that somebody is observing a pile of objects in an attempt to recognize them. If he spots a sharp corner in this scene, he is most likely not going to retrieve smooth spherical objects from his memory in order to check if they appear at this location of the scene. His memory will be accessed directly at all the familiar objects having sharp corners.

The implementation of the above mentioned idea for molecular surface point matching is easier said than done. If the molecular surface points were colored, each by a different color, we could index the ligand points into a table according to their color, and, given the receptor interest point set, check the color of each of its points, access the memory at this color and retrieve the matching ligand point. However, no such symbolic color, which is representative enough is available in general. (It should be noted though, that the (knob/hole) is a set of two symbolic colors, in this sense.) Hence, we shall look for a "geometric color" which describes the geometric attributes of our points. A natural quantitative geometric attribute of a point is its coordinate set. This could be our color. Maybe we can compute the coordinates of the ligand caps and store their identity in a table which is accessed by these coordinates. Then, we could compute the coordinates of the receptor pits, and for each such coordinate access the table and check whether there is a ligand cap matching it. However, as we well know, coordinates are defined relative to some reference frame. How can we ensure that the ligand and receptor interest points are computed in the same reference frame? Moreover, we do know that the ligand geometric representation has to undergo a rotation and translation to bind to the receptor.

Thus, before we adopt the associative memory technique, we must address the following issues:
(i) Rotational and translational invariance.
(ii) Choice of the same reference frame both for the ligand and for the receptor.
(iii) Handling of the fact that only part of the ligand caps will have counterpart receptor pits and vice versa.

These problems are addressed by redundantly representing each molecule in many reference frames, which are naturally attached to this molecule. Now the coordinates of a point will represent not only the point itself, but the point in a given reference frame. We start by unambiguously defining a local 3D reference frame for a pair of points with their associated normals.

6.1.1. Reference frame definition
Let (p_1, n_1) and (p_2, n_2) be two points with their associated normals. A Cartesian

reference frame is unambiguously defined by two points a, b and a direction d, which is different from the ab direction. It can be defined uniquely by picking a as the origin, the ab direction as the x-axis, the cross-product of ab and d as the y-axis, and the cross-product of the new x and y-axes as the z-axis. Our reference set is based on two critical points $a = p_1$ and $b = p_2$, and the d direction can be taken as the average of n_1 and n_2. Note that each such reference frame has a shape signature similar to that defined in section 5, which includes the interpoint distance, the angles between the normals and the interpoint segment, and the resulting torsion angle.

Now, we can compute the coordinates of all the other interest points with their associated normals in this frame. Rotation and translation of the molecule does not change these coordinates. However, representing the molecule in a single reference frame is insufficient, since we cannot ensure *a priori* that the reference interest point pair participates in the binding. Hence, we represent the molecule in all possible interest point pair reference frames. Consequently, each interest point is redundantly represented in many different reference frames.

Below we describe our method for matching of one ligand to one receptor, although a database of ligands could be explored simultaneously. Likewise, a ligand can be docked to a database of receptors.

There are two major phases in the Geometric Hashing algorithm. The first is the Preprocessing phase, where the ligand geometric information is memorized in a look-up table or hash-table and the second is the so-called Recognition phase, where the receptor points are efficiently matched to the ligand points via the precomputed hash-table. One should note that once such a table is built, it can be stored and should not be computed at each run of the algorithm. Also, one should note that the actual runtime of the algorithm is the sum of the times of both phases and not their product. If the ligand table is precomputed in a previous run, the actual execution time is only the time of the Recognition phase.

6.1.2. Ligand preprocessing

- Reference sets based on a pair of ligand critical points with their associated normals are chosen.
- For each ligand reference set $rs(l_1, l_2)$, the coordinates of the other ligand critical points (in a restricted vicinity as explained below) and their normals are computed in the Cartesian frame associated with the reference set. These coordinates serve as an index to a hash-table (memory), where the reference set and the point information are stored along with additional geometric constraints on this triplet of points + normals.

6.1.3. Recognition

For each pair of receptor critical points do the following.

- Build a reference set based on the pair of receptor critical points (and their associated normals).
- The coordinates of the other receptor critical points and their normals are computed in the Cartesian frame associated with this reference set. For each such coordinate the hash-table bins with values close to the coordinate are accessed. The records from the table are retrieved and those ligand reference sets which satisfy the geometric constraints of the receptor reference set receive a vote as matching candidates of this set.

After the voting process is completed for the current receptor reference set, if a ligand reference set receives many votes, the transformation between this set and the receptor reference set is computed, the molecules are superimposed and additional transformation refinement steps are executed. In particular, additional matching points are recovered and the best transformation (in the least squares sense) for the extended set is computed. Transformations causing intermolecule penetration are discarded.

The flow-chart of the algorithm is given in Fig. 1.

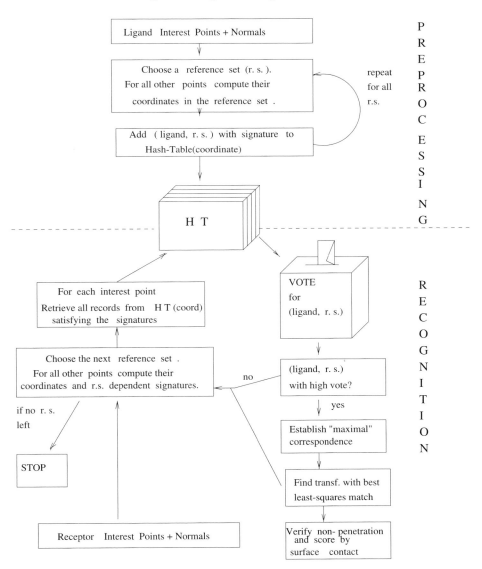

Fig. 1. The flow-chart of the Geometric-Hashing-based docking algorithm.

One should note that the complexity of the Recognition phase is of the order of n^3, where n is the number of receptor interest points.

Following the general description of the algorithm we outline several implementation details which are specific to our application.

6.1.4. Implementation details

We outline the constraints that we have used in our application. To give the reader a feeling of the (in)accuracies involved, we also mention the numerical values of the thresholds used to derive the experimental results presented below. Obviously, these thresholds can be changed according to the application.

6.1.5. Choice of reference sets

We do not use all pairs of points to build reference sets. Some distance and directional constraints are applied to a pair of points p_1, p_2 and their normals n_1 and n_2 in order to build a reference set based on them.

6.1.5.1. Distance constraint. The distance between points p_1 and p_2 should be within a given range:

$$d_{min} < \text{dist}(p_1, p_2) < d_{max}.$$

In our experiments we used $d_{min} = 2.5\,\text{Å}$ and $d_{max} = 9.0\,\text{Å}$, for the non-immunoglobulins. For the immunoglobulins we used $d_{min} = 6.0\,\text{Å}$ and $d_{max} = 13.0\,\text{Å}$.

6.1.5.2. Directional constraints. The orientation of the normals n_1 and n_2 should satisfy the following.
6.1.5.2.1. Angle between normals. The angle formed between n_1 and n_2 should be below a given threshold:

$$\text{acos}(n_1 \cdot n_2) < 1.2\,\text{rad}.$$

6.1.5.2.2. Torsion angle. The torsion angle formed between n_1, p_1, p_2 and n_2 should be smaller than a given threshold. This torsion angle is the angle between two planes. The first is defined by the line segment connecting p_1 and p_2 and by the normal n_1 at point p_1. The second plane is defined by the same line segment and the normal n_2 at point p_2.

$$\text{Torsion}(n_1, p_1, p_2, n_2) < 1.4\,\text{rad}.$$

6.1.5.2.3. Angle between each normal and the line segment p_1, p_2. Let x be the unit vector in the direction of $p_2 - p_1$. Then the normals should satisfy:

$$0.87\,\text{rad} < \text{acos}(n_1 \cdot x), \quad \text{acos}(n_2 \cdot x) < 2.4\,\text{rad}, \quad |\text{acos}(n_1 \cdot x) - \text{acos}(n_2 \cdot x)| < 1.2\,\text{rad}.$$

6.1.6. Choice of points belonging to a reference set

Given a reference set $rs(p_1, p_2)$, built on the critical points p_1 and p_2 (and the average of their normals) we can represent any point in 3D by its coordinates in $rs(p_1, p_2)$. However, we do not take all the points into consideration. We restrict ourselves to those points p which satisfy the constraint

$$\text{dist}(p_1, p), \text{ dist}(p_2, p) < d_{pmax}.$$

The rationale of this proximity constraint is the fact that we are looking for a binding site which is spatially localized, thus far away points are probably not relevant for the given reference set. By applying this constraint we gain in execution time and reduce the hash-table occupancy. Here we used $d_{pmax} = 15.0\,\text{Å}$ for the non-immunoglobulins. For the immunoglobulins we used $d_{pmax} = 18.0\,\text{Å}$.

6.1.7. Voting constraints

Given a receptor reference set $rs(r_1, r_2)$ and a receptor point r which satisfy all the above constraints, the coordinates of r in $rs(r_1, r_2)$ are computed. Let us denote these coordinates as the triplet (x, y, z).

The hash-table is accessed to extract all the records stored at addresses of the form (a, b, c), where (a, b, c) is in the range $(x \pm \epsilon, \ y \pm \epsilon, \ z \pm \epsilon)$. We used a value of $\epsilon = 1.5$. Each of these hash-table triplets (a, b, c) represents the coordinates of a ligand point in a given ligand reference set. Consider a record stored at an address with index (a, b, c). This record was computed for a ligand point l based on a ligand reference set $rs(l_1, l_2)$. It contributes one vote to the match between the current receptor reference set $rs(r_1, r_2)$ and the ligand reference set $rs(l_1, l_2)$, since the coordinates of r in $rs(r_1, r_2)$ are similar to those of l in $rs(l_1, l_2)$.

Before a vote is tallied for such a match we verify that appropriate distance and angular constraints among the participating triplets are met. For details see ref. [9].

6.1.8. Aligning the receptor with the ligand

After the completion of the voting for a given receptor reference set $rs(r_1, r_2)$, all the ligand reference sets $rs(l_1, l_2)$ which have received a number of votes above a certain threshold are retrieved. In our examples this vote threshold is between 7 and 13. For each such $rs(l_1, l_2)$ with a high vote, we compute the rotation and translation of the ligand which corresponds to this vote. In the voting process all the pairs of matching points (r, l) were recorded, and thus, these pairs can be used along with the reference set points (r_1, l_1) and (r_2, l_2) in the computation of the transformation giving the minimal least squares error [28].

6.1.9. Improvement of the transformation

Given a list of matching pairs of atoms and the transformation that best superimposes the matching ligand and receptor points, we verify and "optimize" this match hypothesis. This is done by computing the contact score in a way which is similar to the one described in section 5. Also, the set of corresponding points is verified to satisfy somewhat tighter distance and angular constraints (see ref. [9] for additional details).

6.1.10. Verification of penetration and contact scoring

Finally, the solutions are checked for inter-penetration, and for those passing that filter the contact score is computed. These steps are in essence similar to the same steps described in section 5. The final score indicates the geometric quality of the contact interface among the molecules.

6.2. Experimental results

We have carried out dozens of docking experiments (see ref. [9] for listings of results). Some of the results we present in Table 8. The table also lists the number of surface atoms and the number of interest points of the receptor and the ligand. The number of the receptor active site interest points appear next to the receptor's interest points.

As in the results presented in section 5 our experiments are divided into two classes – bound and unbound. For the first class of bound examples we have examined ligand–receptor complexes appearing in the PDB. We have randomly rotated and translated the ligand, and applied our algorithm to the transformed molecules. The performance of our method is assessed by comparing the proposed docking orientations with the original orientation. To measure the accuracy of a solution, we compute the root-mean-square (rms) distance between the ligand atomic coordinates in their original orientation and the coordinates of the same atoms in the orientation suggested by our algorithm. Commonly an rms below 2 Å is considered to be a successful docking orientation.

The second class of experiments involves docking examples of unbound molecules, where the receptor and ligand were independently crystallized. In these experiments, the correct docking orientation is not known. However, the unbound examples presented here use unbound ligands and unbound receptors that are similar to other ligands and receptors

Table 8
The molecules and their interest points

Complex	Receptor	Ligand	Receptor		Ligand	
			Atoms	Points	Atoms	Points
1cpk	protein kinase	prot. kin. inhibitor	939	544/74	115	119
3dfr	dihydrofolate reductase	methotrexate	576	596/60	214	43
4cpa	carboxypeptidase Aα	inhibitor	761	436/28	166	148
4sgb	serine protease B	potato inhibitor PCI-1	477	296/68	214	163
1cho	α-chymotripsin	ovomucoid 3rd domain	620	449/69	226	167
2mbh	hemoglobin β-sub.	hemoglobin α-sub.	469	273/58	419	249
1fdl	IgG1 Fab fragment	lysozyme	646	53	421	439
Unbound						
	5choe	2ovo	636	449/69	227	167
	2hfl	1lyz	616	62	408	452

Table 9
Results of the docking algorithm for the full ligand and receptor surfaces

Receptor	Ligand	Best rms	Number of solutions	Best rank	Time full (min)
1cpke	1cpki	1.20	37	1	19.6
3dfr	3dfrm	0.65	1140	7	10.5
4cpa	4cpai	1.51	3249	147	16.3
4sgbe	4sgbi	1.09	827	6	52.0
1choe	1choi	0.80	1721	6	27.3
2mhbb	2mhba	0.79	32	1	4.6
1fdll	1fdly	1.97	285	22	9.8
Unbound					
5choe	2ovo	1.32	8490	214	29.7
2hfll	1lyz	1.28	8957	1497	55.8

that have also been crystallized as complexes. These complexes serve as a reference for the expected docking orientation. To assess the performance of our docking method, we compute the rms distance of the unbound ligand in the proposed solution to the ligand in the reference orientation. Here again, rms below 2 Å is considered a correct solution.

The first 7 examples in the tables belong to the first class, while the last two examples are of "unbound" docking. In general, one is interested in docking a ligand onto a receptor for which neither a complex exists nor a reference complex can be found. In these cases, the rms criterion cannot be used to select the correct solution.

Table 9 lists the results obtained in the docking experiments. The rms of the highest ranking solution, which is still below 1.5 Å (2 Å for immunoglobulins) is listed in the first column. The second column lists the number of solutions which scored above the minimal number of votes in the Geometric Hashing stage and have no significant penetration of the ligand into the receptor. These are subsequently ranked by the surface contact score. The third column shows the rank of the solution whose rms is listed in the first column. Table 9 demonstrates that the docked solution that is closest to the crystallographically determined one ranks at the very top for the complexes. The rms in all cases is below 2.0 Å. The times required to complete the docking runs on an SGI Indigo 2 workstation are listed in the next column. The experiments were run using the full receptor surface. One should note that in this experiment we have not applied a clustering procedure on the resulting solutions. Since many candidate solutions tend to cluster, application of such a procedure would have improved the ranking of the "best" solutions.

The docking experiments were repeated using only the critical points of the receptor which belong to the interface (see Table 8). This is motivated by the fact that sometimes we can estimate the location of the active site in advance. Table 10 shows the results for this easier case.

Table 10
Results of the docking algorithm for the full ligand and active site receptor surface

Receptor	Ligand	Best rms	Number of solutions	Best rank	Time (min)
1cpke	1cpki	1.20	37	1	2.2
3dfr	3dfrm	0.65	202	3	0.7
4cpa	4cpai	1.51	730	28	2.1
4sgbe	4sgbi	1.09	645	2	4.2
1choe	1choi	0.80	389	4	4.2
2mhbb	2mhba	0.79	17	1	1.1
1fdll	1fdly	1.97	67	22	1.6
Unbound					
5choe	2ovo	1.32	1490	31	3.4
2hfll	1lyz	1.28	2635	84	3.8

7. Further developments and future tasks

We have discussed two methods for rigid molecule docking which are based only on geometric complementarity search. We have shown that rigid docking of relatively large complexes can be accomplished in short time, and that solutions close to the native score high enough to allow their detection. We have discussed major issues in geometric pattern matching, such as the molecular surface representation, extraction of interest features and efficient matching algorithms. It should be clear that these are not independent issues. Especially the interest features extracted and the matching algorithms used may be tightly coupled in the sense that there are certain feature sets which can trigger the development of one class of algorithms, while for another algorithm class, suitable features may be detected and developed. We have explained the analogy between object recognition tasks in Computer Vision and prediction of biomolecular docking. We believe that this analogy can help to better understand the geometric shape complementarity issues involved in both disciplines.

In our examples the performance on bound cases was truly satisfactory. Although we achieved reasonable results for the unbound case as well, it is obvious that much work has to be done before the performance on these cases is satisfactory. This is mainly due to the fact that in many cases the rigid approximation of the molecule is not accurate enough for docking purposes. Even if the molecules do not undergo major conformational changes, their surfaces often adjust to each other in the docking process thus departing from their rigid shape.

Another major challenge is dealing with major conformational changes in the ligand or receptor. Several efforts for such flexible docking algorithms are currently in progress and the major ones have been mentioned in the Introduction. The challenge there is to develop efficient algorithms, which can effectively search the much bigger parameter space without increasing significantly the complexity with regard to the rigid docking algorithms.

Acknowledgments

We thank our students and colleagues D. Fischer, S.L. Lin, and R. Norel, who have contributed to this research effort and implemented the algorithms discussed in this chapter. The research of R. Nussinov has been sponsored by the National Cancer Institute, DHHS, under Contract No. 1-CO-74102 with SAIC. The contents of this publication do not necessarily reflect the views or policies of the DHHS, nor does mention of trade names, commercial products, or organization imply endorsement by the US Government. The research of H.J. Wolfson and R. Nussinov in Israel has been supported in part by grant No. 95-0028 from the US–Israel Binational Science Foundation (BSF), Jerusalem, Israel, by the Rekanati Foundation, by the Basic Research Fund of Tel Aviv University, and by consortium "Daat" of the Israeli Ministry of Trade and Industry. H.J. Wolfson acknowledges the support of the Hermann Minkowski – Minerva Center for Geometry at Tel Aviv University.

References

[1] Janin, J. and Chothia, C. (1990) J. Mol. Biol. 265, 16027–16030.
[2] Cherfils, J. and Janin, J. (1993) Curr. Opin. Struct. Biol. 3, 265–269.
[3] Kuntz, I.D., Blaney, J.M., Oatley, S.J., Langridge, R. and Ferrin, T.E. (1982) J. Mol. Biol. 161, 269–288.
[4] Shoichet, B.K., Bodian, D.L. and Kuntz, I.D. (1992) J. Comput. Chem. 13, 380–397.
[5] Connolly, M.L. (1986) Biopolymers 25, 1229–1247.
[6] Jiang, F. and Kim, S.H. (1991) J. Mol. Biol. 219, 79–102.
[7] Katchalski-Katzir, E., Shariv, I., Eisenstein, M., Friesem, A.A., Aflalo, C. and Vakser, I.A. (1992) Proc. Natl. Acad. Sci. USA 89, 2195–2199.
[8] Norel, R., Fischer, D., Wolfson, H.J. and Nussinov, R. (1994) Protein Eng. 7, 39–46.
[9] Fischer, D., Lin, S.L., Wolfson, H.J. and Nussinov, R. (1995) J. Mol. Biol. 248, 459–477.
[10] Norel, R., Lin, S.L., Wolfson, H.J. and Nussinov, R. (1994) Biopolymers 34, 933–940.
[11] Norel, R., Lin, S.L., Wolfson, H.J. and Nussinov, R. (1995) J. Mol. Biol. 252, 263–273.
[12] DesJarlais, R.L., Sheridan, R.P., Dixon, J.S., Kuntz, I.D. and Venkataraghavan, R. (1986) J. Med. Chem. 29, 2149–2153.
[13] Leach, A.R. and Kuntz, I.D. (1992) J. Comput. Chem. 13(6), 730–748.
[14] Rarey, M., Kramer, B., Lengauer, T. and Klebe, G. (1996) J. Mol. Biol. 261, 470–489.
[15] Jones, G., Willet, P. and Glen, R. (1995) J. Mol. Biol. 245, 43–53.
[16] Sandak, B., Nussinov, R. and Wolfson, H.J. (1995) Comput. Appl. Biosci. 11, 87–99.
[17] Sandak, B., Nussinov, R. and Wolfson, H.J. (1996) Docking of conformationally flexible molecules. In: 7th Symp. on Combinatorial Pattern Matching, CPM 96, Lecture Notes in Computer Science, Vol. 1075. Springer, Berlin, pp. 271–287.
[18] Lengauer, T. and Rarey, M. (1996) Curr. Opin. Struct. Biol. 6, 402–406.
[19] Lamdan, Y. and Wolfson, H.J. (1988) Geometric hashing: a general and efficient model-based recognition scheme. In: Proc. IEEE Int. Conf. on Computer Vision, Tampa, FL. IEEE Computer Society Press, Los Alamitos, CA, pp. 238–249.
[20] Lamdan, Y., Schwartz, J.T. and Wolfson, H.J. (1990) IEEE Trans. Robotics Automation 6(5), 578–589.
[21] Nussinov, R. and Wolfson, H.J. (1991) Proc. Natl. Acad. Sci. USA 88, 10495–10499.
[22] Richards, F.M. (1977) Ann. Rev. Biophys. Bioeng. 6, 151–176.
[23] Connolly, M.L. (1983) Science 221, 709–713.
[24] Connolly, M.L. (1983) J. Appl. Cryst. 16, 548–558.
[25] Lin, S.L., Nussinov, R., Fischer, D. and Wolfson, H.J. (1994) Prot. Struct. Funct. Genet. 18, 94–101.
[26] Lin, S.L. and Nussinov, R. (1996) J. Mol. Graphics 14, 78–90.

[27] Lin, S.L., Xu, D., Li, A., Rosen, M., Wolfson, H.J. and Nussinov, R. (1997) J. Mol. Biol. 271(5), 838–8452.
[28] Schwartz, J.T. and Sharir, M. (1987) Int. J. Robotics Res. 6(2), 29–44.
[29] Wolfson, H.J. (1990) Model based object recognition by 'geometric hashing'. In: Proc. Eur. Conf. on Computer Vision, Lecture Notes in Computer Science, Vol. 427. Springer, Berlin, pp. 526–536
[30] Bachar, O., Fischer, D., Nussinov, R. and Wolfson, H.J. (1993) Protein Eng. 6(3), 279–288.
[31] Fischer, D., Norel, R., Wolfson, H.J. and Nussinov, R. (1993) Prot. Struct. Funct. Genet. 16, 278–292.

S.L. Salzberg, D.B. Searls, S. Kasif (Eds.), *Computational Methods in Molecular Biology*

CHAPTER 15

Modeling biological data and structure with probabilistic networks[*]

Simon Kasif[1] and Arthur L. Delcher[2]

[1]*Department of Electrical Engineering and Computer Science,
University of Illinois at Chicago, Chicago, IL 60607-7053, USA;
Phone: +1 312-355-0441; Fax: +1 312-413-0024; Email: kasif@eecs.uic.edu;*
[2]*Computer Science Department, Loyola College in Maryland, Baltimore, MD 21210, USA;
Phone: +1 410-617-2740; Fax: +1 410-617-2157; Email: delcher@loyola.edu*

1. Introduction

In this chapter we review the use of graphical probability models (Bayes' networks) for modeling biological data such as proteins and DNA sequences. Bayes' networks (or probabilistic networks) are graphical models of probability distributions that can provide a convenient medium for scientists to:

- Express their knowledge of the domain as a collection of probabilistic or causal rules that describe biological sequences. These rules capture our intuition about the formation of biological sequences.
- Learn the exact structure of the network and the associated probabilities that govern the probabilistic processes generating the sequences by directly observing data.

In particular, causal and statistical independence assumptions are made explicit in these networks thereby allowing biologists to express and study different generative models. Consequently, probabilistic networks provide a convenient medium for scientists to experiment with different empirical models and obtain potentially important insights into the problems being studied.

In general, probabilistic graphical models provide support for the following two important capabilities.

- *Learning,* with a framework for specifying the important probabilistic dependencies that we want to capture in the data. Specifically, in pure graphical models we typically record (learn) tables of conditional probabilities between different variables as specified by the modeler. In particular, the graph structure allows us to reduce the sample complexity (size of biological databases) required to learn biological structures.
- *Inference,* with a computational framework for combining these conditional probabilities using algorithms that take advantage of the graphical properties of the model to simplify and speed-up computation.

Models generated by probabilistic methods have a precise, quantitative semantics that is often easy to interpret and translate to biological rules. In particular, the causal connections in probabilistic network models allow biologists to postulate generative

[*] Research supported by NSF Grants IRI-9529227 and IRI-9616254. Both authors also hold appointments with the Department of Computer Science, Johns Hopkins University, Baltimore, MD 21218, USA.

processes that describe biological structures such as helical regions in proteins or coding regions in DNA. In addition, the probabilistic approach we describe here allows us to perform "virtual experiments" where we can examine (probabilistically) the effect of biological changes on structure. In particular our approach presents the opportunity to perform simulated (virtual) mutagenesis where we perform local substitutions (mutations), and measure (probabilistically) their effect on the biological structure.

2. Probabilistic networks

We assume that the reader has some familiarity with the basics of probability theory and, in particular, Bayes' Law. We refer the reader to Feller[1] for a more thorough treatment of the concepts that we briefly introduce here in an intuitive fashion. We will trade formality for what we hope is a casual and elementary discussion that is accessible to scientists without formal training in probability theory. We note, however, that the material presented here has a fully grounded mathematical basis and the interested reader is referred to Feller[1] for further details.

We first remind the reader that $P(X, Y)$ denotes the joint probability distribution of two random variables X and Y, and that $P(X|Y)$ denotes the conditional probability of X given Y defined as

$$P(X|Y) = \frac{P(X, Y)}{P(Y)} \quad \text{or equivalently} \quad P(X, Y) = P(X|Y)P(Y).$$

This latter expression generalizes to more than two random variables as

$$P(X_1, X_2, \ldots, X_n) = P(X_1|X_2, \ldots, X_n)P(X_2|X_3, \ldots, X_n) \cdots P(X_{n-1}|X_n)P(X_n).$$

From these equations it is easy to derive Bayes' Law:

$$P(X|Y) = \frac{P(Y|X)P(X)}{P(Y)}.$$

Further recall that X and Y are *independent* iff

$$P(X|Y) = P(X) \quad \text{or equivalently} \quad P(X, Y) = P(X)P(Y),$$

and X and Y are *conditionally independent* given Z iff

$$P(X|Y, Z) = P(X|Z) \quad \text{or equivalently} \quad P(X, Y|Z) = P(X|Z)P(Y|Z).$$

The notion of independence can be generalized to sets of random variables in many ways. Let us consider a few illustrative examples to provide some insight into the framework of probabilistic networks. Consider a sequence of random variables $X_1, X_2, \ldots, X_n$. Let us

Fig. 1. Chain model.

assume that each X_i is conditionally independent of $X_1, X_2, \ldots, X_{i-2}$ given X_{i-1}. In other words we are assuming that

$$P(X_i|X_1, \ldots, X_{i-1}) = P(X_i|X_{i-1}).$$

This is often referred to as the Markov assumption, namely, that the future is independent of the entire past given the present event. More importantly this means that we can predict the future event X_{i+1} knowing only the last event in the series X_i.

This also means that the joint probability distribution of the variables $X_1, \ldots, X_n$ has a particularly simple form:

$$P(X_1, X_2, \ldots, X_n) = P(X_1|X_2) P(X_2|X_3) \cdots P(X_{n-1}|X_n) P(X_n).$$

To express these dependencies graphically, we use a chain graph as shown in Fig. 1. By a visual inspection of the chain graph we can observe immediately that every variable X_i in the chain graph separates the chain into two parts, "left" and "right". In particular, every variable X_i separates variable X_{i+1} from variables $X_1, \ldots, X_{i-1}$. This makes it easy to "visualize" the independence assumption expressed by the network.

In this simple chain graph we associate a matrix of conditional probabilities with each edge of the graph. In particular the edge between X_i and X_{i+1} is associated with the matrix of conditional probabilities $P(X_{i+1}|X_i)$. The columns of this matrix correspond to the different values of the random variable X_{i+1} and the rows to the values of X_i. Before we proceed further to discuss more general Bayes' networks, let us consider a simple biological application.

3. Using chain graphs in biology

Consider the following simple problem: we are given a collection of homologous protein sequences $\{D_i\}$ stored in a database $\mathcal{D}$. For simplicity, let us assume the protein sequences are of the same length L (we can easily remove this assumption). We now would like to answer the following question: given a new protein sequence of length L, what is the likelihood that it is a member of the family of sequences stored in $\mathcal{D}$? The standard approach to this problem is to devise a simple probabilistic network model of the sequences in $\mathcal{D}$.

We first associate a random variable X_i with each position $1 \leqslant i \leqslant L$, where the values of X_i range over the alphabet of amino acids. As a first step, we assume the model has the form given in Fig. 2. The distribution of X_i in this case is simply intended to model the

Fig. 2. Simplest possible network assuming independence of the random variables.

relative frequency of each amino acid in position i independently of other positions. With each variable X_i we associate a vector of probabilities (the probability distribution of X_i). By our independence assumption, the joint probability of the variables $X_1, \ldots, X_L$ is defined by

$$P(X_1, \ldots, X_L) = P(X_1) P(X_2) \cdots P(X_L) = \prod_{i=1}^{L} P(X_i).$$

Note that this model is a standard one in biology and is typically expressed as a "consensus matrix" [2].

Let us now discuss the generative interpretation of this simple model. Since the variables are independent we will consider the following *generative process* to create a database ("virtual world") of protein sequences using the model.

(1) For $i = 1$ to L. Using the probability distribution associated with the ith variable generate a random value for it.
(2) Repeat step 1 forever, generating many new sequences.

Now we must ask the question of what is the best network in this very simple class of networks that models the sequences in our database as closely as possible. This is commonly referred to as the *Learning Problem* for Bayes' networks. More formally we want to find a probabilistic model M such that $P(M|\mathcal{D})$ is maximized. Alternatively, (using Bayes' Law) we want to maximize

$$\frac{P(\mathcal{D}|M) P(M)}{P(\mathcal{D})}.$$

Assuming that each model is equally likely this is equivalent to finding a model that maximizes $P(\mathcal{D}|M)$. Again assuming that the sequences are generated independently we can maximize the product of $P(D_i|M)$ where D_i is the ith sequence. Since the model is of the form

$$P(X_1, \ldots, X_L) = P(X_1) P(X_2) \cdots P(X_L)$$

the reader can verify by simple differentiation that the model that maximizes the probability of the database $\mathcal{D}$ (subject to the above assumptions) is given by recording the individual frequencies of each amino acid at each position. That is, learning the best model that generates the data is done simply by recording the relative frequency of each amino acid in each position.

Now given a new protein sequence $y = y_1, y_2, \ldots, y_L$ we evaluate its probability of being generated by the model as

$$P(y|M) = P(y_1) P(y_2) \cdots P(y_L) = \prod_{i=1}^{L} P(y_i).$$

To score the new sequence we typically compute the ratio of $P(y|M)/P_r(y)$, where P_r is the probability y has been generated "at random", i.e., by some model to generate all possible sequences. We then compare this ratio to some carefully chosen threshold. This is exactly what the consensus matrix approach for protein modeling achieves. Note that the network that is comprised of L unconnected nodes is equivalent to a 0th-order Markov chain.

Notice that this model is described using a relatively small number of parameters that are necessary to describe the local probability distribution of a segment of a biological structure. Typically, as the models become more complex the number of parameters required for learning the model increases. However, the simple learning methodology generalizes to more complex probabilistic networks. As a result we obtain a very useful toolkit for experimenting with different models of biological structure in an easy fashion (a key point of this chapter). In fact, this learning methodology is valid for arbitrary fixed(predetermined)-structure Bayes' networks without hidden variables.

Let us now examine a slight generalization of the above models, namely a chain network for protein modeling which is equivalent to a first-order Markov model (see Fig. 1). Again we describe the model as an ordered collection of nodes each associated with random variable X_i. We draw an edge from node X_{i-1} to node X_i. With each such edge we associate a matrix of conditional probabilities of the form $P(X_i|X_{i-1})$. Note that the number of parameters has increased compared to the zeroth-order Markov chain. For the protein problem presented here, we simply need to "learn" a collection of conditional probabilities of the form $P(X_i|X_{i-1})$. This is done simply by recording frequencies of amino-acid pairs for each position in the protein sequence.

With this model, given a new protein sequence $y = y_1, y_2, \ldots, y_L$ we evaluate the probability that y was generated by the model as the product

$$P(y_1|y_2) \cdots P(y_{L-1}|y_L)P(y_L).$$

This generalizes the consensus matrix approach for biological modeling. This suggestion was followed by Salzberg[3] and produced a statistically significant improvement on the false positive rate of detecting donor/acceptor sites in DNA sequences.

4. Modeling hidden state systems with probabilistic networks

Now consider a very simple hidden Markov model (HMM) for protein sequence modeling (see chapter 4 of the present volume for more details on hidden Markov models). That is, we assume each amino acid in the protein can be in one of three states: α-helix, β-sheet, or coil. With each state s_i we associate a set of transition probabilities. In particular, we use $A_{i,j}$ to denote the probability of moving from state s_i to state s_j. With each state we also associate a matrix of "output" probabilities. That is, in each state we "output" an amino acid according to some probability distribution. Specifically, we let $B_{i,k}$ be the probability of generating the kth amino acid in state s_i. Figure 3 illustrates this model.

This formulation naturally leads to a generative model of proteins. We simply use the transition probabilities and the output probabilities to create protein sequences according to the following algorithm. Let L be the length of the protein sequence.
(1) Set the current state $S = \text{coil}$.

340

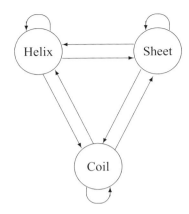

Fig. 3. Simple HMM for generating protein sequences.

(2) Output an amino acid according to the output probability distribution of S, $B_{S,k}$.
(3) Transition to a new state S according to the probability distribution associated with the current state S, $A_{S,j}$.
(4) Repeat steps 2–3 $L-1$ more times.
We can repeat the entire algorithm forever to generate arbitrarily many sequences.

   Let us now consider a probabilistic network analog of this model. First we will fix the length of the sequences we are considering (L). Then we associate a random variable PS_i with the state in position i. Then we associate a random variable E_i with the amino acid "generated" in position i. The network is depicted in Fig. 4. The reader can easily verify that the network represents exactly the same probabilistic model as given by the HMM, provided we are modeling sequences of length L and less. In fact, every HMM has an equivalent probabilistic network that is a two-layer tree network. The model here is similar to the network used in some of the experiments described by Delcher et al. [4].

   Note that during learning all the variables are fully observable. During inference, however, (e.g., when predicting the likely structure of a protein) the values of the state variables are hidden. They indeed correspond to the hidden states of a hidden Markov

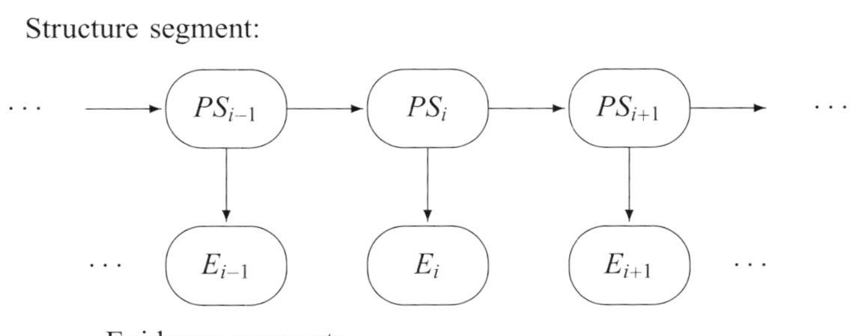

Structure segment:

$\cdots \longrightarrow PS_{i-1} \longrightarrow PS_i \longrightarrow PS_{i+1} \longrightarrow \cdots$

$\cdots \quad E_{i-1} \quad E_i \quad E_{i+1} \quad \cdots$

Evidence segment:

Fig. 4. Causal tree model for protein structure.

model. In more complex models the state variables might not be observable even during learning the network from data (see section 6).

5. *Tree networks*

Let us now generalize further beyond simple Markov models and consider probability distributions in the form of general trees. We use the example of a causal tree network defined on seven binary random variables as shown in Fig. 5. This model implies the following:

- The joint probability distribution on seven variables can be "factored" as a product of seven terms:

$$P(X_0, X_1, X_2, X_3, X_4, X_5, X_6)$$
$$= P(X_0)\, P(X_1|X_0)\, P(X_2|X_0)\, P(X_3|X_1)\, P(X_4|X_1)\, P(X_5|X_2)\, P(X_6|X_2).$$

- Each internal node X_i separates variables above it from variables below it. For example,

$$P(X_2|X_0, X_1, X_3, X_4) = P(X_2|X_0).$$

It is useful to review the generative process associated with the tree network above. We first generate a random value according to the probability distribution of the root variable X_0. Given that value we can now rely on the conditional probabilities $P(X_1|X_0)$ and $P(X_2|X_0)$ to generate values for X_1 and X_2, respectively, and recursively for the remaining variables in the network.

There are three very important properties of tree-probability distributions that are useful for biological modeling.

- It is possible to learn the best tree network that has a specific predetermined structure and maximizes the probability of the data. All we need to do is to learn (estimate) frequencies of the form $P(X_i|X_j)$ where X_j is the parent of X_i in the tree.

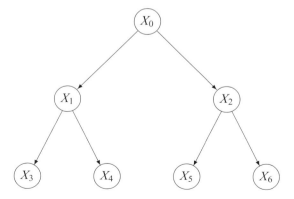

Fig. 5. Tree model.

- It is possible to learn the best probability distribution in the form of any tree that models a collection of biological sequences. This is done using a very efficient algorithm that first estimates the mutual information between any pair of variables in the model, and then constructs the maximal spanning tree using the pairwise mutual information as edge-weights (see Pearl [5] for more details).
- It is possible to perform probabilistic inference in tree networks using simple and efficient algorithms (see next section).

Note that trees allow us to define more flexible generative models for proteins than standard Markov processes. In particular, since a Markov process is a special case of a tree we can ask the question what is the best tree model that fits the data. We refer the reader to a useful application of probabilistic tree models for donor/acceptor sites in DNA sequences by Delcher et al. [6].

6. Probabilistic networks: learning and inference

As mentioned above there are two basic computational processes that must be performed in probabilistic networks: learning and inference. Learning involves the estimation of network parameters or structure. For fixed-structure networks without hidden variables these parameters are conditional probabilities which can be estimated from observed frequencies in a data set, as illustrated for chain and tree networks above.

There are several inference processes associated with networks.

- Compute the probability of all N variables in the network each taking a particular value. In other words we compute a particular event in the joint probability distribution that the network models. This computation is typically very efficient and simply involves the computation of a product of terms of the form

$$P(X = x | Z_1 = z_1, \ldots, Z_k = z_k),$$

where x and z_i are the values taken by node X and its parents Z_i, respectively. One example of this process is illustrated above in section 3 when we compute the probability that a particular protein is produced by a chain model. If we assume the tree model for proteins in Fig. 5, then to compute the probability that a particular protein is generated by the tree, we just compute a product of conditional probabilities:

$$P(x_0) P(x_1 | x_0) P(x_2 | x_0) P(x_3 | x_1) P(x_4 | x_1) P(x_5 | x_2) P(x_6 | x_2),$$

where x_i is the particular amino acid found in position X_i.

- Compute the conditional probability distribution of a single random variable given evidence in the form of specific values associated with other variables. This allows us to formulate and compute answers to biologically interesting queries. For example, given a family of proteins we can ask and obtain answers to a query about the likelihood of a particular amino acid in position seven given that positions two, three and four contain a particular protein structure.

- Compute the most likely joint set of values $\{z_i\}$ for a given set of variables $\{Z_i\}$ given evidence in the form of specific values associated with other variables. (See the application of this in section 7.)

We will not describe the actual computation of probabilistic inference in general Bayes' networks here. The reader is referred to Pearl [5] for a complete description of these procedures in networks. In particular, the tree networks above are generalized to general graphical representations of probability networks in the form of directed acyclic graphs (DAGs). We just note that the problem of computing the probability distribution of a single random variable given a collection of others in Bayes' networks is computationally intractable (NP-hard), and we often do not have accurate estimates of all the probabilities. However, it is known that when the structure of the network has a special form (e.g., a tree) it is possible to perform a complete probabilistic analysis efficiently.

In the following sections we describe the application of tree networks to the problem of performing secondary structure prediction in protein sequences. This description is given primarily for illustrative purposes in order to motivate additional applications of graphical models in computational biology.

7. Application of probabilistic networks for protein secondary structure prediction

In this section we discuss several simple experiments with probabilistic networks for predicting the secondary structure of proteins. A number of methods have been applied to this problem with various degrees of success [7–14]. In addition to obtaining experimental results comparable to other methods, there are several theoretically and practically interesting observations that we have made in experimenting with our systems. The most important aspect of this approach is that the results obtained have a precise probabilistic semantics. Conditional independence assumptions are represented explicitly in our networks by means of causal links, and we can associate a generative explanation of the underlying model.

The majority of previously exploited methods for protein folding were based on a windowing approach. That is, the learning algorithm attempted to predict the structure of the central amino acid in a "window" of k amino-acid residues. It is well recognized that in the context of protein folding, very minimal mutations (amino-acid substitutions) often cause significant changes in the secondary structure located far from the mutation site. Our approach is aimed at capturing this behavior.

In the next few sections we describe initial experiments, for which we have chosen the simplest possible models. We first describe a causal-tree model using Pearl's belief updating [5]. Then we describe the application of the Viterbi algorithm to this model. We then illustrate the utility of probabilistic models in the context of modeling the effect of mutations on secondary structure. Finally, we describe an application of hidden Markov models to modeling protein segments with uniform secondary structure (i.e., runs of helices, sheets or coils).

In what follows the set of proteins is assumed to be a set of sequences (strings) over an alphabet of twenty characters (different capital letters) that correspond to different amino acids. With each protein sequence of length n we associate a sequence of secondary structure descriptors of the same length. The structure descriptors take three values, h, e and c, that correspond to α-helix, β-sheet and coil. That is, if we have a subsequence of $hh \ldots h$ in positions $i, i + 1, \ldots, i + k$ it is assumed that the protein sequence in those positions folded as a helix. Many previous approaches to secondary structure prediction in proteins (e.g., using neural networks or memory-based approaches) have used the following method: the classifier receives a window of length $2K + 1$ (typically $K < 12$) of amino acids. The classifier then predicts the secondary structure of the central amino acid (i.e., the amino acid in position K) in the window. The prediction problem in this paper is stated as follows. Given a protein sequence of length n, generate a sequence of structure predictions of length n which describes the corresponding secondary structure of the protein.

8. A probabilistic framework for protein analysis

In this section we demonstrate one approach for modeling the secondary structure of protein sequences using tree networks. The tree networks we describe correspond to hidden Markov models (HMMs) described in chapter 4 of the present volume.

The general schema we advocate has the following form. The set of nodes in the networks are either protein-structure nodes (PS nodes) or evidence nodes (E nodes). Each PS node in the network is a discrete random variable X_i that can take values which correspond to descriptors of secondary structure, i.e., segments of hs, es and cs. With each such node we associate an evidence node that again can assume any of a set of discrete values. Typically, an evidence node would correspond to an occurrence of a particular subsequence of amino acids at a particular location in the protein. With each edge in the network we will associate a matrix of conditional probabilities. The simplest possible example of a network is given in Fig. 4 (above).

We assume that all conditional dependencies are represented by a causal tree. This assumption violates some of our knowledge of the real-world problem, but provides an approximation that allows us to perform an efficient computation. For an exact definition of a causal tree see Pearl [5]. Causal belief networks can be considered as a generalization of classical Markov chains that have found many useful applications in modeling.

9. Protein modeling using causal networks

As mentioned above, the network is comprised of a set of protein-structure nodes and a set of evidence nodes. Protein-structure nodes are finite strings over the alphabet $\{h, e, c\}$. For example the string $hhhhhh$ is a string of six residues in an α-helical conformation, while $eecc$ is a string of two residues in a β-sheet conformation followed by two residues folded as a coil. Evidence nodes are nodes that contain information about a particular region of the protein. Thus, the main idea is to represent physical and statistical rules in the form

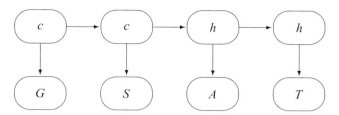

Fig. 6. Example of causal tree model showing protein segment *GSAT* with corresponding secondary structure *cchh*.

of a probabilistic network. We note that the main point of this chapter is to describe the framework of causal networks as an experimental tool for molecular biology applications rather than focusing on a particular network. The framework allows us flexibility to test causal theories by orienting edges in the causal network.

In our first set of experiments we converged on the following model that seems to match the performance of many existing approaches. The network looks like a set of *PS* nodes connected as a chain. To each such node we connect a single evidence node. In our experiments the *PS* nodes are strings of length two or three over the alphabet $\{h, e, c\}$ and the evidence nodes are strings of the same length over the set of amino acids. The following example clarifies our representation. Assume we have a string of amino acids *GSAT*. We model the string as a network comprised of four evidence nodes G, S, A, T and four *PS* nodes. The network is shown in Fig. 6. A correct prediction will assign the values c, c, h and h to the *PS* nodes as shown in the figure.

The generative model associated with this network is as follows. We assume we first generate (with the appropriate probability) the first structure node of the network. Then we independently generate the next structure node of the network and in parallel the amino acid associated with the first structure node. Biologically, it is not totally unreasonable to generate the ith structure node based on the previous structure node. However, this might be a very interesting and also controversial point of whether the structure of the ith position in the protein is independent of the actual amino-acid sequence that precedes it given the knowledge of the preceding structure.

Let $PS_0, PS_1, \ldots, PS_n$ be a set of *PS* nodes connected as in Fig. 4. The conditional distribution for the variable PS_i given values of the evidence nodes (i.e., given an amino-acid sequence) in a causal tree network can be computed using the following formulae. Let $E_{PS_i}^- = E_i, E_{i+1}, \ldots, E_n$ denote the set of evidence nodes to the right of PS_i, and let $E_{PS_i}^+ = E_1, E_2, \ldots, E_{i-1}$ be the set of evidence nodes to the left of PS_i. By the assumption of independence explicit in the network we have

$$P(PS_i|PS_{i-1}, E_{PS_i}^+) = P(PS_i|PS_{i-1}).$$

Thus,

$$P(PS_i|E_{PS_i}^+, E_{PS_i}^-) = \alpha\, P(E_{PS_i}^-|PS_i)\, P(PS_i|E_{PS_i}^+),$$

where α is a normalizing constant. For length consideration we will not describe the algorithm to compute the probabilities. The reader is referred to Pearl for a detailed

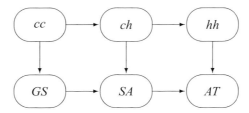

Fig. 7. Example of causal DAG model using pairs of amino acids.

description [5]. Pearl gives an efficient procedure to compute the belief distribution of every node in such a tree. Most importantly, this procedure operates by a simple efficient propagation mechanism that operates in linear time.

A more accurate model of protein structure is obtained if we create networks in which nodes represent more than a single amino-acid position in the protein. In Fig. 7 we show the model for overlapping pairs of amino acids. Note that in this model it is necessary to add connections between evidence nodes to ensure that the pairs of amino acids are consistent on the overlapped position. As a result, the model is no longer a tree, but a direct acyclic graph instead.

10. Protein modeling using the Viterbi algorithm

In this section we describe a simple scheme to compute the most likely set of values for the state nodes of the network (secondary structure) given evidence in the form of the primary structure (the amino-acid sequence). This procedure (Viterbi) has been heavily used in speech understanding systems, where it achieves remarkable performance on speaker-independent continuous speech understanding.

We have implemented the Viterbi algorithm and compared its performance to the method outlined above. The method is briefly discussed here. We follow the discussion by Forney [15].

As mentioned above, the above network models a simple probabilistic process which at time k can be described by a random variable X_k that assumes a discrete number of values (protein states) $1, \ldots, M$. The process is Markov, i.e., the probability $P(X_{k+1}|X_0, \ldots X_k) = P(X_{k+1}|X_k)$. We denote the entire process by the sequence $X = X_0, \ldots, X_n$. We are given a set of observations $Z = Z_0, \ldots, Z_n$ such that Z_i depends only on the transition $T_i = X_i \rightarrow X_{i+1}$. Specifically,

$$P(Z|X) = \prod_{k=0}^{n-1} (Z_k|T_k).$$

Given the amino-acid sequence Z, the Viterbi algorithm computes the sequence of values for the state variables X that maximizes $P(Z|X)$.

An intuitive way to understand the problem is in graph theoretic terms. We build an n-level graph that contains nM nodes (see Fig. 8). With each transition we associate an

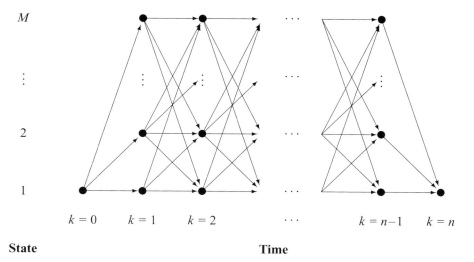

$$M$$

$$\vdots$$

$$2$$

$$1$$

$$k = 0 \qquad k = 1 \qquad k = 2 \qquad \cdots \qquad k = n-1 \qquad k = n$$

State **Time**

Fig. 8. Modeling the Viterbi algorithm as a shortest path problem.

edge. Thus, any sequence of states has a corresponding path in the graph. Given the set of observations Z with any path in the graph we associate a length $L = -\log P(X,Z)$. We are seeking a shortest path in the graph. However, since

$$P(X,Z) = P(X)P(Z|X)$$
$$= \left(\prod_{k=0}^{n-1} P(X_{k+1}|X_k) \right) \left(\prod_{k=0}^{n-1} P(Z_k|T_k) \right),$$

if we define $\lambda(T_k) = -\log P(X_{k+1}|X_k) - \log P(Z_k|T_k)$ we obtain that

$$-\log P(X,Z) = \sum_{k=0}^{n-1} \lambda(T_k).$$

Now we can compute the shortest path through this graph by a standard application of dynamic programming to computing the shortest path in directed acyclic graphs. For each time step i we merely maintain M paths which are the shortest paths to each of the possible states we could be in at time i. To extend the path to time step $i+1$ we simply compute the lengths of all the paths extended by one time unit and maintain the shortest path to each one of the M possible states at time $i+1$.

Our experimentation with the Viterbi algorithm was completed in Spring 1992 [16] and is described in ref. [4]. A number of other applications of Viterbi for modeling protein structure are described in the chapter by Krogh who originally proposed the application of HMMs (using Viterbi) for protein modeling. As mentioned above, the Viterbi algorithm predicts the most likely complete sequence of structure elements, whereas other causal-tree inference methods make separate predictions about individual PS nodes.

The results of our experiments are reported in ref. [4]. We reported prediction accuracies both on individual residues and on runs of helices and sheets. The most accurate model was the chain-pair model shown in Fig. 7.

11. Towards automated site-specific mutagenesis

An experiment which is commonly done in biology laboratories is a procedure where a particular site in a protein is changed (i.e., a single amino-acid mutation) and then it is tested whether the protein settles into a different conformation. In many cases, with overwhelming probability the protein does not change its secondary structure outside the mutated region. One experiment that is easy to do using our method is the following procedure. We assume the structure of a protein is known anywhere outside a window of length l, for $l = 1, 2, 3, \ldots$ and try to predict the structure inside the unknown window. Table 1 shows results from one such experiment.

Table 1

Accuracy of prediction of a subsegment of amino acids, given the correct secondary structure information for the remainder of the protein; results are averaged over all possible segments of the given length in all proteins

Length of predicted segment	Amino-acid positions predicted correctly
1	90.38%
2	87.29%
3	85.18%
4	82.99%
6	79.32%
8	76.49%
12	72.39%
16	69.85%
20	68.08%
24	66.94%

The results above are conservative estimates of the accuracy of prediction for this type of an experiment and can be easily improved. One simple explanation for this result is that the high accuracy of prediction over small windows is just a result of momentum effects and the prediction accuracy for transitions from coil-regions to helices and sheets remains low.

12. Using the EM algorithm

We now briefly mention one more set of experiments that can be performed with a probabilistic model of the type discussed in this chapter. The idea is very simple and is strongly influenced by the methodology used in speech recognition. Our goal in this

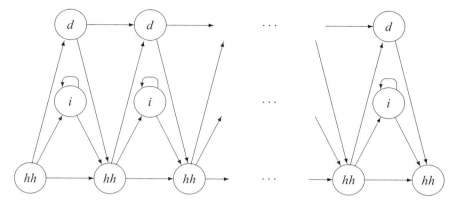

Fig. 9. Hidden Markov Model used to recognize a sequence of helices. With each edge out of a node, there is an associated probability of taking that transition together with a set of probabilities of generating each of the 20 possible amino acids while making that transition. Nodes labelled i allow for the insertion of extra amino acids for long chains. Nodes labelled d represent deletion of amino acids thereby permitting the model to generate short chains. Edges to d-nodes generate no amino acid.

experiment is to create a simple probabilistic model that recognizes runs of helices. We use the framework of the Viterbi algorithm described above. We previously defined the notion of the most likely path in the probabilistic network given all the evidence. This path can be described as a sequence of nodes (states) in the network, i.e., given a particular sequence of amino acids, we want to find a sequence of states which has the highest probability of being followed given the evidence. Alternatively, we can regard the network as a probabilistic finite state machine that generates amino-acid outputs as transitions are made.

In this experiment we would like to create the most likely model that recognizes/generates sequences of helices. Intuitively (and oversimplifying somewhat), we would like to find a network for which the probabilities of traversing a path from initial to final state given helical sequences of amino acids are greater than the probabilities for non-helical sequences. Figure 9 shows the network that we used. This particular model was inspired by the work of Haussler et al. [17,18]. (See more details in Krogh's chapter 4 of the present volume). An equivalent causal tree for this HMM is shown in Fig. 10.

Initially we assigned equal probabilities to every transition from a given node, and for each transition we set the probabilities of outputting amino acids to the relative frequencies of those amino acids in the training data. We then use the Baum–Welch method (or EM, expectation maximization) [19] to adjust the probabilities in the network to increase its probability of recognizing the input sequences.

We constructed three networks (for helix, sheet and coil runs) and trained them to recognize their respective runs. All the networks were of the form shown in the figure, but were of different lengths, corresponding to the average length of the respective type of run. The helix network had nine nodes on the bottom level, the sheet network has four, and the coil network six nodes. We then tested the networks by giving each one the same run sequence and computed its probability (using the Viterbi algorithm) of generating that sequence. Table 2 shows the relative frequency with which each of the three networks had

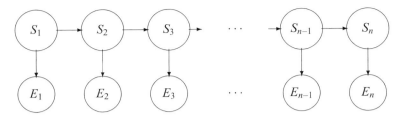

Fig. 10. Causal tree that is equivalent to the HMM in Fig. 9. Each state node is a random variable whose values are the different states in the HMM. The evidence nodes simply represent the amino acid sequence. The conditional probabilities connecting state nodes are the state transition probabilities in the HMM, modified to incorporate all paths through deletion nodes, which have no output. The conditional probabilities connecting state and evidence nodes are the probabilities of generating each amino acid on a transition out of the corresponding state.

Table 2
Relative frequencies with which HMM networks had highest probabilities of generating sequences of a particular type

Input type	Network trained to recognize:					
	Helices		Sheets		Coils	
Helix	469	(91.1%)	34	(6.6%)	12	(2.3%)
Sheet	231	(28.2%)	344	(42.0%)	244	(29.8%)
Coil	433	(33.0%)	114	(8.7%)	766	(58.3%)

the highest probability of generating each type of input sequence. The fact that helices are predicted far more accurately than sheets is in part attributable to the fact that the helix network is much larger.

By way of comparison, we used the causal tree model of Fig. 7 to predict the same segments, it predicted only about 20% of helix-run sequences correctly. This is not surprising when we consider that most of the sequence examples were coils, which strongly biased the model to predict coils.

13. Discussion

In this chapter we have described the application of probabilistic networks to modeling biological data. One of the main advantages of the probabilistic approach we described here is our ability to perform detailed experiments where we can experiment with changes to the data and its effect on structure. In particular, we can easily perform local substitutions (mutations) and measure (probabilistically) their effect on the global structure. Window-based methods do not support such experimentation as readily. Our method is efficient both during training and during prediction, which is important in order to be able to perform many experiments with different networks.

Our initial experiments have been done on the simplest possible models where we ignore many known dependencies. For example, it is known that in α-helices hydrogen bonds are formed between every ith and $(i + 4)$th residue in a chain. This can be incorporated in our model without losing efficiency. We also can improve our method by incorporating additional correlations among particular amino acids as in Gibrat et al. [20]. We achieve prediction accuracy similar to many other methods such as neural networks. Typically, the current best prediction methods involve complex hybrid methods that compute a weighted vote among several methods using a combiner that learns the weights. For example, the hybrid method described by Zhang et al. [13] combines neural networks, a statistical method and memory-based reasoning in a single system and achieves an overall accuracy of 66.4%.

We also have used a more sophisticated model based on hidden Markov models or, equivalently, on probabilistic tree models with hidden variables. Our networks are trained to recognize runs of α-helix, β-sheets and coil. Thus, the helix network is designed to generate sequences of amino acids that are likely to generate runs of helices. HMMs were originally proposed in the paper by Haussler et al. [17,18] to recognize specific families of proteins, namely globins. We independently suggested the use of probabilistic networks that are equivalent to a class of HMMs, and reported some preliminary results in using such networks for predicting secondary structure. The network that was trained to generate runs of helices did relatively well on identifying such runs during testing on new sequences of helices.

Bayesian classification is a well-studied area and has been applied frequently to many domains such as pattern recognition, speech understanding and others. Statistical methods also have been used for protein-structure prediction. What characterizes our approach is its simplicity and the explicit modeling of causal links (conditional independence assumptions). We also believe that for scientific data analysis it is particularly important to develop tools that clearly display such assumptions. We believe that probabilistic networks provide a very convenient medium for scientists to experiment with different empirical models which may yield important insights into problems.

To summarize, data mining of biological sequences is an important potential application of probabilistic networks. We believe that the ultimate data analysis system using probabilistic techniques will have a wide range of tools at its disposal and will adaptively choose various methods. It will be able to generate simulations automatically and verify the model it constructed with the data generated during these simulations. When the model does not fit the observed results the system will try to explain the source of error, conduct additional experiments, and choose a different model by modifying system parameters. If it needs user assistance, it will produce a simple low-dimensional view of the constructed model and the data. This will allow the user to guide the system toward constructing a new model and/or generating the next set of experiments. We believe that flexibility, efficiency and direct representation of causality in probabilistic networks are important and desirable features that make them very strong candidates as a framework for biological modeling systems. In recent work we have applied this methodology to problems such as biological motif finding in DNA and protein sequences and modeling splice junctions in DNA sequences.

352

References

[1] Feller, W. (1968) An Introduction to Probability Theory and Its Applications, Vol I. Wiley, New York.

[2] Mount, S., Burks, C., Hertz, G., Stormo, G., White, O. and Fields, C. (1992) Splicing signals in *Drosophila*: intron size, information content, and consensus sequences. Nucleic Acids Res. 20, 4255–4262.

[3] Salzberg, S. (1997) A method for identifying splice sites and translational start sites in eukaryotic mRNA. Comput. Appl. Biosci. (CABIOS) 13, 365–376.

[4] Delcher, A., Kasif, S., Goldberg, H. and Hsu, W. (1993) Probabilistic prediction of protein secondary structure using causal networks. In: L. Hunter, D.B. Searls and J. Shavlik (Eds.), Proc. 1993 Int. Conf. on Intelligent Systems for Computational Biology. AAAI Press, Menlo Park, CA, pp. 316–321.

[5] Pearl, J. (1988) Probabilistic Reasoning in Intelligent Systems. Morgan Kaufmann, San Francisco, CA.

[6] Cai, D., Delcher, A., Kao, B. and Kasif, S. (1998) Modeling DNA sequences with Bayes networks. In review.

[7] Chou, P. and Fasman, G. (1978) Prediction of the secondary structure of proteins from their amino acid sequence. Adv. Enzymol. 47, 45–148.

[8] Garnier, J., Osguthorpe, D. and Robson, B. (1978) Analysis of the accuracy and implication of simple methods for predicting the secondary structure of globular proteins. J. Mol. Biol. 120, 97–120.

[9] Holley, L. and Karplus, M. (1989) Protein secondary structure prediction with a neural network. Proc. Natl. Acad. Sci. USA 86, 152–156.

[10] Cost, S. and Salzberg, S. (1993) A weighted nearest neighbor algorithm for learning with symbolic features. Mach. Learning 10(1), 57–78.

[11] Qian, N. and Sejnowski, T. (1988) Predicting the secondary structure of globular proteins using neural network models. J. Mol. Biol. 202, 865–884.

[12] Maclin, R. and Shavlik, J.W. (1993) Using knowledge-based neural networks to improve algorithms: Refining the Chou–Fasman algorithm for protein folding. Mach. Learning 11, 195–215.

[13] Zhang, X., Mesirov, J. and Waltz, D. (1992) A hybrid system for protein secondary structure prediction. J. Mol. Biol. 225, 1049–1063.

[14] Muggleton, S. and King, R. (1991) Predicting protein secondary structure using inductive logic programming. Technical report, Turing Institute, University of Glasgow, Scotland.

[15] Forney, G.D. (1973) The Viterbi algorithm. Proc. IEEE 61(3), 268–278.

[16] Delcher, A., Kasif, S., Goldberg, H.R. and Hsu, B. (1992) Probabilistic prediction of protein secondary structure using causal networks. Technical report JHU/CS-92-23, Computer Science Dept., Johns Hopkins University, Baltimore, MD.

[17] Haussler, D., Krogh, A., Mian, S. and Sjolander, K. (1992) Protein modeling using Hidden Markov Models. Technical Report UCSC-CRL-92-23, University of California, Santa Cruz, CA.

[18] Krogh, A., Brown, M., Mian, S., Sjolander, K. and Haussler, D. (1994) Hidden Markov models in computational biology: Applications to protein modeling. J. Mol. Biol. 235, 1501–1531.

[19] Rabiner, L.R. (1989) A tutorial on hidden Markov models and selected applications in speech recognition. Proc. IEEE 77(2), 257–286.

[20] Gibrat, J.-F., Garnier, J. and Robson, B. (1987) Further developments in protein secondary structure prediction using information theory. J. Mol. Biol. 198, 425–443.

PART IV

Reference Materials

S.L. Salzberg, D.B. Searls, S. Kasif (Eds.), *Computational Methods in Molecular Biology*

Software and databases for computational biology on the Internet

The on-line version of this appendix, in which all the systems below are hot links, can be found at
`http://www.cs.jhu.edu/~salzberg/compbio-book.html`

1. Sequence Analysis Programs

- VEIL (the Viterbi Exon–Intron Locator) uses a custom-designed hidden Markov model (HMM) to find genes in eukaryotic DNA. The training of the current version of VEIL used Burset and Guigo's database of 570 vertebrate sequences, so VEIL will work best on sequences from vertebrates. The VEIL site is at Johns Hopkins University.
- MORGAN is an integrated system for finding genes in vertebrate DNA sequences. MORGAN uses a variety of techniques to accomplish this task, including a decision tree classifier, Markov chains to recognize splice sites, and a frame-dependent dynamic programming algorithm. MORGAN has been trained and tested primarily on vertebrate sequence data. Results showing MORGAN's accuracy and the source code to the system can be obtained from this site. The MORGAN site is at Johns Hopkins University.
- GENSCAN is a program designed to predict complete gene structures, including exons, introns, promoter and poly-adenylation signals, in genomic sequences. It differs from the majority of existing gene finding algorithms in that it allows for partial genes as well as complete genes and for the occurrence of multiple genes in a single sequence, on either or both DNA strands. Program versions suitable for vertebrate, nematode (experimental), maize and Arabidopsis sequences are currently available. The vertebrate version also works fairly well for Drosophila sequences. Sequences can be submitted on a web-based form at this site. The GENSCAN Web site is at Stanford University.
- GRAIL and GenQuest provide analysis and putative annotation of DNA sequences both interactively and through the use of automated computation. This site provides access both to GRAIL, which finds genes in eukaryotic DNA sequences, and to GENQUEST, which is a DNA sequence comparison and alignment server. These systems are at the Oak Ridge National Laboratory in Tennessee.
- The BCM Gene Finder finds splice sites, genes, promoters, and poly-A recognition regions in eukaryotic sequence data. The underlying technology uses linear discriminant analysis. You can submit sequences to BCM using a Web interface found here. The site is located is at the Baylor college of Medicine.
- The GeneID server contains the GeneID system for finding genes in eukaryotes. GeneID is a hierarchical rule-based system, with scoring matrices to identify signals and rules to score coding regions. You can use this page to submit a genomic DNA sequence to the GeneID program. The GeneID site is at Boston University.

- Genie, a gene finder based on generalized hidden Markov models, is at the Lawrence Berkley National Laboratory. It was developed in collaboration with the Computational Biology Group at the University of California, Santa Cruz. Genie uses a statistical model of genes called a Generalized Hidden Markov Model (GHMM) to find genes in vertebrate and human DNA. In a GHMM, probabilities are assigned to transitions between states and to the generation of each nucleotide base given a particular state. Machine learning techniques are applied to optimize these probabilities using a standardized gene data set, which is available on this site. The page has a link to the Genie Web server, to which sequences may be submitted.
- GeneParser identifies protein coding regions in eukaryotic DNA sequences. The home page at the University of Colorado includes various documents describing GeneParser's theory and performance as well as some sample output screens. The complete system is available here.
- GenLang is a syntactic pattern recognition system that uses the tools and techniques of computational linguistics to find genes and other higher-order features in biological sequnce data. Patterns are specified by means of rule sets called grammars, and a general purpose parser, implemented in the logic programming language Prolog, then performs the search. This system is at the University of Pennsylvania.
- THREADER2 is a program for predicting protein tertiary structure by recognizing the correct fold from a library of alternatives. Of course, if a fold similar to the native fold of the protein being predicted is not in the library, then this approach will not succeed. Fortunately, certain folds crop up time and time again, and so fold recognition methods for predicting protein structure can be very effective. In the first prediction contest held at Asilomar, organized by John Moult and colleagues, THREADER correctly identified 8 out of 11 target structures which either globally or locally resembled a previously observed fold. Preliminary analysis of the results from the second competition (CASP2) show that THREADER 2 has shown clear improvement in both fold recognition sensitivity AND sequence-structure alignment accuracy. In CASP2, the new version of THREADER recognized 4 folds correctly out of 6 targets with recognizable structures (including the difficult task of assigning a jelly-roll fold rather than other beta-sandwich topologies for one target). THREADER 2 produced more correct fold predictions (i.e. correct folds ranked at No. 1) than any other method.
- MarFinder uses statistical patterns to deduce the presence of MARs (Matrix Association Regions) in DNA sequences. MARs constitute a significant functional block and have been shown to facilitate the processes of differential gene expression and DNA replication. This tool and Web site are at the National Center for Genome Resources.
- NetPlantGene is at the Technical University of Denmark. The NetPlantGene Web server uses neural networks to predict splice sites in Arabidopsis thaliana DNA. This site also contains programs for other sequence analysis problems as well, such as the recognition of signal peptides.
- MZEF and Pombe. This page contains software tools designed to predict putative internal protein coding exons in genomic DNA sequences. Human, mouse and arabidopsis exons are predicted by a program called MZEF, and fission yeast exons are predicted by a program called Pombe. The site is located at the Cold Spring Harbor Laboratory.

- PROCRUSTES finds the multi-exon structure of a gene by aligning it with the protein databases. PROCRUSTES uses an algorithm called spliced alignment, which explores all possible exon assemblies and finds the multi-exon structure with the best fit to a related protein. If a database sequence exists that is closely similar to the query. PROCRUSTES will produce a highly accurate prediction. This program and Web page are at the University of Southern California.
- Promoter Prediction by Neural Network (NNPP) is a method that finds eukaryotic and prokaryotic promoters in a DNA sequence. The basis of the NNPP program is a time-delay neural network. The time-delay network consists mainly of two feature layers, one for recognizing the TATA-box and one for recognizing the "Initiator", which is the region spanning the transcription start site. Both feature layers are combined into one output unit, which gives output scores between 0 and 1. This site is at the Lawrence Berkley National Laboratory. Also available at this site is the splice site predictor used by the Genie system. The output of this neural network is a score between 0 and 1 indicating a potential splice site.
- Repeat Pattern Toolkit (RPT) consists of tools for analyzing repetitive sequences in a genome. RPT takes as input a single sequence in GenBank format, and attempts to find both coding (possible gene duplications, pseudogenes, homologous genes) and non-coding repeats. RPT locates all repeats using a fast Senstive Search Tool (SST). These repeats are evaluated for statistical significance utilizing a sensitive All-PAM search, and their evolutionary distance is estimated. The repeats are classified into families of similar sequences. The classification output is tabulated using perl scripts and plotted using gnuplot. RPT is at the Institute for Biomedical Computing at Washington University in St. Louis.
- SorFind, RepFind, and PromFind are at RabbitHutch Biotechnology Corporation. The three programs are currently available without charge to interested parties. SorFind (current version 2.8) identifies and annotates putative individual coding exons in genomic DNA sequence, RepFind (current version 1.7) identifies common repetitive elements, and PromFind (current version 1.1) identifies vertebrate promoter regions.
- SplicePredictor is a program designed to predict donor and acceptor splice sites in maize and Arabidopsis sequences. Sequences can be submitted on a web-based form at this site. The system is at Stanford University.
- The TIGR Software Tool Collection is at The Institute for Genomic Research. A number of software tools are freely available for download. Tools currently available include:
  - ADE (Analysis, Display, Edit Suite): a set of relational database schemas and tools for management of cDNA sequencing projects, including database searching and analysis of results
  - autoseq_tools: a set of utilities for DNA sequence analysis
  - btab: a BLAST output parser
  - Glimmer: a bacterial gene finding system (with its own separate page; see elsewhere on this page)
  - grasta: Modified FastA code that searches both strands and outputs btab format files
  - hbqcm (Hexamer Based Quality Control Method): a quality control algorithm for DNA sequencing projects

358

- TIGR Assembler: a tool for assembly of large sets of overlapping sequence data such as ESTs, BACs, or small genomes
- TIGR-MSA: a multiple sequence alignment algorithm for the MasPar massively parallel computer.

This page is at The Institute for Genomic Research in Rockville, Maryland.

- TESS (Transcription Element Search Software) is a set of software routines for locating and displaying transcription factor binding sites in DNA sequence. TESS uses the Transfac database as its store of transcription factors and their binding sites. This page is at the University of Pennsylvania's Computational Biology and Informatics Laboratory.
- WebGene (GenView, ORFGene, SpliceView) is a Web interface for several coding region recognition programs, including:
  - GenView: a system for protein-coding gene prediction
  - ORFGene: gene structure prediction using information on homologous protein sequences
  - SpliceView: prediction of splicing signals
  - HCpolya: a hamming Clustering Method for Poly-A prediction in eukaryotic genes

This page is at the Istituto Tecnologie Biomediche Avanzate in Italy.

- Glimmer is a system that uses Interpolated Markov Models (IMMs) to identify coding regions in microbial DNA. IMMs are a generalization of Markov models that allow great flexibility in the choice of the "context"; i.e., how many previous bases to use in predicting the next base. Glimmer has been tested on the complete genomes of *H. influenzae, E. coli, H. pylori, M. genitalium, A. fulgidus, B. burgdorferi, M. pneumoniae,* and other genomes, and results to date have proven it to be highly accurate. Annotation for some of these genomes, as well as the system source code, is available from this site.
- GeneMark is a system for finding genes in bacterial DNA sequences. The algorithm is based on non-homogeneous 5th-order Markov chains, and it was used to locate the genes in the complete genomes of H. influenzae, M. genitalium, and several other complete genomes. The site includes documentation and a Web interface to which sequences can be submitted. This system is at the Georgia Institute of Technology in Atlanta, GA.
- The Staden Package contains a wealth of useful programs for sequence assembly, DNA sequence comparison and analysis, protein sequence analysis, and sequencing project quality control. The site is mirrored in several locations around the world.

2. Databases

- The NCBI WWW Entrez PubMed Browser, at the National Center for Biotechnology Information (NCBI), is one of the most important resources for searching the NCBI protein, nucleotide, 3-D structures, and genomes databases. You can also browse NCBI's taxonomy and search for bibliographic entries in Entrez PubMed.
- NCBI dbEST at the National Center for Biotechnology Information is is a division of GenBank that contains sequence data and other information on "single-pass" cDNA sequences, or Expressed Sequence Tags, from a number of organisms.

- HHS Sequence Classification. HHS is a database of sequences that have been clustered based on a variety of criteria. The database and clustering algorithms are described in chapter 6. This Web page, at the Insitute for Biomedical Computing at Washington University in St. Louis, allows one to access classifications by sequence, group listing, structure, and alignment.
- The Chromosome 22 Sequence Database is at the University of Pennsylvania's Computational Biology and Informatics Laboratory. It allows queries on a wide variety of features associated with the sequence of Chromosome 22. Also on this site is a map of Chromosome 22, which allows you to search for loci and yacs.
- The EpoDB (Erythropoiesis Database) is a database of genes that relate to vertebrate red blood cells. A detailed description of EpoDB can be found in chapter 5. The database includes DNA sequence, structural features and potential transcription factor binding sites. This Web site is at the University of Pennsylvania's CBIL.
- The LENS (Linking ESTs and their associated Name Space) database links and resolves the names and identifiers of clones and ESTs generated in the I.M.A.G.E. Consortium/WashU/Merck EST project. The name space includes library and clone IDs and names from IMAGE Consortium, EST sequence IDs from Washington University, sequence entry accession numbers from dbEST/NCBI, and library and clone IDs from GDB. LENS allows for querying of IMAGE Consortium data via all the different IDs.
- PDD, the NIMH–NCI Protein–Disease Database, is at the Laboratory of Experimental and Computational Biology at the National Cancer Institute. This server is part of the NIMH-NCI Protein–Disease Database project for correlating diseases with proteins observable in serum, CSF, urine and other common human body fluids based on biomedical literature.
- The Genome Database (GDB), at the Johns Hopkins University School of Medicine, comprises descriptions of the following types of objects: regions of the human genome, including genes, clones, amplimers (PCR markers), breakpoints, cytogenetic markers, fragile sites, ESTs, syndromic regions, contigs and repeats; maps of the human genome, including cytogenetic maps, linkage maps, radiation hybrid maps, content contig maps, and integrated maps. These maps can be displayed graphically via the Web; variations within the human genome including mutations and polymorphisms, plus allele frequency data.
- The Johns Hopkins University BioInformatics Web Server. This page includes Prot-Web, a collection of protein databases, and links to other biological databases at Hopkins. It also has an excellent page of links to other biological web servers around the world.
- The TRANSFAC Database is at the Gesellschaft für Biotechnologische Forschung mbH (Germany). TRANSFAC is a transcription factor database. It compiles data about gene regulatory DNA sequences and protein factors binding to them. On this basis, programs are developed that help to identify putative promoter or enhancer structures and to suggest their features.
- TransTerm – Translational Signal Database – is a database at the University of Otago (New Zealand). TransTerm contains sequence contexts about the stop and start codons of many species found in GenBank. TransTerm also contains codon usage data for these same species and summary statistics for the sequences analysed.

3. Other software and information sources

- The Banbury Cross Site is a web page for benchmarking gene identification software. Banbury Cross is at the Centre National De La Recherche Scientifique. This Benchmark site is intended to be a forum for scientists working in the field of gene identification and anonymous genomic sequence annotation, with the goal of improving current methods in the context of very large (in particular) vertebrate genomic sequences.
- CBIL bioWidgets, at the University of Pennsylvania, is a collection of software libraries used for rapid development of graphical molecular biological applications. It includes:
  - bioWidgets for Java^{TM}, a toolkit of biology-specific user interface widgets useful for rapid application development in Java^{TM}
  - bioTK, a toolkit of biology-specific user interface widgets useful for rapid application development in Tcl/Tk
  - RSVP, a PostScript tool which lets your printer do nucleic acid sequence analysis; it generates very nice color diagrams of the results.
- Human Genome Project Information at Oak Ridge National Laboratory contains many interesting and useful items about the U.S. Human Genome Project. They also have a more technical Research site.
- FAKtory: A software environment for DNA Sequencing is at the University of Arizona. It is a prototype software environment in support of DNA sequencing. The environment consists of
  - (1) their software library, FAK, for the core combinatorial problem of assembling fragments
  - (2) a Tcl/Tk based interface
  - (3) a software suite supporting a database of fragments and a processing pipeline that includes clipping, tagging, and vector removal modules.
  A key feature of FAKtory is that it is highly customizable: the structure of the fragment database, the processing pipeline, and the operation of each phase of the pipeline may be specified by the user.
- Computational Analysis and Annotation of Sequence Data. This is a tutorial by A. Baxevanis, M. Boguski, and B.F. Ouellette on how to use alignment programs and databases for sequence comparison. It is a review that will appear in the forthcoming book Genome Analysis: A Laboratory Manual (Bruce Birren, Eric Green, Phil Hieter, Sue Klapholz and Rick Myers, eds) to be published by Cold Spring Harbor Laboratory Press. The hypertext version of the review is linked to Medline records, software repositories, sequences, structures, and taxonomies via the Entrez system of the National Center for Biotechnology Information.

S.L. Salzberg, D.B. Searls, S. Kasif (Eds.), *Computational Methods in Molecular Biology*

Suggestions for further reading in computational biology

This appendix contains a selected list of articles that provide additional technical details or background to many of the chapters in the text. We have annotated each selection very briefly, usually with just a sentence or two. Included are many papers written by the authors of the preceding chapters. The purpose of including these is to provide a guide for the reader who wants to find out more about the technical work behind any of the chapters. The brief descriptions should give the reader an advance sense of what is contained in each paper, which we hope will make searching the literature for additional reading materials more effective. We have mixed in a selection of influential papers that contributed to the thinking behind some of the chapters. This selection is by no means comprehensive, and it is of necessity heavily biased towards the topics covered in this book. It should be interpreted as a good place to start learning more about computational biology. The selections that follow are listed alphabetically by first author.

Altman, R.B. (1993) Probabilistic structure calculations: a three-dimensional tRNA structure from sequence correlation data. In: Proc. 1st Int. Conf. on Intelligent Systems for Molecular Biology (ISMB), pp. 12–20. This describes a probabilistic algorithm to calculate tRNA structures.

Altschul, S.F., Gish, W., Miller, W., Myers, E.W. and Lipman, D.J. (1990) Basic local alignment search tool. J. Mol. Biol. 215(3), 403–410. This describes the BLAST algorithm, one of the most widely used (and fastest) tools for sequence alignment.

Bagley, S.C. and Altman, R.B. (1995) Characterizing the microenvironment surrounding protein sites. Prot. Sci. 4, 622–635. Here the authors develop the FEATURE system that represents microenvironments by spatial distributions of physico-chemical features. The microenvironments of protein sites and control background nonsites are compared statistically to characterize the features of sites.

Borodovsky, M. and Mcininch, J.D. (1993) Genemark: Parallel gene recognition for both DNA strands. Comput. Chem. 17(2), 123–133. This is the original paper describing the Genemark system, a Markov chain method for finding genes in bacterial DNA.

Bowie, J.U., Luthy, R. and Eisenberg, D. (1991) A method to identify protein sequences that fold into a known three-dimensional structure. Science 253, 164–170. This is the pioneering paper that originated the idea of protein threading. Their method aligns an amino acid sequence to a "profile", which is a sequence of environments from a known protein structure.

Brendel, V., Bucher, P., Nourbakhsh, I.R., Blaisdell, B.E. and Karlin, S. (1992) Methods and algorithms for statistical analysis of protein sequences. Proc. Natl. Acad. Sci. USA 89, 2002–2006. Describes the computer program SAPS, which implements a number of statistical methods for analyzing protein sequence composition.

Brunak, S., Engelbrecht, J. and Knudsen, S. (1991) Prediction of human mRNA donor and acceptor sites from the DNA sequence. J. Mol. Biol. 220, 49–65. An analysis of human splice sites and a description of NetGene, which is a neural network based system for splice site prediction.

Bryant, S.H. and Lawrence, C.E. (1993) An empirical energy function for threading protein sequence through the folding motif. Proteins: Struct. Func. Genet. 16, 92–112. This describes one of the first threading algorithms.

Burge, C. and Karlin, S. (1997) Prediction of complete gene structures in human genomic DNA. J. Mol. Biol. 268, 78–94. A description of the GENSCAN system for finding genes in eukaryotic DNA. GENSCAN is a semi-Markov HMM, which allows it to take account of exon length distributions as well as local dependencies between bases. GENSCAN is currently the leading gene-finding system for eukaryotic DNA.

Burset, M. and Guigo, R. (1996) Evaluation of gene structure prediction programs. Genomics 34(3), 353–367. A thorough comparison of all the available programs (as of early 1996) for finding genes in vertebrate DNA sequences. Also introduced a data set of 570 vertebrate sequences that became a standard benchmark.

Churchill, G. (1992) Hidden Markov Chains and the analysis of genome structure. Comput. Chem. 16(2), 107–115. One of the first papers to describe the use of HMMs for the analysis of genomic sequence data.

Dayhoff, M.O., Schwartz, R.M. and Orcutt, B.C. (1978) A model of evolutionary change in proteins. Atlas Prot. Seq. Struct. 5(suppl. 3), 345–352. This describes the construction of the first amino acid substitution matrix, the PAM matrix, based on evolutionary data. PAM matrices are widely used by protein sequence alignment programs.

Dahiyat, B.I. and Mayo, S.L. (1997) De novo protein design: fully automated sequence selection. Science 278, 82–87. The first-ever design of a complete protein sequence by computer in which the protein was designed and synthesized, its structure was solved, and the structure was found to match the intended design.

Durbin, R.M., Eddy, S.R., Krogh, A. and Mitchison, G. (1998) Biological Sequence Analysis. Cambridge University Press. A book covering sequence alignment and search, hidden Markov models, phylogeny and general grammars. It has an emphasis on probabilistic methods.

Eddy, S.R. (1996) Hidden Markov models. Curr. Opin. Struct. Biol. 6, 361–365. A short review of hidden Markov models for sequence families.

Eisenhaber, F., Persson, B. and Argos, P. (1995) Protein structure prediction: recognition of primary, secondary, and tertiary structural features from amino acid sequence. Crit. Rev. Biochem. Mol. Biol. 30(1), 1–94. A comprehensive review of algorithms for protein structure prediction.

Fickett, J. and Tung, C.-S. (1992) Assessment of protein coding measures. Nucleic Acids Res. 20(24), 6441–6450. A survey and comparative evaluation of 21 different measures used to distinguish coding from noncoding DNA sequences.

Fickett, J.W. (1996) Finding genes by computer: the state of the art. Trends Genet. 12, 316–320. Short review of methods for finding genes.

Fischer, D., Lin, S.L., Wolfson, H.J. and Nussinov, R. (1995) A geometry-based suite of molecular docking processes. J. Mol. Biol. 248, 459–477. Presents the application of the Geometric Hashing method to protein-ligand docking.

Fischer, D., Tsai, C.J., Nussinov, R. and Wolfson, H.J. (1995) A 3-D Sequence-independent representation of the Protein Databank. Prot. Eng. 8(10), 981–997. An automated classification of the PDB chains into representative folds.

Fischer, D., Rice, D., Bowie, J.U. and Eisenberg, D. (1996) Assigning amino acid sequences to 3-dimensional protein folds. FASEB J. 10(1), 126–36. A comparison of several leading threading algorithms that discusses the key components of each algorithm.

Green, P., Lipman, D., Hillier, L., Waterston, R., States, D. and Claverie, J.-M. (1993) Ancient conserved regions in new gene sequences and the protein databases. Science 259, 1711–1716. Describes a large-scale computational comparison that located regions of homology between distantly related organisms. These are denoted "ancient conserved regions" because of the evolutionay distance between the organisms in which they were found.

Grundy, W., Bailey, T., Elkan, C. and Baker, M. (1997) Meta-MEME: Motif-based hidden Markov models of protein families. Comput. Appl. Biosci. 13(4), 397–403. Describes the MEME system, an HMM for finding one or more conserved motifs in a set of protein sequences.

Guigo, R., Knudsen, S., Drake, N. and Smith, T. (1992) Prediction of gene structure. J. Mol. Biol. 226, 141–157. Describes the GeneID program, a hierarchical rule-based system for finding genes in eukaryotic sequences.

Henderson, J., Salzberg, S. and Fasman, K. (1997) Finding genes in human DNA with a Hidden Markov Model. J. Comput. Biol. 4(2), 127–141. This describes the VEIL system, an HMM that predicts gene structure in DNA sequences. VEIL is a "pure" HMM for gene finding (as is Krogh's HMMgene), in that it uses a single HMM architecture for all its processing.

Henikoff, S. and Henikoff, J.G. (1992) Amino acid substitution matrices from protein blocks. Proc. Natl. Acad. Sci. USA 89, 10915–10919. This describes the construction of the BLOSUM substitution matrix using aligned blocks of protein sequences. The highly sensitive BLOSUM matrix is the current default matrix in the BLAST online server.

Hinds, D.A. and Levitt, M. (1992) A lattice model for protein structure prediction at low resolution. Proc. Natl. Acad. Sci. USA 89(7), 2536–40. Describes a model for protein structure prediction in which the locations of the molecules are restricted to grid points on a lattice.

Jones, D., Taylor, W. and Thornton, J. (1992) The rapid generation of mutation data matrices from protein sequences. Comput. Appl. Biosci. 8(3), 275–282. A detailed description of how to construct an amino acid substitution matrix. Included is an updated version of the PAM matrices using recent data.

Karlin, S. and Altschul, S. (1990) Methods for assessing the statistical significance of molecular sequence features by using general scoring schemes. Proc. Natl. Acad. Sci. USA 87, 2264–2268. A mathematical analysis of the statistical significance of results from sequence alignment algorithms.

Krogh, A., Mian, I. and Haussler, D. (1994) A Hidden Markov Model that finds genes in E. coli DNA. Nucleic Acids Res. 22, 4768–4778. Description of the design and performance of an HMM for finding genes in E. coli sequence data.

Krogh, A., Brown, M., Mian, I.S., Sjolander, K. and Haussler, D. (1994) Hidden Markov models in computational biology: Applications to protein modeling. J. Mol. Biol. 235(5), 1501–1531. Introduction of a profile-like HMM architecture which can be used for making multiple alignments of protein families and to search databases for new family members.

Lander, E. and Waterman, M. (1988) Genomic mapping by fingerprinting random clones: a mathematical analysis. Genomics 2, 231–239. An analysis that includes the now-classic curves showing the relationship between the length of a genome, the number of clones sequenced, and the number of separate "islands" or contigs in a sequencing project.

Lathrop, R.H. (1994) The protein threading problem with sequence amino acid interaction preferences is NP-complete. Protein Eng. 7(9), 1059–68. A proof that the unconstrained protein threading problem is computationally hard.

Lathrop, R.H. and Smith, T.F. (1996) Global optimum protein threading with gapped alignment and empirical pair score functions. J. Mol. Biol. 255, 641–665. A technical description of the branch and bound algorithm for protein threading that appears in chapter 12 of this volume.

Lawrence, C. and Reilly, A. (1990) An expectation maximization (EM) algorithm for the identification and characterization of common sites in unaligned biopolymer sequences. Prot. Struct. Funct. Genet. 7, 41–51. Describes an algorithm for finding and grouping together homologous subsequences from a set of unaligned protein sequences. This is the basis for an algorithm for finding protein motifs.

Lawrence, C.E., Altschul, S.F., Boguski, M.S., Liu, J.S., Neuwald, A.F., and Wootton, J.C. (1993) Detecting subtle sequence signals: a Gibbs sampling strategy for multiple alignment. Science 262, 208–214. This paper describes the influential Gibbs sampling method for detecting subtle residue (or nucleotide) patterns common to a set of sequences.

Moult, J., Pedersen, J.T., Judson, R. and Fidelis, K. (1995) A large-scale experiment to assess protein structure prediction methods. Proteins 23(3), ii–v. This is the introduction to CASP1, the first "competition" that compared various protein structure prediction algorithms. It includes evaluations of homology modeling, threading, and ab initio prediction algorithms.

Mount, S. (1996) AT-AC Introns: An ATtACk on Dogma, Science 271(5256), 1690–1692. A nice description of non-standard splice sites in eukaryotic genes, important information for anyone designing gene-finders or splice site identification systems.

Murthy, S.K., Kasif, S. and Salzberg, S. (1994) A system for induction of oblique decision trees. J. Artif. Intell. Res. 2, 1–33. Describes the decision tree classification system that is used in the Morgan gene-finding system. The source code is available.

Needleman, S.B. and Wunsch, C.D. (1970) A general method applicable to the search for similarities in the amino acid sequence of two proteins. J. Mol. Biol. 48, 443–453. This is the pioneering paper on sequence alignment using dynamic programming.

Nevill-Manning, C.G., Sethi, K.S., Wu, T.D. and Brutlag, D.L. (1997) Enumerating and ranking discrete motifs. In: Proc. 5th Int. Conf. Intelligent Systems for Mol. Biol. (ISMB), 202–209. Describes methods for efficiently and exhaustively evaluating discrete motifs in protein sequences.

Norel, R., Lin, S.L., Wolfson, H.J. and Nussinov, R. (1995) Molecular surface complementarity at protein–protein interfaces: the critical role played by surface normals at well placed, sparse points in docking. J. Mol. Biol. 252, 263–273. A method which has been especially successful in large protein-protein docking both in bound and unbound cases.

Nussinov, R. and Wolfson, H.J. (1991) Efficient detection of three-dimensional motifs in biological macromolecules by computer vision techniques. Proc. Natl. Acad. Sci. USA 88, 10495–10499. This paper explains the analogy between object recognition problems in Computer Vision and the task of structural comparison of proteins.

Reese, M., Eeckman, F., Kulp, D. and Haussler, D. (1997) Improved splice site detection in Genie. In: RECOMB '97. ACM Press, pp. 232–240. Describes how the Genie gene finding system was improved by adding conditional probabilities to its splice site recognition modules.

Salzberg., S. (1995) Locating protein coding regions in human dna using a decision tree algorithm. J. Comput. Biol. 2(3), 473–485. This demonstrates how to use decision tree classifiers to distinguish coding and noncoding regions in human DNA sequences. It includes a comparison of decision trees to linear discriminant classifiers.

Salzberg, S., Delcher, A., Kasif, S. and White, O. (1997) Microbial gene identification using interpolated Markov models, Nucleic Acids Res. 26, 544–548. This paper describes Glimmer, an Interpolated Markov Model for finding genes in microbial DNA (bacteria and archaea). The Glimmer system can be obtained from the Web page given in Appendix A.

Sandak, B., Nussinov, R. and Wolfson, H.J. (1995) An automated computer vision and robotics-based technique for 3-D flexible biomolecular docking and matching. Comput. Appl. Biosci. (CABIOS) 11(1), 87–99. In this paper a flexible docking method motivated by a Computer Vision technique is presented. Its major advantage is that its matching complexity does not increase compared with the rigid docking counterpart.

Sankoff, D. and Kruskal, J.B. (1983) Time Warps, String Edits, and Macromolecules: The Theory and Practice of Sequence Comparison. Addison-Wesley, Reading, MA. A classic book describing basic sequence alignment methods and many other topics.

Schaffer, A.A., Gupta, S., Shriram, K. and Cottingham Jr., R., (1994) Avoiding recomputation in linkage analysis. Human Heredity 44, 225–237. An elegant algorithm to compute genetic linkage between inherited human traits. The algorithm is implemented in a system that can handle substantially larger linkage problems than were previously possible.

Searls, D.B. (1992) The linguistics of DNA. Am. Sci. 80(6), 579–591. A gentle introduction to a view of macromolecules based on computational linguistics.

Sippl, M.J. and Weitckus, S. (1992) Detection of native-like models for amino acid sequences of unknown three-dimensional structure in a data base of known protein conformations. Prot. Struct. Funct. Genet. 13, 258–271. This describes a threading method that uses knowledge-based potentials.

Smith, T.F. and Waterman, M.S. (1981) Identification of common molecular subsequences. J. Mol. Biol. 147(1), 195–7. This paper is the original description of the classic Smith–Waterman sequence alignment algorithm.

Snyder, E.E. and Stormo, G.D. (1995) Identification of coding regions in genomic DNA. J. Mol. Biol. 248, 1–18. This describes the GeneParser system, a combination of neural nets and dynamic programming for finding genes in eukaryotic sequence data.

Solovyev, V.V., Salamov, A.A. and Lawrence, C.B. (1994) Predicting internal exons by oligonucleotide composition and discriminant analysis of spliceable open reading frames. Nucleic Acids Res. 22, 5156-5163. Description of a gene finder for human DNA that uses hexamer coding statistics and discriminant analysis. The basis for the FGENEH gene finding system, which is among the best.

Srinivasan, R. and Rose, G. (1995) LINUS: A Hierarchic procedure to predict the fold of a protein. Prot. Struct. Funct. Genet. 22, 81–99. The LINUS system predicts the 3-D shape of a protein by using a hierachical procedure that first folds small local environments into stable shapes, and then gradually folds larger and larger environments. The inter-molecular forces are modeled with rough approximations.

Staden, R. and McLachlan, A.D. (1982) Codon preference and its use in identifying protein coding regions in long DNA sequences. Nucleic Acids Res. 10, 141–156. A seminal paper on the compositional analysis of exons.

Stoltzfus, A., Spencer, D.F., Zuker, M., Logsdon Jr., J.M. and Doolittle, W.F. (1994) Testing the exon theory of genes: the evidence from protein structure. Science 265, 202–207. This paper describes the development and application of objective methods to test the exon theory of genes, i.e. the theory that protein-coding genes arose from combinations of primordial mini-genes (exons), perhaps corresponding to individual protein domains, separated by spacers (introns). They find no significant evidence that exons correspond to distinct units of protein structure.

Sutton, G., White, O., Adams, M. and Kerlavage, A. (1995) TIGR Assembler: A new tool for assembling large shotgun sequencing projects. Genome Sci. Technol. 1, 9–19. Describes the system that was used to assemble over 20 000 sequence fragments into the first complete genome of a free-living organism, the bacteria *H. influenzae*.

Tsai, C.J., Lin, S.L., Wolfson, H.J. and Nussinov, R. (1996) A dataset of protein–protein interfaces generated with a sequence order independent comparison technique. J. Mol. Biol. 260(4), 604–620. An automatic classification of the interfaces appearing in the PDB.

Uberbacher, E.C. and Mural, R.J. (1991) Locating protein-coding regions in human DNA sequences by a multiple sensors-neural network approach. Proc. Natl. Acad. Sci. USA 88, 11261–11265. This paper describes the computational approach used by GRAIL for locating protein-coding portions of genes in anonymous DNA sequence, by combining a number of coding-related features using a neural network.

Waterman, M.S. (1995) Introduction to Computational Biology. Chapman & Hall, London. Textbook covering the mathematics of sequence alignment, multiple alignment, and several other topics central to computational biology.

Xu, Y. and Uberbacher, E.C. (1997) Automated gene structure identification in large-scale genomic sequences. J. Comput. Biol. 4(3), 325–338. This paper describes a computational method for parsing predicted exons into (multiple) gene structures, based on information extracted from database similarity searches.

Xu, Y., Mural, R.J., Einstein, J.R., Shah, M.B. and Uberbacher, E.C. (1996) GRAIL: A multi-agent neural network system for gene identification. Proc. IEEE 84, 1544–1552. This summary paper describes a number of general techniques used in GRAIL's gene identification algorithm, which include neural-net based coding region recognition, exon prediction in erroneous DNA sequences, and dynamic programming for gene structure prediction.

Zuker, M. (1994) Prediction of RNA secondary structure by energy minimization. Methods Mol. Biol. 25, 267–294. Describes the popular mfold program for predicting RNA secondary structure using energy minimization.

Subject Index

370